AF847745

Topics in Biomedical Gerontology

Pramod C. Rath · Ramesh Sharma
S. Prasad
Editors

Topics in Biomedical Gerontology

Springer

Editors
Pramod C. Rath
Molecular Biology Laboratory, School of Life Sciences
Jawaharlal Nehru University
New Delhi, Delhi
India

S. Prasad
Biochemistry and Molecular Biology Laboratory, Department of Zoology
Banaras Hindu University
Varanasi, Uttar Pradesh
India

Ramesh Sharma
Department of Biochemistry
North Eastern Hill University
Shillong, Meghalaya
India

ISBN 978-981-10-2154-1 ISBN 978-981-10-2155-8 (eBook)
DOI 10.1007/978-981-10-2155-8

Library of Congress Control Number: 2016950884

© Springer Science+Business Media Singapore 2017
This work is subject to copyright. All rights are reserved by the Publisher, whether the whole or part of the material is concerned, specifically the rights of translation, reprinting, reuse of illustrations, recitation, broadcasting, reproduction on microfilms or in any other physical way, and transmission or information storage and retrieval, electronic adaptation, computer software, or by similar or dissimilar methodology now known or hereafter developed.
The use of general descriptive names, registered names, trademarks, service marks, etc. in this publication does not imply, even in the absence of a specific statement, that such names are exempt from the relevant protective laws and regulations and therefore free for general use.
The publisher, the authors and the editors are safe to assume that the advice and information in this book are believed to be true and accurate at the date of publication. Neither the publisher nor the authors or the editors give a warranty, express or implied, with respect to the material contained herein or for any errors or omissions that may have been made.

Printed on acid-free paper

This Springer imprint is published by Springer Nature
The registered company is Springer Nature Singapore Pte Ltd.
The registered company address is: 152 Beach Road, #22-06/08 Gateway East, Singapore 189721, Singapore

Dedicated to our teacher Prof. Madhu Sudan Kanungo (1927–2011)

A world-class Indian Biochemist and Gerontologist who devoted his entire career of teaching and research to Biochemistry of Aging at the Banaras Hindu University (BHU), Varanasi, India for 49 years. After completing his Ph.D. under Prof. C.L. Prosser at the University of Illinois, USA, he joined the Department of Zoology, BHU in early 1960s and continued until the end. He was always actively involved in teaching and research and developed an excellent teaching and research program of Animal Biochemistry and produced several Ph.D. students as next generation of teachers and scientists in many universities and institutes of India and abroad. He published 150 research papers and authored two books on aging: Biochemistry of Aging (Academic Press, 1980) and Genes and Aging (Cambridge University Press, 1994). He received numerous prestigious scientific recognitions for research (notably S.S. Bhatnagar Prize, Fellows of all the Indian Science Academies, Jawaharlal Nehru Fellowship, Emeritus Professorship of BHU) and civilian award (Padmashree). His dedication to teaching and research has been exemplary. He founded and was President and Patron of the Association of Gerontology of India (AGI). He is regarded as the father of aging research in India. He represented many national and international bodies concerned with aging and policies for the elderly.

Foreword I

Aging is an intrinsic biological process from which there is no escape (Arking 2003, Kanungo 2003, Kirkwood 2002). Now biological investigations of aging have grown into cutting-edge science (Morley 2004). Understanding why aging occurs may help devise means to counter or modulate our aging by altering the aging process and senescence. How to live longer and add life to years so that the elderly live in good health is the main aim of aging research.

Experimental research on aging in laboratory animals continues to provide information about multiple deleterious biological and molecular events that accumulate in different tissues over time and gradually reduce an organism's physiological effectiveness, state of maintenance, and health. Such studies give information about physiological ages—measurement of physiological processes involved in aging (Wolow et al. 2010). Furthermore, insight into the mechanisms underlying the cellular and molecular changes that can contribute to senescence is of great interest.

It can be argued that an adult's overall health will remain stable if gene expression remains stable (Rao et al. 1996). Microarray determination of the gene expression patterns between young and old subjects has shown relationships between alterations in gene expressions and aging.

There exists, however, no specific gene that may cause aging as no gene has been evolutionarily selected for aging (Kirkwood 2002). However, experimental manipulation of specific genes has been reported to extend life span of some organisms. Organisms are said to be programmed for survival and not for death. Genes influencing aging are those that affect the durability and maintenance of the soma. That is why; the subject of genomics/gene action in aging is of paramount importance. Study of model organisms for aging research, e.g., *Drosophila, Caenorhabditis, and Saccharomyces* have shown that there are genes that alter the rate of aging. Thus, organisms are genetically programmed to survive although the program is not efficient enough to make them last indefinitely. Efficient mechanisms for somatic maintenance and repair (genetic mechanisms of longevity assurance) can secure long and healthy life span. So aging occurs because organisms fail to make required energy allocations for somatic maintenance and repair (Arking 2003). Therefore, some interventions should be possible to retard/slow aging. The field of healthy aging medicine has also emerged (Ohlansky et al. 2004; Disterhoft and Oh 2006).

In experimental studies, calorie restriction has been found to reduce the development of many normal and pathological markers of aging and to slow the intrinsic aging rates (Sharma 2004). It brings about changes that help maintaining the optimal function of tissues. It has been found to lengthen primate lives. It, however, remains to be seen how it will benefit humans. Undernutrition mimicking dietary caloric restriction was shown to be actually beneficial in human subjects in maintaining good DNA repair capacity (Rao et al. 1996). Oxidative stress is seen as a major contributor to the aging process. Thus anti-oxidative measures pharmacologically have been found to reduce aging. Increasing the level of genetic defense mechanisms against oxidative stress is also of significance.

This volume "Topics in Biomedical Gerontology" presents aging research data at molecular, cellular, and organismal levels on a wide spectrum of subjects derived from animal models such as *Drosophila, Dictyostelium*, mouse, rat, and also humans.

Data on human aging are reported in three papers. Frailty is an important expression of aging—like process in humans, and down regulation of sirtuins is shown to be a biomarker of this process. How morphology of the human pancreas is altered during development and aging is described in an interesting paper. The aging nervous system encompasses several research areas. A paper has shown how aging affects the population of neurons and glia in human cochlear nucleus.

The lysosomal and autophagy pathways are both involved during normal aging. Autophagy tends to decline during normal aging, and it is likely to be important for promoting healthy aging. Described in an experimental paper are data from studies on *Dictyostelium discoideum*, which show how autophagy promotes life extension.

Neurons are post-mitotic cells, which cannot divide to replace damaged or deteriorated cells. Thus, accumulated damage from oxidative stress has been considered a key factor contributing to normal aging of the nervous system (Poon et al. 2004). The paper on oxidative stress in the brain highlights the role of oxidative stress in brain's cognitive aging. How oxidative stress influences other tissues such as kidney is also described.

Electrical activity is the functional basis of the nervous system. Aging of the brain involves changes in its electrophysiological activity and age-related disorganization of electrical activity causes derangement of neural and cognitive functions (Disterhoft and Oh 2006). There is thus a potential link between electrophysiological and behavioral consequences of aging.

That the anti-aging neurosteroid dehydroepiandrosterone may have antiepileptic influence on the nervous system of elderly is shown in an experimental study. A number of papers included in this volume also present information concerning anti-aging influence of calorie restriction, e.g., down-regulation of insulin-like growth factor-1 and upregulation of cardiac and skeletal muscle inorganic pyrophosphatase. Experimental data on the role of noncoding RNA and Pax-6 transcription factor are described during aging. Also, a review on chromatin and aging is of interest.

A major neurodegenerative disease associated with aging is Alzheimer's disease in which amyloid-β protein pathology plays an important role (Mishra et al. 2016). The paper on amyloid pathology attempts to show how this may cause behavioral changes in the context of human-specific behavior such as empathy, sympathy and pro-social tendency in rats.

Disorders such as diabetes, depression as well as dietary factors can enhance decline of the nervous system function. A paper examines how experimentally-induced diabetes mellitus type 2 induces brain aging and memory impairment and how these two effects are prevented by *Bacopa monnieri*. In another paper how recovery from age-related memory loss may be possible is described. Healthy aging through the principles of *Ayurveda* has also been described.

Rameshwar Singh
Formerly Professor of Neurobiology
School of Life Sciences, Jawaharlal Nehru University
New Delhi

References

Arking R (2003) Aging: a biological perspective. Am Sci 91:508–515

Disterhoft JF, Oh MM (2006) Pharmacological and molecular enhancement of learning in aging and Alzheimer's disease. J Physiol Paris 99:180–192

Kanungo MS (2003) Mechanism of aging. Proc Indian Nat Sci Acad B69:121–140

Kirkwood TBL (2002) Evolution of aging. Mech Aging Dev 123:737–745

Mishra N, Singh R, Sharma D (2016) Differential response of amygdala and hippocampus consequent to Aβ 40 and Aβ 42—induced toxicity in the rat brain: a comparative study. Int J Adv Res 4:959–994

Morley JE (2004) The top 10 hot topics in aging. J Gerontol Med Sci 59A:24–33

Ohlansky SJ, Hayflick L, Perls TT (2004) Anti-aging medicine: the hype and the reality—Part 1. J Gerontol Biol Sci 59A:513–514

Poon HF, Calabrese V, Scapagnini G, Butterfield DA (2004) Free radicals: key to brain aging and heme oxygenase as a cellular response to oxidative stress. J Gerontol Med Sci 59A:478–493

Rao KS, Ayyagiri S, Raji NS, Murthy KJR (1996) Under-nutrition and ageing: effects on DNA repair in human peripheral lymphocytes. Curr Sci 71:464–469

Sharma R (2004) Dietary restriction and its multifaceted effects. Curr Sci 87:1203–1210

Wolow CA, Zou S, Mattson MP (2010) Aging of the nervous system: In: Wolf NS (ed) The comparative biology of aging. Springer Science+Business Media BV, pp 319–352

Foreword II

The graying of the world population is one of the most serious issues in the twenty-first century. The life span of people in developed as well as developing countries has been increasing during the past several decades and appears to increase further (Vaupel 2010). This is a good news, but it has also caused increase in the number of frail and/or diseased elderly people, putting a heavy social, economical. and family burden on younger generations in many countries.

It is often stated by geriatric physicians that the major risk factor in the development of age-associated diseases such as dementia and cardiovascular disorders is aging itself. This implies that biological changes with aging are likely to underlie the etiology and progress of the diseases in addition to causing the physiological impairments in many tissues although aging is not a disease.

Biomedical Gerontology is therefore important not only as a basic science to unravel the mechanisms of biological aging, but also to extend health span of people, reducing risks of age-related diseases, and improving the activities of daily living to retard the decline of quality of life with aging. Aging is very complex biological phenomena as represented in drawings of the famous Indian legend "The blind men and the elephant," symbolizing that a big complex object may be recognized as different things if examined in part depending on from which side it is viewed. In fact, more than three hundred theories have been proposed for mechanisms of biological aging already many years ago (Medvedev 1990), and the number is still increasing and the theories are being modified (Goto 2015).

This comprehensive book covers a wide range of major current topics on aging in both basic and applied fields by expert Indian scientists including disciples of late Prof. Madhu Sudan Kanungo, my old friend, the founder of Indian biomedical gerontology and the leader of aging research in Asian countries. He was already well known internationally when I started the research on aging forty years ago.

I do expect that "Topics in Biomedical Gerontology" will make significant contributions to the advancement of gerontological research, stimulating the investigators in the world, and help the realization of healthy living of elderly people.

Sataro Goto, Ph.D.
Visiting Professor
Juntendo University Graduate School
Professor Emeritus, Toho University, Japan
Emeritus Member, Japan Society for Biomedical Gerontology

References

Goto S (2015) The biological mechanisms of aging: a historical and critical overview. In: Mori N, Mook-Jung I (eds) Aging mechanisms-longevity, metabolism and brain aging. Springer

Medvedev ZA (1990) An attempt at a rational classification of theories of ageing. Biol Rev Camb Phylos Soc 65:375–398

Vaupel JW (2010) Biodemography of human ageing. Nature 464:536–542

Preface

Human race has been experiencing an explosion in the middle and old age population over the last century. One does desire to live long, but not many years in disability. Hence, the goal of biomedical gerontology is to achieve quality healthy life rather than merely adding years to one's life. With advancing age, senior citizens suffer from loss of muscle strength and function that causes hardship in walking and climbing and often requires living assistance. Aging per se is not a disease, however, it is a potential risk factor for various diseases including neurodegenerative (Alzheimer's and Parkinson's), cancer, cardiovascular, ocular, pulmonary, and osteoporosis. Hence, understanding the mechanisms involved in the causation of aging process will help devise interventions to postpone age-related diseases and provide healthy aging. A pivotal goal of aging research is not only to maximize human life span, but also to improve the quality of life with advancing age.

Aging results from multiple interactions between genes and environment. Hence, it is under the strong influence of nature and nurture. Studies in human and other experimental animals indicate that right nutrition and regular physical exercise contribute remarkably to enhance health span and prolong life span as well. Preventive and rehabilitative strategies might help elderly maintain active lifestyle and preserve intellectual and functional competence in old age. One of the well-studied strategies is the dietary restriction that has been shown to increase both mean and maximum life span together with reduction in morbidity and increased quality of life.

Advancement in the field of biology with revised central dogma of molecular biology and RNA world has lead to newer concepts and approaches to explore the interventional strategies in increasing healthy quality life. In the era of "omics," one can use metabolomic, functional genomic, and proteomic approaches to understand the mechanisms of aging process and design ways to modulate such processes to achieve healthy aging. Several strategies are being employed to achieve this goal by biomedical gerontologists such as macronutrients and micronutrients, vitamins, phytoceuticals, and dietary restriction.

Our venture to produce this book is to put together various topics which are useful in finding causation and mechanisms of aging in various experimental animals including humans. Attempts have also been made to include the work on various interventional strategies to achieve healthy aging. It all originated from the International conference on Engaging and Empowering the Elderly and 17th Biennial Conference of Association of Gerontology (India) held at Thiruvananthapuram during September 15–16, 2014. In addition, we have tried to include some additional relevant topics for larger group of readers in the field of biomedical gerontology to add intellectual understanding among researchers of healthful aging.

New Delhi, India — Pramod C. Rath
Shillong, India — Ramesh Sharma
Varanasi, India — S. Prasad

Contents

About the Editors

Pramod C. Rath is a Professor of Molecular Biology and has been Acting Dean of the School of Life Sciences, Jawaharlal Nehru University, New Delhi. He received his Ph.D. in 1988 from the Biochemistry Section of the Zoology Department, Banaras Hindu University, Varanasi, India on "Expression and Modulation of Genes during Aging of the Rat" under the supervision of Prof. M.S. Kanungo. He did his post-doctoral research at the Institute for Molecular Biology I, University of Zurich, ETH-Hoenggerberg from 1989 to 1990 on "Regulation of Interferon Gene Expression in Mammalian Cells" with Prof. Charles Weissmann, Director. He joined the School of Life Sciences, Jawaharlal Nehru University, New Delhi as an Assistant Professor (Molecular Biology) in December 1990. He went to Cold Spring Harbor Laboratory, New York, as a Short-term Scholar in March, 1998 to work on "Advanced Genome Sequencing and Analysis" with Prof. Richard MacCombie, Chief, Genome Sequencing Research Group. He was a Visiting Research Scientist during 1998–1999 at the University of Texas MD Anderson Cancer Center, Houston, Texas and worked on "Cytokine Signalling and Cancer" with Prof. Bharat B. Aggarwal, Chief, Cytokine Research Laboratory. He also went to the Stem Cell Institute, Medical School, University of Minnesota, Minneapolis as a Short-term Scholar during November–December, 2005 and worked on "Multipotent Adult Progenitor Cells (MAPCs)" with Prof. Catherine Verfaillie, Director. He has 25 years of teaching and research experience, having mentored 17 Ph.D. and three M.Phil. students. He teaches Molecular Biology, Molecular Genetics and Cell Signalling courses at M.Sc. and Pre-Ph.D. levels. His research areas include (i) Genomic Biology of Repetitive DNA and Noncoding RNA; (ii) Cytokines, Transcription Factors, Cell Signalling, and Disease and (iii) Mesenchymal Stem Cells from Bone Marrow and Regenerative Medicine. He has many research publications to his credit in reputed international journals, such as, Ageing Research Reviews, Molecular Neurobiology, Journal of Molecular Neuroscience, International Journal of Developmental Neuroscience, International Journal of Biological Macromolecules, RNA Biology, PLos One, Journal of Clinical Immunology, Journal of Interferon and Cytokine Research, Biochim. Biophys. Acta (Molecular Biology of Disease), Biochem. Biophys. Res. Commun.,

FEBS Letts., Mol. Biol. Rep. etc. He has also contributed chapters to books including "Gene and its Engineering," "Textbook of Biotechnology," "Textbook of Geriatric Medicine," and "Quality Aging." He has been the recipient of several awards including the A.B. Mishra award for being the gold-medalist during his M. Sc.; the Council of Scientific and Industrial Research fellowship for completing his Ph.D.; Swiss National Fund post-doctoral fellowship; Young Scientist research project of the Department of Science and Technology, Government of India, 1993–1995; Advanced Genome Sequencing and Analysis course fellowship, Cold Spring Harbor Laboratory, New York, 1998; Clayton Research Foundation Visiting Scientist fellowship, University of Texas MD Anderson Cancer Center, Houston, 1998–1999 etc. He is a life member of many scientific bodies, including Vice-President of the Association of Gerontology, India (AGI), Indian Immunology Society (IIS), Indian Society of Cell Biology (ISCB), Association of Biotechnologists of India (ABI), Indian Association for Cancer Research (IACR), Society of Biological Chemists (India), and the Society for Translational Cancer Research (STCR). He has been a part of many scientific meetings, conferences and committees. He is a reviewer for many reputed journals, such as Cancer Investigations, DNA and Cell Biology, Indian Journal of Biochemistry and Biophysics, J Genetics, Scientific Reports (NPG), FEBS Letts, BMC Cancer (Biomed Central), Molecular Neurobiology, Neuroscience Letts, Molecular Biology Reports, e-CAM, Biomed. Res. Int., J. Cell. Biochem., Biogerontology etc. He is an Editorial Board member for Frontiers in Cell and Developmental Biology, Switzerland.

Ramesh Sharma (Ph.D., Banaras Hindu University, 1981) is a Professor at the Department of Biochemistry, North-Eastern Hill University, Shillong, India. He did his post-doctoral research at the University of California, Berkeley, USA (1986–1988), and was a Visiting Professor at the Institute of Enzyme Research at Tokushima University, Japan (1989–1990) and at the Department of Biochemistry at the School of Pharmaceutical Sciences, Toho University, Chiba, Japan (2003–2004). He has been teaching various disciplines of physiology and biochemistry for the past 33 years, mostly at the postgraduate/research level. He has been on the editorial board of international journals such as "Biogerontology" from Springer, the Netherlands; "International Journal of Developmental Neurosciences" from Elsevier, USA; "Okinawa Wellness & Longevity Journal" Okinawa, Japan; and "Frontiers in Genetics of Aging," Science Park, Switzerland. He has produced more than 80 publications in international and national journals of repute in the field of aging research and has produced 16 Ph.D.s. In 2008, he was elected as a Fellow of the National Academy of Sciences (India). He has been the Head of the Department of Biochemistry for three terms (1993–1996; 1999–2002; 2011–2014) and is presently Coordinator of the UGC-SAP programme in Biochemistry and the Dean, School of Life Sciences. He has also been Coordinator of the University with Potential for Excellence (UPE) given by UGC to NEHU. He has been a member of the board of management of NERIST and a research council member of CSIR-RRL (now NEIST), Jorhat, Assam. Professor Sharma has also been

Chairman of the Board for undergraduate and postgraduate studies in Biochemistry, Home Science and Nursing and a member of the Academic Council of NEHU and Mizoram University. He has been the President of Association of Gerontology (India) and is a member of the Task Force on Neuro-Disease Biology, DBT, GOI (2014–2017). Professor Sharma visited Semmelweis University, Budapest, Hungary (2014) under INSA-Senior Scientist Exchange Fellowship and was a Visiting Researcher at Toho University, Chiba, Japan (January 01–March 31, 2016).

S. Prasad did his B.Sc. (1978, Bot., Zool., Chem.); M.Sc. (1980, Zoology with Biochemistry Specialization); Ph.D. (1988, Zoology) from the Banaras Hindu University (BHU), Varanasi, India and joined as a Lecturer in 1991 and is serving at present as a Professor at the Department of Zoology, Institute of Science, BHU since 2008. He worked on "Age-associated Studies on HMG Non-Histone Proteins of the Rat Liver" for his doctoral degree under the mentorship of Prof. M.K. Thakur, F.N.A.Sc., F.I.A.N. His current research interests are focused on molecular, behavioral and neurobiological mechanisms of aging as well as learning and memory during aging and diseases. He visited Royal Cancer Centre Research Institute, Tokyo under AMBO International Fellowship to work on "Genomics of Cancer" in 1991 and availed MRC Visiting Fellowship to work on "Estrogen-induced Expression of Genes in Aging Brain" with Prof. J.A. Edwardson at MRC Neurochemical Pathology Unit, New Castle General Hospital, New Castle upon Tyne, UK during 1995–1996. He is associated with teaching of Biochemistry and Molecular Biology, Physiology, Neuroscience, Molecular Techniques at undergraduate, postgraduate and doctoral levels in the Department of Zoology since last 26 years and Biochemistry and Neurobiology at the Department of Molecular Human Genetics until 2011. He has produced three Ph.D. students and four to submit their Ph.D. theses. He has guided and trained several postgraduate students from BHU and other universities for their dissertation work. He has been on several academic and administrative committees of the Institute of Sciences and University. He has served as member on Board of Studies for Biotechnology and Bioinformatics and Zoology at the North Eastern Hill University, Shillong, Molecular Human Genetics, Biophysics, Forensic Science and Interdisciplinary School of Life Sciences at BHU. He has published 34 research publications in journals of high impact on Aging, Molecular Biology and Neuroscience. He has written book chapters and review articles. He has edited a book on "Cellular and Molecular Neurobiology" (Narosa Publishing House, New Delhi). He is in the reviewers' panel of many journals, e.g., Biogerontology (Springer), Neurochemistry International (Elsevier), World Journal of diabetes and ECAM (Hindawi). He has been elected as a member of the National Academy Sciences (India), Allahabad in 2010 and is associated with Association of Gerontology (India), Society of Biological Chemists, Indian society of Cell Biology, Indian Academy of Neurosciences and Society of Neurochemistry (India) as a life member and currently serving as the Secretary of Association of Gerontology (India).

Part I
Models and Markers in Aging

Drosophila melanogaster: A Prime Experimental Model System for Aging Studies

Renu Yadav, Soram Idiyasan Chanu, Kritika Raj, Nisha and Surajit Sarkar

Abstract Aging, the process of growing old is largely characterized by gradual deterioration of normal cellular functions, leading to progressive and steady decline in the biological, physical and psychological abilities. The phenomenon of aging is genetically determined and environmentally modulated. This is one of the most common yet mysterious aspects of biological studies, even after being a subject of interest to humans since the beginning of recorded history. Moreover, precise molecular basis of aging remains poorly understood, in part, because we lack a large number of molecular markers which could be used to measure the aging process in specific tissues. Moreover, limitations of human genetics and associated ethical issues further make it difficult to identify or analyze candidate gene(s) and pathways in greater details, and with the fact that the basic biological processes remain conserved throughout phylogeny; model organisms from bacteria to mammals have been utilized to resolve different aspects of aging. Classical model system such as *Drosophila melanogaster* has emerged as an excellent system to elucidate essential genetic/cellular pathways of human aging, due to its short generation time, availability of powerful genetic tools and functionally conserved physiology. Several key cellular events and signaling cascades have been deciphered by utilizing *Drosophila* as system of aging research and continues to add novel insights into this complex process. Present article attempts to introduce *Drosophila* as a model system for aging studies and also provides a brief overview of its decades of contribution in aging research.

Keywords *Drosophila* · Aging · Molecular chaperone · Oxidative stress · Insulin signalling · Neurodegeneration

R. Yadav · S.I. Chanu · K. Raj · Nisha · S. Sarkar (✉)
Department of Genetics, University of Delhi South Campus,
Benito Juarez Road, New Delhi 110021, India
e-mail: sarkar@south.du.ac.in

© Springer Science+Business Media Singapore 2017

P.C. Rath et al. (eds.), *Topics in Biomedical Gerontology*,
DOI 10.1007/978-981-10-2155-8_1

1 Introduction

Aging is not a passive activity, but an actively regulated complex process or collection of gradual senescence processes at both physiological and cellular levels. Some of the most prominent characteristics of aging include progressive decrease in physiological capacity, reduced ability to respond adaptively to environmental stimuli, increased vulnerability to infection and complex diseases and, increased mortality. Aging at large, is genetically determined and environmentally modulated. Aging activates some irreversible series of biological changes that inevitably result in death of the organism. Although, the causes of these changes may be entirely unrelated in different cases implying no common mechanism, yet they often imply a mutual element of descent. Therefore, aging is one of the most common yet mysterious aspects of biological studies, even after being a subject of interest to human race since the beginning of recorded history.

Decades of research on aging has found several genes and many biological processes those are associated with them; however, several fundamental questions continue to be intensely debated. Some of such unanswered questions are: (i) How many biological processes contribute to aging? What are they? (ii) Is it possible to reverse the phenomenon of aging? (iii) Can a single gene mutation recapitulate all the aging induced consequences? Also, the molecular basis of aging remains poorly understood, in part, because we lack a large number of molecular markers of aging which can be used to measure the aging process in specific tissues. Thus, unravelling the mysteries of aging is still on the frontier of biomedical research.

The last two decades have witnessed a tremendous upsurge in the genetic analyses of aging, with a greater emphasis towards the elucidation of the molecular mechanisms, pathways, and physiological processes implicated in longevity. Since the limitations associated with human genetic studies make it difficult to identify or analyze candidate gene(s) and pathway(s) in greater details, and with the fact that the basic biological processes remain conserved throughout phylogeny, model organisms from bacteria to mammals have been utilized to resolve different aspects of aging. However, classical model systems such as *Caenorhabditis elegans* and *Drosophila melanogaster* have emerged as excellent systems to elucidate essential genetic/cellular pathways of human aging. *Drosophila* particularly, holds tremendous promise for identifying genes and to decipher other mechanisms which influence age-related functional declines. Some of the major advantages associated with *Drosophila* have been discussed below:

2 *Drosophila melanogaster* as a Model Organism for Aging Research

D. melanogaster, commonly known as “fruit fly” is one of the most studied organisms in biology, particularly in genetics and developmental biology (Fig. 1a). Some of the major advantages of using *Drosophila* for aging related studies include

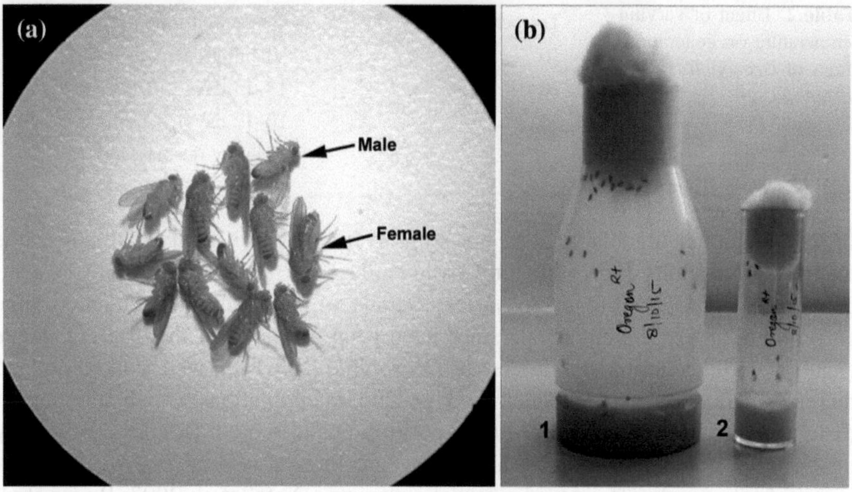

Fig. 1 **a** Mixed population (male and female) of wild type (*Oregon R*$^+$) *Drosophila melanogaster* as appears under stereozoom binocular microscope. **b** Wild type (*Oregon R*$^+$) *Drosophila melanogaster* raised on cornmeal in culture bottle (1) and vial (2)

its short life span of 50–70 days, high fecundity (female lay up to 100 eggs per day), availability of powerful genetic tools, accessibility of stocks with many different alterations, knowledge of the complete genomic sequence and large homogeneous populations. In addition, ease of culturing and affordability of maintaining large populations within the confines of a laboratory further makes flies a remarkable model organism (Fig. 1b). Besides, absence of meiotic recombination in males and presence of balancer chromosomes allow populations of flies carrying heterozygous mutations to be maintained without undergoing any constant screening for the mutations. Moreover, completely sequenced and annotated genome distributed on four chromosomes makes *Drosophila* a well acceptable system to perform large-scale genetic screens for identification of potential modifiers of aging and disease related phenotype(s). One of the striking features of *Drosophila* is the existence of morphologically distinct developmental stages which includes embryonic, larval, pupal and adult phase (Table 1); thus, the sexually matured "aging" adults phase could be easily distinguished in the developing population.

Table 1 Different developmental stages in life cycle of *Drosophila melanogaster* at 25 °C

Stages	Time (in h)	Duration (in h)
Egg	0	24
First instar larvae	24	24
Second instar larvae	48	24
Third instar larvae	72	48
Pupa	120	72
Adult fly	192	–

Table 2 Effect of varying temperature on generation times of *Drosophila melanogaster*

Temperature (°C)	Generation time (days)
10	Viability decrease
20	13–15 days
25	10–12 days
35	Viability decrease

In several model organisms, it is not so conventional to visually distinguish the mature aging adults form immature or juvenile stage. Depending upon the temperature, *Drosophila* life cycle varies. Details of the different generation time corresponding to different temperature have been provided in the Table 2. Since anatomy and developmental process of *Drosophila* have been well worked out and therefore, creating environmental and genetic manipulations which alter aging dynamics and life span could be easily performed and scored. Besides, availability of the large number of mutants and transgenic lines at several *Drosophila* stock centers further makes it a popular model organism (Dietzl et al. 2007; Ryder et al. 2007).

Similarity of different genes and families which are structurally and functionally related in both *Drosophila* and mammals, makes flies a good model in human based research. It is increasingly clear now that *Drosophila* genome has approximately 75 % of known human disease genes and ~50 % of proteins have mammalian homologs (Reiter et al. 2001). Moreover, the adult fly harbors a well-coordinated sophisticated brain and nervous system, which makes it capable of exhibiting complex behaviors such as learning and memory, much like the human brain (Jones and Grotewiel 2011). Disruption of such well-coordinated motor behaviors leads to neuronal death and dysfunction. Mammalian aging related phenotypes such as locomotory and sensory impairments, learning disabilities, sleep like behavior etc. are well manifested in *Drosophila* (Jones and Grotewiel 2011). *Drosophila* lack a functional blood brain barrier which could otherwise prevent access of drugs to the tissues of central nervous system; as a result flies become extremely useful for pharmacological screening for identification of novel therapeutic drug targets (Jones and Grotewiel 2011). Interestingly, the response towards many drugs that has shown effects within the *Drosophila* CNS is quite similar as observed in mammalian systems (Wolf and Heberlein 2003; Pandey and Nichols 2011).

Drosophila provides powerful genetic tools which can easily manipulate gene expression in a tissue specific manner during various stages of life cycle. UAS-Gal4 system is a commonly used genetic tool to achieve ectopic expression of desired genes or to suppress the expression of a target gene by UAS-RNAi transgene (Brand and Perrimon 1993). Additionally, FLP-FRT system, a site- directed recombination technology, has been progressively used to manipulate the fly genome in vivo, under controlled condition (Theodosiou and Xu 1998). Utilizing this technology loss-of-function of any lethal gene can be easily studied in a given target organ in a spatially controlled manner, in the cases where model organism would not survive as a result of loss of this gene in other organs. The effect of altered gene can also be studied over time, by using an inducible promoter to trigger the recombination activity late in development. This prevents the genetic alteration from affecting overall development of the organ, and also allows single cell comparison of the one lacking the gene to normal neighboring cells in the same environment.

In comparison to other model systems, a few additional advantages offered by *Drosophila* for aging studies include presence of almost fully differentiated post-mitotic cells throughout the adult fly, representing synchronized aging (Arking 1991). Enlightening the first aspect, the instigation of adulthood in *Drosophila* is said to occur only after the fly ecloses out of the pupal case. During this stage of its life, it becomes sexually mature and competent to reproduce and thus, aging is thought to be initiated (Shaw et al. 2008). This is in great disparity with other model systems where it is often difficult to find out when the organism has attained maturity (Helfand and Rogina 2003a). The second aspect has been focused on the rarely dividing neurons of the brain which makes the *Drosophila* brain an excellent model for the cytological studies and to relate with human aging (Herman et al. 1971). Hence, aging related structural changes could be easily and conclusively deduced by observing a set of synchronously aging cells. Moreover, due to the absence of blood vessels in insect brain, the pathological changes due to blood vasculature can be debarred. In view of above noted advantages, *Drosophila* has been widely utilized to decipher various aspects of aging. A brief overview of the history of *Drosophila* aging research has been provided below.

3 History of *Drosophila* Aging Research

For the first time Thomas H. Morgan used the small invertebrate, *Drosophila melanogaster*, to write the purpose of his research and this marks the beginning of an era of groundbreaking research utilizing this system in his “fly room” at Columbia, USA. This led to the discovery of the ‘chromosomal theory of inheritance’ and he was eventually awarded Nobel Prize in 1933 for his excellent finding. Following this, the researchers have come a long way in terms of exploiting the powerful genetics offered by this tiny fruit fly. Remarkably, Loeb and Northrop in 1916 reported the first use of *Drosophila* as a model system to study aging. They performed several experiments to demonstrate the effects of temperature and food on fly longevity (Loeb and Northrop 1916). They concluded that longevity of flies as poikilothermic organisms depends on the temperature of the environment (Loeb and Northrop 1916). In addition, they also examined the effect of starvation and sugar concentration on fly longevity (Loeb and Northrop 1917). Subsequently, Pearl and co-workers demonstrated that longevity in flies is hereditable (Pearl and Parker 1921, 1922). Consistent to Pearl’s finding, the significance of genetic influence in regulation of life span of adult flies was further reported by Clark and Gould in 1970. By utilizing *Drosophila* as a model system, several small compounds such as biotin, pyridoxine and pantothenic acid were identified which extend the life span upon regulated feeding (Gardner 1948). The effect of reproductive behavior on aging has been a topic of aging research since middle of 20th century when J. Maynard Smith and colleagues reported that longevity of flies could be affected by changing their reproductive behavior (Smith 1958). Their studies had established *Drosophila* as a good model system to study the fitness trade-offs and life span (Smith 1958). Since then, the mechanistic correlation between reproduction and longevity has been a topic of great interest in the aging research. Consistently, the plasticity behaviors between fly longevity and reproductive output was further confirmed by the selection experiments performed in the 1980s, which showed that longevity could be significantly extended when female flies were selected for late-life fertility (Rose and Charlesworth 1980, 1981; Luckinbill et al. 1984; Luckinbill and Clare 1985). Michael Rose has reviewed the history of laboratory-based evolution experiments and the use of different genomic technologies to comprehend genetics of longevity in *Drosophila* (Rose and Charlesworth 1980, 1981). Interestingly, independent studies performed during end of 20th century led to identification of two different life extending mutations, the *Methuselah (mth)* and *I'm not dead yet (Indy)* by performing random genetic alterations. It was demonstrated that partial loss-of-function mutation in either *mth* or *indy* extend the life span in both male and female flies, without loss of fertility (Lin et al. 1998; Rogina et al. 2000). In modern era of aging research, in addition to classical approaches several contemporary approaches and novel strategies are being adopted to decipher the mechanistic in-depth of aging and longevity. Some of the popular genetic approaches include selective breeding, mutagenesis followed by forward genetic analysis, cellular and molecular genetics and QTL analysis (Jazwinski 2000). These methods, so far, have allowed identification of numerous genes involved in diverse cellular functions including aging and longevity in *Drosophila*. Table 3 provides a brief collection of some genes and their assigned function(s) which have been found to be associated with longevity in *Drosophila*. An overview of various methods and approaches related to *Drosophila* aging research have been provided below.

4 Evaluating Aging in *Drosophila*: Methods and Approaches

Over the past decades understanding the complex mechanisms underlying the process of aging has emerged as a great frontier of biomedical research considering not only the welfare of humankind but also to overcome the challenges associated with this complex biological phenomenon. As discussed above, aging is a process of progressive, irreversible changes at the molecular and cellular level, which results in the decline of organismal performances. The stereotypic/phenotypic changes which are associated with aging in most of the organisms are the result of the changes at molecular, physiological and cellular levels. Therefore, due to the

Table 3 A brief collection of some genes found to expand life span in *Drosophila melanogaster*

Gene	Function in the cell	Longevity increases due to	References
14-3-3ε	Antagonist to dFoxo	Knockdown	Nielsen (2008)
chico	Insulin receptor substrate	Knockdown	Clancy et al. (2001)
dFoxo	*Drosophila* forkhead transcription factor	Over-expression	Giannakou et al. (2004)
dilps	*Drosophila* insulin-like peptides	Knockdown	Grönke et al. (2010)
dInR	*Drosophila* Insulin receptor	Knockdown	Tatar et al. (2001)
dPTEN	*Drosophila* phosphatase and tensin homolog, controls cell growth and proliferation	Over-expression	Hwangbo et al. (2004)
dS6K	Major downstream kinase of the TOR pathway	Knockdown	Kapahi et al. (2004)
dsir2	Histone and non-histone, NAD-dependent deacetylase	Over-expression	Rogina and Helfand (2004), Bauer et al. (2009)
dTOR	Serine/threonine protein kinase that regulates growth, proliferation, survival, transcription, etc. of the cell	Knockdown	Kapahi et al. (2004)
dTsc1, *dTsc2*	Act synergistically to inhibit TOR	Over-expression	Kapahi et al. (2004)
GSH	Antioxidant enzyme involved in formation of reduced glutathione	Over-expression	Mockett et al. (1999)
Hep	JNK kinase	Over-expression	Wang et al. (2003)
Hsp22	Stress response	Overexpression	Morrow et al. (2004b)
Hsp23	Stress response	Overexpression	Morrow and Tanguay (2003)
Hsp26	Stress response	Overexpression	Liao et al. (2008)
Hsp27	Stress response	Overexpression	Liao et al. (2008)
Hsp68	Stress response	Overexpression	Wang et al. (2003)
Hsp70	Stress response	Overexpression	Tatar et al. (1997)
Indy	Succinate and citrate transmembrane transporter	Knockdown	Rogina et al. (2000)
mth	G-protein coupled receptor	Knockdown	Lin et al. (1998)
puc	Inhibits JNK by its specific phosphatase activity	Knockdown	Zeitlinger and Bohmann (1999), Wang et al. (2003)
SOD	Antioxidant enzyme involved in degeneration of superoxide radicals to molecular oxygen	Over-expression	Parkes et al. (1998), Sun and Tower (1999)

fact that aging follows the normal laws of chemical, physical and several of the complex biological phenomenon; combined efforts of molecular, genetics, physiological, anatomical and behavioral approaches have been used to assess the mysteries behind aging. In the following text, details of the different approaches which have been used to assay the process of aging in *Drosophila* have been discussed.

4.1 Assessing Life Expectancy

It is difficult to measure how an individual changes with age, but demographic assay such as age of the dead individuals in a cohort can be easily measured. Even though the age of the dead individual does not provide any direct information on what causes death, it does signify some important aspects of the aging process including the stochastic nature of the life span and relationship of mortality to age. In contemporary *Drosophila* aging research, determination of life span and progression of aging is performed and compared by analyzing survivorship curves. Figure 2 provides a representative survivorship graph of aging over time in wild type and a symbolic mutant strain of *Drosophila*. Assuming the fact that shortening or lengthening of life span of an organism is the result of relative aging, comparative analyses among mean, median and maximum life span of different populations under different conditions could be treated as one of the factors to measure aging process. Considering the primary potential role of life span assay in aging research, it is also important to consider the interventions such as genetic and environmental factors during the analysis because in *Drosophila* a mild change will affect the age of the individuals.

4.2 Behavioral Assay

Several of the behaviors including locomotor activities, circadian rhythm, sleep patterns and even cognitive functions can be quantitatively assessed in *Drosophila* and functional deficits could be clearly observed and recorded in aging adults. Behavioral activities of *Drosophila* could be studied with two widely used simple methods, i.e. Rapid Iterative Negative Geotaxis (RING) and *Drosophila* activity monitoring (DAM) system (Nichols et al. 2012; Sun et al. 2013). RING is one of

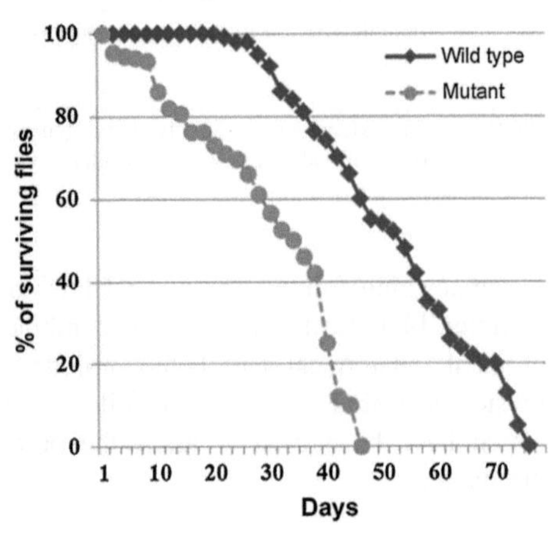

Fig. 2 Comparative survival curve of wild type and a symbolic mutant line of *D. melanogaster*. With advancing age, increased mortality and drastic decrease in the life span of the mutant population could be observed

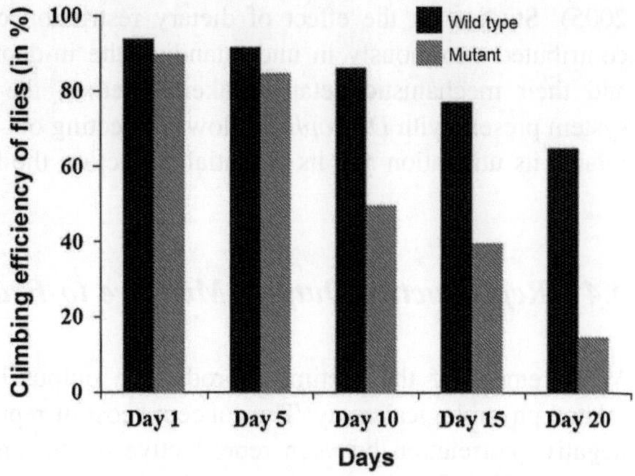

Fig. 3 Comparative climbing behavior of wild type and a symbolic mutant line of *D. melanogaster* with age. A decline in climbing activity in wild type population could be observed with increasing age, which is radically affected in mutant population

the most commonly used systems to assess the locomotor behavior of flies. Taking the advantage of inherent negative geotaxis response of the flies, this assay records the climbing ability of the flies against the gravity on the wall of empty vial after being tapped on the bottom of the container. Aging studies in *Drosophila* had reported gradual decline in the locomotor activities in almost all the species that have been studied (Iliadi and Boulianne 2010). The functional decline in the locomotor activity of the flies with age in wild type and in a symbolic mutant strain is depicted in Fig. 3. In the case of DAM based analysis, flies are kept individually in sealed activity tubes of DAM system and activity of the fly is measured based on the frequency of the event recorded each time a fly breaks an infrared light beam across the middle of the activity tube (Pfeiffenberger et al. 2010). It is mostly used to study circadian rhythm, sleep patterns, hypo and hyperactivity of flies. Moreover, more sophisticated video based tracking systems have been developed to analyze various fly behaviors including movement pattern, courting behavior etc. (Branson et al. 2009).

4.3 Assessing Aging on the Basis of Dietary Composition

Similar to other organisms, major environmental factors such as diet or food has a huge impact on the lifespan in *Drosophila* (Tatar et al. 2014). Thus, calorie or dietary restriction based studies is also among the important methods used to study aging in flies. Dietary restriction, by diluting all or specific components of food ingredients has two important impacts on physiology of flies: life span extension and reduction in the reproductive ability (Partridge et al. 2005). Intriguingly, studies based on dietary restriction allowed discovery of some fundamental regulators of aging (Partridge et al. 2005). It has been found that dietary restriction mediated life span extension is primarily controlled by some major metabolic pathways such as insulin/IGF-1 signaling, TOR (Target of Rapamycin) pathway etc. (Partridge et al.

2005). Studies on the effect of dietary restriction on aging and longevity have contributed enormously in understanding the in-depth of aging related pathways and their mechanistic details. Taken together, the powerful molecular genetic system present with *Drosophila* allows dissecting out the relationship between food intake, its utilization and its potential impact on the life span of the organism.

4.4 Reproductive Output: Measure to Evaluate Aging

Measurement of the lifetime reproductive output is another aspect of lifespan related physiological assay. The concept "cost of reproduction" in aging signifies a negative correlation between reproductive output and longevity of the organism (Tatar 2010). To measure the reproductive output of the flies, lifetime egg production in once mated female or number of progeny from the mating of male and female are measured. Selection experiments in *Drosophila* has resulted selection of long lived flies with decreased early reproduction and selection of the late life reproduction leads to the identification of lines with increased life span; moreover, long lived mutant females have reduced fecundity or fertility (Iliadi et al. 2012). Considering the cost of reproduction in *Drosophila* system, virgin/sterile females live longer than fertile control ones and fertile flies with increased reproduction results in increased susceptibility to stress (Salmon et al. 2001). The reason behind the extended life span with reduced reproduction may be probably the energy cost from lower or delayed egg production, as well as reduced cost of mating.

4.5 Stress as a Measure to Study Aging

Certain environmental stresses such as oxidative stress, starvation, crowded culture condition, heat or cold shock etc. have profound effect on aging and can be evaluated in *Drosophila*. According to the free radical theory of aging, an accumulative damage to the major biological macromolecules is the result of increasing level of cellular Reactive Oxygen species (ROS) (Harman 1992). In *Drosophila* system also, measurement of resistance against different stresses is another widely used method to study aging. Survival in the presence of a strong oxidizing agent like Paraquat (N,N′-dimethyl-4,4′-bipyridinium dichloride), an organic compound widely used as herbicide has been used to assess resistance against stress (Vermeulen et al. 2005). Paraquat feeding in flies induces various ROS and consequently, due to increased oxidative damage survival of flies declines. Starvation resistance is another interesting aspect which has been found to extend life span in *Drosophila* as it deals with their ability to manage with energy shortage (Minois and Le Bourg 1999). Moreover, stresses such as extremes of temperature have significant impact on *Drosophila* aging as both adversely affect their survival and life span (Minois and Le Bourg 1999).

4.6 Aging Analysis Utilizing Genetic Approaches

Since 1920 when Pearl's studies demonstrated for the first time that longevity in flies is heritable, genetic approaches remain as invaluable method for identifying the physiological mechanism of aging process. It includes alternation of single genes and careful analysis of the resultant phenotypes which affect the longevity and behavioral response of the flies. This method can be adapted to confirm any of the existing hypotheses based on the candidate gene approach or to explore new genes using the random gene alternation approach (Helfand and Rogina 2003b). In *Drosophila* system, a number of genetic approaches have been developed to generate mutation and to manipulate gene expression for aging studies. Some of such popular approaches include insertional mutagenesis by P-element, gene expression alternation by UAS-Gal4 system, inducible gene expression by Gene-switch Gal4 (GSG-UAS system) and gene knockdown by RNA interference (RNAi) strategy (Sun et al. 2013).

5 Contribution of *Drosophila* in Excavating Molecular and Genetic Mechanisms of Aging

As discussed earlier, *Drosophila* has been extensively utilized to unravel the molecular and genetic aspects of aging and longevity. In addition to genetic factors, environmental stresses which deteriorate cellular functions are largely known to be instrumental in instigating the process of aging. Therefore, several approaches have been undertaken to modify the genetic makeup of flies to promote extension of life span by modulating the cellular response to environmental stresses. Some of them have been briefly addressed in following texts:

5.1 Oxidative Stress

About a century ago the observation that animals with higher metabolic rates generally exhibit shorter life span led to the foundation of "Rate-of-living Hypothesis"; though the mechanistic association between metabolic rate and life expectancy was unknown during that period. Interestingly, in contrast to this theory, some species don't exhibit any strict inverse correlation between metabolic rate and longevity, particularly in birds and primates (Finkel and Holbrook 2000). In

1956, Denham Harman proposed mechanistically stronger theory of aging known as "Free-radical theory of aging"; according to which cumulative oxidative damage to biological macromolecules, brought about by ROS over the time results in deterioration of cellular function and stability, which ultimately act as a driving force for progression of aging (Harman 1956; Yadav et al. 2015). It was a decade later when enzyme superoxide dismutase (SOD) (enzyme with sole function of degeneration of superoxide anions) was discovered and first compelling evidence in the support of Harman's theory was presented (McCord and Fridovich 1969). Later in 1985 extensive research in redox biology concept of oxidative stress was used to symbolize the damage incurred in biological systems due to excessive ROS production and/or inadequate antioxidant defense (Sies and Cadenas 1985). Subsequently, the free-radical theory of aging was revised to the Oxidative stress theory of aging which subsequently emerged as the most persuasive theory in aging research (Pérez et al. 2009). A great deal of research work was performed to substantiate this theory but the results were inconsistent and partially challenging as well (Lapointe and Hekimi 2010). However, large number of findings from various organisms including *Drosophila* is reminiscent that decline in oxidative stress level is directly associated with increased life expectancy (Bokov et al. 2004). Therefore, intricate balance in the production of oxidants along with the capability of the organism to counteract the oxidative stress is critically linked to the progression of aging.

D. melanogaster has been widely used at the forefront to examine the oxidative stress hypothesis. The elementary idea behind such studies reside on the assumption that factors which aid in decreasing oxidative stress should have beneficial effects against aging and hence should result in enhancement of life expectancy. In support to this claim, linear correlation between oxidative stress resistance and longevity has been found in *Drosophila* utilizing various strains (Dudas and Arking 1995). In such cases, strains with extended life span exhibited either higher resistance to oxidative stress or had enhanced level of antioxidant enzymes (Dudas and Arking 1995; Harshman and Haberer 2000). For instance, reduced function of *Methuselah* (*mth*) gene which is a G-protein coupled receptor results in increase in life span. P-element insertion line of *mth* enhances longevity of the flies by approximately 35 % (Lin et al. 1998). In addition to increase in life span, this gene also provides tolerance against several stresses including high temperature, dietary paraquat (intracellular free radical generator) and starvation (Lin et al. 1998). Though, the explicit function of the *mth* is still unknown but it has been proposed to be involved in transmitting cues for regulating stress response pathways (Lin et al. 1998). Figure 4 attempts to provide a schematic representation of various signaling cascades which are known to modulate aging and longevity in *D. melanogaster*.

Relationship between oxidative stress tolerance and longevity has been tested in *Drosophila* by overexpressing antioxidant genes, utilizing transgenic approaches. Increase in the expression of glutathione reductase (GSH) antioxidant enzyme (involved in formation of reduced glutathione) results in high level of oxidative stress tolerance and prolonged life span in flies exposed to hyperoxic conditions, though no effect on the longevity was observed when the flies were reared at

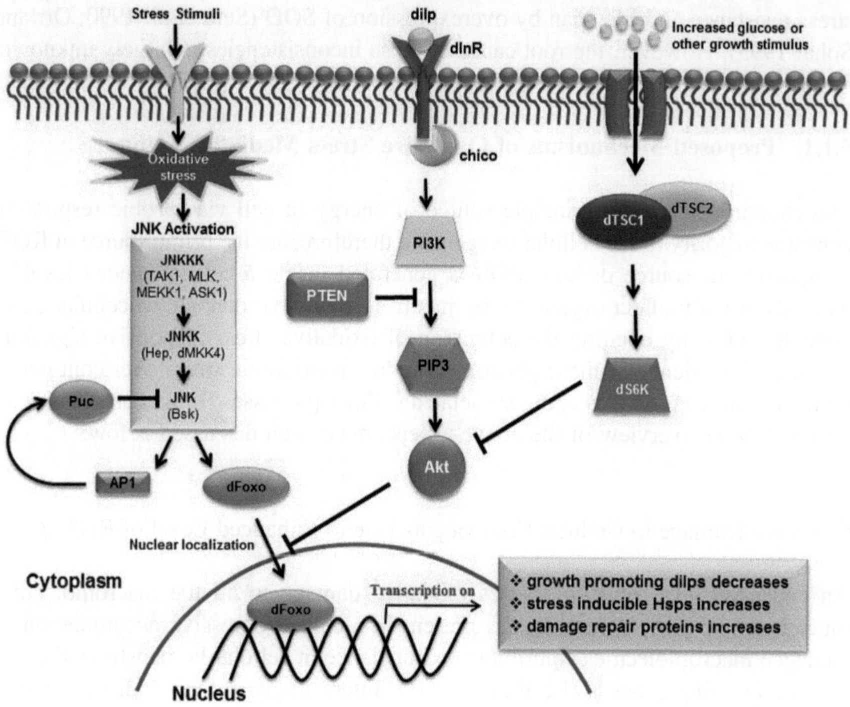

Fig. 4 Schematic representation of various stimulus and signaling pathways which modulate progression of aging and longevity in *D. melanogaster* (please refer text for detail)

normoxic condition (Mockett et al. 1999). In addition, it has also been demonstrated that decrease in expression of antioxidant enzymes such as superoxide dismutase (SOD) (scavenges superoxide anion radicals) and catalase (involved in eradication of H_2O_2), shortens the life span indicating the significance of ROS detoxification on life expectancy (Phillips et al. 1989; Phillips and Hilliker 1990; Missirlis et al. 2001; Kirby et al. 2002). However, in this context it is also important to note that since the mutation was prevalent even during the fly development, and therefore, a decrease in life span could also be partially attributed to the damage accumulated during development and might not be solely due to oxidative stress. To overcome this discrepancy, studies were focused on increasing the fly expectancy by overexpressing the antioxidant enzymes. In this respect numerous studies displayed higher oxidative stress resistance and modest enhancement in life span, either by combined overexpression of SOD and catalase or SOD alone (Orr and Sohal 1994; Parkes et al. 1998; Sun and Tower 1999). A noteworthy report in these findings was 40–50 % increase in life span by overexpressing human SOD in motor neurons of fly (Parkes et al. 1998). Achieving modest increase in life span by overexpression of antioxidant enzymes supports oxidative stress theory of aging. However, on the other side some studies reported slight or insignificant enhancement of oxidative stress resistance and life span by overexpression of SOD (Seto et al. 1990; Orr and Sohal 1993). However, the root cause of these inconsistencies is largely unknown.

5.1.1 Proposed Mechanisms of Oxidative Stress Mediated Aging

Mitochondria being the principle source of energy in cell via aerobic respiration consume majority of the cellular oxygen and therefore, are the prime source of ROS. Irrespective of source or how ROS is generated inside a cell, enhanced level of oxidants broadly effect organisms by incurring oxidative damage to cellular components and/or by eliciting the activation of oxidative stress responsive signaling cascades. Prevalence of these phenomenon due to oxidative stress over continuous periods of time stimulates aging associated cellular processes (Finkel and Holbrook 2000). A brief overview of the above aspects have been discussed below:

Oxidative Damage to Cellular Components Due to Enhanced Level of ROS

An enhanced level of ROS causes oxidative damage to all the macromolecules (nucleic acids, proteins and lipids) present in a cell. Progressive accumulation of damaged macromolecules contributes to imbalance in cellular homeostasis, thereby instigating aging process (Le Bourg 2001). Interestingly amongst all the cellular organelles, mitochondria in spite of being the major source of ROS, are also the key targets of oxidants. Moreover, due to the close proximity of mitochondrial elements to ROS production site, they are more susceptible to the damage by ROS. Further, lack of histone protection and repair mechanism in mitochondrial DNA aggravates their susceptibility to ROS mediated damages. All this cumulatively results in mitochondrial dysfunctioning and has been profoundly linked to manifestation of aging progression (Sohal 2002; Wallace 2005; Yadav et al. 2013).

There have been several studies carried out in *Drosophila* where correlative data on age associated changes in the structure and functions of mitochondria, are suggestive of the idea that gradual mitochondrial dysfunctioning is associated with aging process (Wallace 2005). One such study investigating the effect of aging on *Drosophila* flight muscles reported a specific reorganization of mitochondrial cristae under oxidative stress, with aging (Walker and Benzer 2004). Aging induces local rearrangement of the cristae in a “swirl” like pattern in individual fly mitochondria (Walker and Benzer 2004). Rapid and extensive accrual of the same pathological condition was witnessed even in young flies under the condition of severe oxidative stress. From functional aspect of this pathological condition, cristae associated with swirling pattern were found to have reduced enzymatic activity of cytochrome c (COX) or complex IV, which is an important respiratory enzyme present in mitochondria. Furthermore, occurrence of swirls is accompanied by alteration in the structural conformation of the cytochrome c and extensive apoptosis of the cells present in the tissue of flight muscles in *Drosophila* (Wallace 2005; Cho et al. 2011).

Electron transport chain (ETC) occurring in mitochondria is one of the most vital processes which is essential for cellular homeostasis; primarily because this process is associated with energy production in the cell. A comprehensive study examining the ETC functioning with aging in *Drosophila* reported a decrease in several aspects of ETC such as electron transport and respiration with gradual increase in aging (Ferguson et al. 2005). Interestingly, compared to the other mitochondrial ETC enzymes which were examined, age-associated reduction was predominantly found in the activity of COX (Ferguson et al. 2005). Also, drug mediated inactivation of COX in mitochondria obtained from young flies result in enhanced ROS production. These observations suggest that ROS induced mitochondrial impairment results in further enhanced production of ROS which exaggerates the mitochondrial damages, forming a "vicious cycle" and thereby acting as driving force in aging and age associated impairments (McCarroll et al. 2004).

ROS Mediated Activation of Oxidative Stress Response Signaling Cascades

Oxidative stress triggers activation of several signaling cascades such as extracellular signal-regulated kinase (ERK), p38 mitogen-activated protein kinase (MAPK), p53 activation, nuclear factor (NF)-kB signaling cascade, c-Jun amino-terminal kinase (JNK), the phosphoinositide 3-kinase (PI(3)K)/Akt pathway etc. (Fig. 4) as a mechanism to combat stress (Finkel and Holbrook 2000). Amongst them, JNK pathway activated by ROS or other stimuli has been recognized as an evolutionarily conserved cascade which can potentially increase life span in flies by triggering protective gene expression program to alleviate toxic effects of oxidative stress (Wang et al. 2003; 2005).

In vertebrates, each components of JNK pathway is represented by huge gene families; however in *Drosophila,* JNK signaling is significantly less complicated thereby making its genetic analysis much simpler than other model organisms (Fig. 4) (Johnson and Nakamura 2007; Igaki 2009). JNK pathway is a part of MAPK signaling cascade which in *Drosophila* constitutes several JNK kinase kinase (JNKKK) [i.e. TGF-β Activated Kinase 1 (TAK1), Mixed Lineage Protein Kinase 2/Slipper (MLK), MEK Kinase1 (MEKK1) and Apoptotic signal-regulating Kinase 1 (ASK1)], two JNK kinase (JNKK) [i.e. Hemipterous (Hep) and dMKK4)] and one JNK [Basket(Bsk)] (Boutros et al. 2002; Chen et al. 2002; Geuking et al. 2009; Biteau et al. 2011). Activation of the pathway by stress stimuli including ROS results in activation of transcription factors AP-1 and dFoxo (*Drosophila* forkhead transcription factor) by Bsk phosphorylation which instigates changes in gene expression resulting in stress specific cellular response. JNK pathway is regulated by negative feedback loop where puckered (*puc*), one of the target gene of AP-1, dampens JNK signaling by its specific JNK phosphatase activity (Wang et al. 2005; Biteau et al. 2011). Studies in *Drosophila* revealed that either reduction in the dosage of *puc* gene product or overexpression of JNKK/Hep in neuronal tissues enhances basal JNK signaling levels resulting in heightened oxidative stress tolerance and increased life span (Zeitlinger and Bohmann 1999; Wang et al. 2003).

Along the same lines, mutant flies for JNKK/Hep gene displayed higher sensitivity towards oxidative stress and were observed incapable of eliciting JNK signaling dependent transcriptional factor induced stress response (Fig. 4) (Wang et al. 2003).

It has been demonstrated that availability of dFoxo transcription factor is essential to achieve JNK signaling mediated increased longevity in *Drosophila*. There is an antagonistic relationship between JNK and insulin/insulin-like growth factor (IGF)-like signaling (IIS) pathways (Wang et al. 2005). JNK inhibits IIS both cell autonomously and systemically (endocrine mechanism) to control life expectancy in *Drosophila*. Cell autonomously, JNK inhibits IIS by promoting nuclear localization of dFoxo, inducing transcription of genes involved in growth control, stress defense and damage repair (Wang et al. 2005; Hotamisligil 2006; Biteau et al. 2011). JNK inhibits IIS signaling systemically by repressing the expression of IIS ligand, *Drosophila* insulin-like peptide (dilp2) in insulin-producing neuroendocrine cells present in the fly brain (Karpac and Jasper 2009; Wang et al. 2005). Therefore, dFoxo and dilp2 dependent antagonistic relationship between JNK and IIS fairly explains the effect of oxidative stress on aging phenomenon.

5.2 *Molecular Chaperones*

As discussed earlier, aging is a complex process involving both genetic and non-genetic factors. In natural populations, several of the environmental factors including extreme temperatures, starvation, oxidative stress and other stresses etc. influence the survival capacity of organisms. Life span of an individual is the ability to withstand against these stresses which result in irreversible cellular damages; therefore, longevity of the organism is largely determined by the stress response of the individual. According to several proposed theories, aging is the result of an imbalance between damage and repair of cellular macromolecules (Vijg 2008), and moreover, with increasing age the response of the organism and its cells towards such damage tends to decline (Campisi and Vijg 2009). Generally, proteins are particularly more subjected to aging-related damages, including cleavage of the polypeptide chain, covalent modification of amino acid side chains, oxidative lesions, crosslinking, and denaturation (Stadtman 2006). Therefore, correct synthesis, proper folding of nascent and denatured protein, and turnover of proteins become one of the most crucial functions in cellular physiology, and failure of the stringent regulation of the cellular protein quality control system results in proteotoxicity, which is a key component of aging and aging related diseases (Morimoto and Cuervo 2009). Cellular protein quality control system is a combined network of molecular chaperones and their regulators, the ubiquitin-proteasome system (UPS) and autophagy system, allowing the proper folding, timely removal of the misfolded and aggregated proteins (Chen et al. 2011; Amm et al. 2014). Interestingly, despite of the unclear molecular mechanism(s) of aging process, increase in the damage of cellular macromolecules including proteins, nucleic acids, lipid etc. due to the upsurge of cellular oxidative stress stands one of the most accepted theories of aging. Aging dependent progressive accumulation of abnormal mitochondria in several *Drosophila* tissues further supports the view of increased production of ROS with advancing age (Walker and Benzer 2004; Chistiakov et al. 2014). It appears that with increased oxidative stress and reduction in ATP synthesis, mitochondrial dysfunction stands as one of the core reasons behind aging related abnormal protein creation and accumulation.

Since the time heat shock response and genes encoding for molecular chaperones were discovered for the first time in *Drosophila*, the functions of these proteins, their regulation in heat shock response and their potential correlation with aging and longevity has been a topic of immense interest (Ritossa 1962, 1996). Despite their constitutive expression during normal homeostasis, molecular chaperones are also known as stress proteins or Heat shock proteins (Hsps) because of their induced expression during stress condition(s). Based on the sequence conservation and molecular weight, Hsps have been divided into several families. Conventionally, principal Hsps range in molecular mass from 15 to 110 kDa which are grouped into 5 major families, viz. Hsp100 (100–104 kDa), Hsp90 (82–90 kDa), Hsp70 (68–75 kDa), Hsp60 (58–65 kDa) and the small Hsp (15–30 kDa) (sHsps) families (Sarkar et al. 2011). Similar to other organisms, *Drosophila* also harbors homologs of the Hsp families which include Hsp83 (Hsp90), Hsp/Hsc70 complex (Hsp70 family), Hsp60, Hsp40 and sHsp (Hsp22, Hsp23, Hsp26 etc.) (Morrow and Tanguay 2003). In addition to its role in assisting denovo folding of the nascent polypeptide chains, Hsps are also defined by their ability to bind and refold the denatured proteins (Morrow et al. 2006). Therefore, induced expressions of Hsps are found in response to stresses that cause protein denaturation, such as heat and oxidative stresses (Morimoto 2008). Expression of the Hsps is mediated by binding of heat shock transcription factor (HSF) to heat shock response elements (HSEs) localizing at promoters of the heat shock genes, and activates their high-level transcription (Voellmy 2004). Interestingly, subsets of heat shock genes are also induced by oxidative stress through the JNK pathway and the transcription factor dFoxo (Wang et al. 2005).

During development and throughout the life span of *Drosophila*; Hsps, especially sHsps exhibit well-regulated, distinct and stage specific expression dynamics, though, upon exposure to environmental stresses like heat, Hsps shows upregulated expression (Morrow and Tanguay 2003). Interestingly, definite role of Hsps in aging and increased sensitivity of the aged flies to environmental stress have emerged from the comparative analysis of the heat shock response between young and old flies (Fleming et al. 1988). Comparative analysis of the expression profile of old and young flies revealed a greater abundance of damaged proteins in the old flies. Interestingly, detection of the same set of induced proteins in young flies fed with canavanine (an arginine analogue used to mimic accumulation of damaged proteins) as otherwise found only in old flies suggests increased sensitivity due to accumulation of aging mediated damaged proteins (Fleming et al. 1988; Niedzwiecki et al. 1991). Consistent to the above conclusion, even in unstressed flies, enhanced expression of Hsps has been found during normal fly aging in tissue-specific patterns (Morrow and Tanguay 2003). For instance, up-regulation of *hsp22* and *hsp70* at both, RNA and protein level while hsp23 at RNA level could be observed during normal *Drosophila* aging (Morrow and Tanguay 2003). With the aim of elucidating transcriptional changes accompanying the aging process, studies based on genome wide gene profiling in *Drosophila* have revealed aging associated upregulation of Hsps in aged flies (Curtis et al. 2007; Pletcher et al. 2002; Zou et al. 2000). Interestingly, dramatic upregulation of subset of Hsps including Hsp70 and sHsps and the genes for innate immune response was reported in old flies, in contrast, down-regulation of genes involved in energy synthesis and electron transport chain was found in same set of flies (Curtis et al. 2007; Pletcher et al. 2002; Zou et al. 2000). Moreover, extensive overlap between the gene expression profile of aged flies and young flies exposed to oxidative stress, further suggests the potential relationship between aging and oxidative stress (Zou et al. 2000).

The beneficial effect of Hsps on longevity is also evident from mild stress experiments known as "hormesis" in *Drosophila*, which activates the stress response without causing cellular damages (Minois 2000). Exposing the organisms to sub-lethal stress, induces hormetic effect through the modulation of heat shock response and helps the animals to live longer by counteracting negative effects of aging (Minois 2000). Besides, young adults of *Drosophila* strains with increased life span also exhibit intrinsic increased expression of sHsps, further suggesting that enhanced expression of Hsp might have a role in favoring longevity (Kurapati et al. 2000). Consistent to the above observation, mutation in *hsp70* or *hsp22* shows reduction in the adult fly survival, and these mutants with decreased lifespan also become more sensitive to stress. The role of Hsps in longevity was further confirmed by HDAC inhibitors mediated enhanced expression of hsp70 and sHsp, which in turn increase the life span of adult flies (Zhao et al. 2005). Remarkably, several of the independent studies have revealed decreased survival of the flies against heat and other stresses in the *Drosophila* mutants of *hsp22* (Morrow et al. 2004a) and all six copies of the *hsp70* (Gong and Golic 2006). In addition, *hsp83* mutant flies become more sensitive to the toxic effects of stresses like sleep deprivation (Shaw et al. 2002).

Interestingly, unlike the sHsps, major Hsps like Hsp70, Hsp60 etc. have failed to demonstrate any substantial effect on longevity, except reduced mortality rates upon mild stress, enhanced heat tolerance and a small increase in overall life span (Tatar et al. 1997; Minois et al. 2001). Among several of the sHsp in *Drosophila*, four of the sHsps i.e. Hsp27, Hsp26, Hsp23 and Hsp22 are well characterized and result in substantial life span extension upon tissue specific over-expression (Morrow and Tanguay 2003; Wang et al. 2003; Liao et al. 2008; Tower 2011). For instance, ubiquitous expression of Hsp22 in motor neuron increases the life span by 30 % and these flies exhibit increase resistance against stress and improved locomotor activity (Morrow et al. 2004b). Therefore, because of the ubiquitous nature of Hsps and its crucial role in variety of cellular processes by interacting with many different proteins, it can be concluded that the widespread outcome of aging is the consequence of the aging associated chaperone failure, and therefore, molecular chaperones itself represent one of the vital intrinsic components to govern the aging process in the living system.

5.3 Insulin/Insulin-like Growth Factor (IGF)-like Signaling (IIS)/TOR Pathway in Regulation of Longevity and Aging

The Insulin/insulin-like growth factor (IGF)-like signaling (IIS) pathway has been long known to serve an established role of regulating somatic growth and development (Butler and Le Roith 2001), reproduction (Netchine et al. 2011), stress resistance (Holzenberger et al. 2003), metabolic homeostasis (Vowels and Thomas 1992; Saltiel and Kahn 2001) and even in aging and longevity (Partridge and Gems 2002; Tatar 2003; Kenyon 2005) in most organisms. There has been substantial evidences which suggest that compromised IIS signaling by introducing mutation (s) in the component(s) of the IIS pathway increase lifespan. On contrary, mutations that tend to shorten lifespan, have been proposed to do so by introducing pathological changes in the cell rather than by speeding up the process of normal aging (Giannakou and Partridge 2007).

The IIS pathway was first elucidated in *Drosophila* as one of the major pathways regulating growth and size of cells (Leevers et al. 1996). However, a plausible link between IIS pathway and longevity originated from studies on *C. elegans* when *daf-2* (a homolog of the insulin/IGF-1 receptor) mutants were found to extend lifespan (Kimura et al. 1997). Later, similar findings were reported in *Drosophila* when null mutants of insulin receptor substrate gene *chico* were found to be responsible for lifespan extension to as much as 48 % in homozygous female flies (Clancy et al. 2001). Interestingly, *chico* heterozygous female flies, though, exhibit an increase of 31 % in median lifespan, but their capacity to resist paraquat-induced acute oxidative stress was found more than their homozygous counterparts. As opposed to females, homozygous *chico* males are short-lived as compared to heterozygous males. Notably, long-lived homozygous mutants displayed higher levels of lipid and SOD activity (Böhni et al. 1999; Clancy et al. 2001; Kabil et al. 2007). Such contradictory observations, therefore, suggest that the trait of stress resistance may not contribute to the phenomenon of longevity via IIS signaling in flies. This can be justified by the fact that free radical generation and oxidative stress responses could be associated with a host of other reasons than just IIS signaling.

Another noteworthy IIS-linked mutation found to increase life span in *Drosophila* was that of the insulin like receptor (dInR). Adult flies with a mutated copy of the dInR gene tend to live longer than their wild type counterparts (Tatar et al. 2001). In this case as well, long-lived *inr* mutants exhibit higher triglyceride content and SOD activity. However, level of lipid content and SOD activity has also been found to be raised in some short lived mutants. In view of above, it may be postulated that in addition to increase in SOD activity and lipid levels, other pleiotropic effects are necessarily involved in fly longevity. Furthermore, reduced expression of *Drosophila* insulin-like peptides (*dilps*), the ligands for dInR (Grönke et al. 2010) or increased expression of dPTEN (*Drosophila* phosphatase and tensin homolog), the negative regulator of insulin pathway (Hwangbo et al. 2004), also results in lifespan extension. dPTEN was shown to be doing so by antagonizing the action of the signal transducer PI3K (phosphatidylinositol-3-kinase) leading to nuclear localization of dFoxo, which in turn downregulates the expression of tissue specific chaperones and *dilps*, thereby completing the loop. Interestingly, increase in lifespan by activation of JNK signaling in response to various stresses as discussed above, also mediates its effect via dFoxo. The mechanism that dFoxo follows in this case comprises at-least in part of reduced IIS, since upregulation of JNK signaling in brain median neurosecretory cells (MNCs) has been shown to be linked with reduced transcript levels of *dilps* 2 and 5 (Wang et al. 2005). This is an interesting finding owing to the fact that MNCs are the site of *dilp* 2, 3 and 5 in the brain. Moreover, low levels of *dilp* 5 and subsequent lifespan extension had also been demonstrated in flies subjected to dietary restrictions (Min et al. 2008). Moreover, upregulation of dFoxo itself has been found to increase lifespan in *Drosophila* (Giannakou et al. 2004).

Furthermore, *dilp*-producing MNCs in adult *Drosophila* brain that integrate external signals to the IIS have also been implicated in influencing longevity. Flies carrying ablated MNCs exhibited up to 33.5 % increase in lifespan which was however accompanied by an age-related reduction in egg laying capacity (Broughton et al. 2005). These flies also demonstrated enhanced levels of circulating glucose along with stored carbohydrates and lipids. They could also resist paraquat- and starvation-induced stresses more efficiently as compared to wild type, though such flies are more sensitive to heat and cold stresses (Broughton et al. 2005). In view of above findings, it may be postulated that some compensatory alterations of related pathways which interact with IIS might be needed in order to balance out the undesirable effects of reduced IIS, so that longevity can be increased without any fitness cost.

One of the major pathways that interact with IIS to regulate growth and longevity in *Drosophila* is TOR pathway (Oldham and Hafen 2003). Two major complexes instigate the TOR pathway—TOR complex 1 (TORC1) and TORC2. TORC1 is sensitive to rapamycin and is implicated in controlling the temporal aspects of growth within a cell (Um et al. 2006) whereas TORC2 is insensitive to rapamycin and is involved in controlling the spatial facets of cellular growth (Jacinto et al. 2004). Reduction in TOR signaling by ubiquitously upregulating dTsc1 and dTsc2, or expression of a dominant negative variant of TOR or expression of a mutated dS6K, a major downstream kinase of the TOR pathway, led to substantial increases in *Drosophila* lifespan (Kapahi et al. 2004). Moreover, rapamycin-mediated inhibition of TOR signaling was also shown to prolong lifespan in flies by as much as 10 % (Bjedov et al. 2010). Rapamycin has been suggested to do so by inactivating TORC1, and by lowering the rate of protein translation in the cell and inducing autophagy (Bjedov et al. 2010). Notably, the long lifespan of *dtsc2* mutants cannot be further extended by subjecting the flies to caloric restriction (Kapahi et al. 2004). Thus, the mechanisms of life extension by inhibited TOR signaling and dietary restriction could be overlapping in nature.

Some of the most apparent evidences of interaction between the IIS and TOR pathway have been elucidated in *Drosophila* models of neurodegenerative disorders. Reduced activity of the IIS/TOR pathway has been found to suppress mutant proteins mediated neurotoxicity in a variety of neurodegenerative disease models (Hirth 2010). Though, the precise modulations required for IIS/TOR signaling to bring about neuroprotection remain elusive. It also remains uncertain whether specific modulations protect against specific forms of neurotoxicity or there is a common link between neuroprotection and IIS/TOR pathways.

5.4 Dietary Restriction

As mentioned earlier, dietary restriction is a phenomenon linked to increase in life expectancy by limiting the nutrient intake. The process of dietary restriction controlling aging is conserved across the species (Piper and Partridge 2007). Influence of dietary restriction on aging has been center of curiosity among the researchers for deciphering the underlying genetic and molecular mechanism(s) involved. One of the hypotheses to explain the role of dietary restriction on aging states that it reduces the body metabolic rate thereby decreasing ROS generation which in turn slows down the aging process. Though this ideology is consistent with the existing relationship between oxidative damage and aging but experimental validation is still awaited. However, several nutrient sensing pathways such as Sirtuin (Sir2) and TOR signaling which operate under indirect control of IIS signaling have been identified to be crucial for dietary restriction mediated life span enhancement.

Sir2 are the members of highly conserved protein family which act as NAD-dependent deacetylases and target both, histone as well as non-histone proteins. They have been implicated as one of the key mediators in dietary restriction triggered increased life expectancy (Dali-Youcef et al. 2007). Subsequently several studies have demonstrated that flies overexpressing dSir2 proteins have higher life expectancy (Rogina and Helfand 2004; Bauer et al. 2009). The maximum increase in mean life span in flies was 57 %, which was achieved by ubiquitous overexpression of dSir2 under the influence of tubulin-Gal4 driver (Rogina and Helfand 2004). Above studies highlight the conserved role of dsir2 in facilitating the favorable effect of dietary restriction on fly life expectancy.

Similar to Sir2, IIS and TOR signaling pathway have been well characterized in demonstrating their noteworthy contribution in fly aging process by coupling growth to nutrition (Tatar et al. 2001; Kapahi et al. 2004; Broughton et al. 2005). It has been proposed that mutants of these signaling pathways extend the life span primarily by slowing down the growth and rate of metabolism (Tatar et al. 2001;

Broughton et al. 2005). Remarkably, it has been shown that reduction in life expectancy due to dFoxo mutation in flies can be compensated by dietary restriction, which further highlights the crosstalk between IIS and TOR pathway in regulating aging (Giannakou et al. 2008). However, despite availability of large information, the accurate role of TOR and dietary restriction in aging is still illusive and further investigations are expected to generate novel insights.

6 Aging and Neurodegeneration

Aging is one of the major risk factors for onset of brain related neurodegenerative disorders such as Alzheimer's disease (AD), Parkinson's disease (PD), Huntington's disease (HD) etc. Neuronal loss, shrinkage of cell bodies and axons of neuronal cells and loss of synapse collectively leads to reduced brain volume and weight in aging individuals (Reiter et al. 2001). Subsequently, progressive deterioration of brain function leads to cognitive decline, memory loss, movement disorders and finally to functional decline and death. With a hastily increasing aging population and due to lack of effective treatment measures, these disorders have emerged as major economic and social burden. Therefore, in view of the fact that these disorders show substantial interference with aging; in-depth investigation on age-related molecular mechanisms or pathways may potentially help in developing novel therapeutic strategies.

Although, it appears quite rational to hypothesize that disease related proteins enhance disease toxicity by accelerating the aging process, however, it is still unclear whether aging related changes are responsible for driving neuronal pathology or both aging and disease associated proteins act synergistically to develop neuronal dysfunction. For instance, in *C. elegans*, mutation that extends longevity in poly(Q) disease reveals age dependent reduction in protein aggregate formation and toxicity, consequently testifying the effect of aging in poly(Q) mediated cellular dysfunction (Morley et al. 2002). Several reports including our own findings demonstrate progressive aggravation of poly(Q) mediated neurotoxicity in age dependent manner. Targeted expression of Htt-93(Q) in *Drosophila* eye exhibits cellular degeneration characterized by retinal depigmentation and cellular toxicity. Our studies on individual flies expressing Htt-93(Q) transgene during aging suggest that the magnitude of retinal depigmentation and cellular toxicity progressively increases with age (Fig. 5). Moreover, involvement of common signaling networks in longevity and mitigation of poly(Q) toxicity raises the prospect that slowing down aging may act as a neuroprotective measure. Therefore, in order to cultivate novel strategies to prevent onset and progression of such deadly disorders, it will be interesting to explore how aging dysfunction and poly (Q)mediated neuropathology are interlinked and how they interact during disease pathogenesis.

As stated earlier, all eukaryotic life forms have well evolved protein quality control machinery, which includes chaperone network, ubiquitin-proteosome and

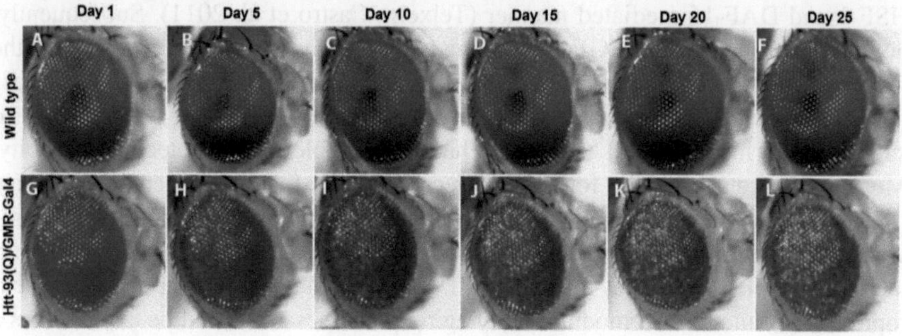

Fig. 5 Progressive increase in poly(Q) mediated neurodegeneration in *Drosophila* eye with advancing age. Adult wild type and poly(Q) diseased fly have been chased from day 1 to day 25. With aging, while wild type eye appears continuously normal (A-F), *GMR-Gal4* driven eye specific expression of Htt-93(Q) exhibits progressive increase in eye depigmentation and cellular degeneration (G-L)

lysosome-mediated autophagy system. Stringency of these systems is essential for post translational modifications, protein folding, stress response and clearance/translocation of damaged proteins (Soti and Csermely 2003; Arslan et al. 2006). Induction and functional capacity of chaperones and cellular proteasome system gets distorted during aging and disease stress condition; therefore, the post mitotic neurons become susceptible to toxic protein aggregates and ultimately, leads to neurodegeneration. Therefore, it is not surprising that overexpression of Hsps ameliorates the neurotoxicity and age related cellular impairments. Overexpression of Hsp70 in *Drosophila* poly(Q) disease models suppresses neurotoxicity by restoring axonal transport, cell death and ultimately extends the life span (Muchowski and Wacker 2005). In addition, role of Hsp70 and Hsp40 in regulating poly(Q) aggregation and toxicity has also been demonstrated in poly(Q) models of *S. cerevisiae*, *C. elegans* and mouse (Muchowski et al. 2000; Cummings et al. 2001). Several mechanisms have been proposed to explain progressive decline in the level of Hsps in neurodegenerative diseases, including transcriptional deficit of Hsps expression via the toxic misfolded protein and sequestration of cellular soluble Hsps along with the toxic aggregates to form IBs. Evidences like CBP mediated transcriptional impairment of Hsp70 in *Drosophila* via reduction of HSF-1 activity further support the transcriptional deficit hypothesis (Hands et al. 2008). Therefore, it appears that mis-regulation of molecular pathways and several factors which are responsible for cellular protein quality control might be the risk factor for disease occurrence, which could be considered while designing novel therapeutic strategies.

In addition to molecular chaperones, potential involvement of insulin/IGF-1 signaling in protein aggregation and toxicity has also been reported. Studies on *C. elegans* suggested a direct link of insulin/IGF-1 signaling in protein aggregation for the first time, when it was demonstrated that insulin/IGF-1 also protects the worms from motility impairment by neutralizing the poly(Q) aggregation and toxicity in

HSF-1 and DAF-16 mediated manner (Teixeira-Castro et al. 2011). Subsequently, downregulation of insulin/IGF-1 signaling pathway was demonstrated to reduce the level of toxic aggregates in poly(Q) mediated Machado-Joseph disease (MJD) (Cohen 2012). Several studies performed on mouse HD and AD models also suggest that insulin/IGF-1 signaling has remarkable neuroprotective capacity. Mouse knockout models for IGF-1 receptor and Insulin Receptor Substrate (IRS) have shown rescue the animals from poly(Q) induced behavioral impairments along with learning and memory deficit (Raj et al. 2012). Collectively, it is increasingly clear now that insulin/IGF-1 signaling plays an essential role in neuroprotective function via modulation of aging processes and could be exploited as a novel pathway to develop new therapeutic strategies.

7 Concluding Remarks

Though it is increasingly clear now that aging is regulated by explicit signaling pathways, however, whether the influence of these signals is applicable to an organism "as whole" or operating at tissue specific manner, which then affects aging systemically remains to be determined. In this context it is also interesting to note that a number of genetic manipulations which extend life span in *Drosophila* and other species have sex-specific preferences. Also, dietary restriction results in a greater extension of life span in female versus male flies. Therefore, exactly how these various pathways/factors control life span and influence the phenomenon of aging is still a "great scientific mystery". The dramatic progress made in recent years utilizing various model organisms has demonstrated the feasibility of decoding this mystery and further studies are expected to reveal the insights of the biological aging and longevity.

Acknowledgments Research programmes in the laboratory have been supported by grants from the Department of Science and Technology (DST), Department of Biotechnology (DBT), Government of India, New Delhi; DU/DST-PURSE scheme and Delhi University R & D fund to SS. RY, SIC, KR and Nisha are supported by DST-INSPIRE, UGC-SRF, UGC-JRF and DBT-JRF fellowships respectively.

References

Alpatov WW, Pearl R (1929) Experimental studies on the duration of life. XII. Influence of temperature during the larval period and adult life on the duration of the life of the imago of *Drosophila melanogaster*. Am Nat 63:37–67

Amm I, Sommer T, Wolf DH (2013) Protein quality control and elimination of protein waste: the role of the ubiquitin-proteasome system. Biochim Biophys Acta 1843:182–196

Arking R (1991) Biology of ageing: observations and principles. Prentice Hall, Englewood Cliffs, NJ

Arslan MA, Csermely P, Soti C (2006) Protein homeostasis and molecular chaperones in aging. Biogerontology 7:383–389

Bauer JH, Morris SNS, Chang C, Flatt T, Wood JG, Helfand SL (2009) dSir2 and Dmp53 interact to mediate aspects of CR-dependent life span extension in *D. melanogaster*. Aging 1:38–49

Bishop NA, Lu T, Yankner BA (2010) Neural mechanisms of ageing and cognitive decline. Nature 464:529–535

Biteau B, Karpac J, Hwangbo D, Jasper H (2011) Regulation of *Drosophila* lifespan by JNK signaling. Exp Gerontol 46:349–354

Bjedov I, Toivonen JM, Kerr F, Slack C, Jacobson J, Foley A, Partridge L (2010) Mechanisms of life span extension by rapamycin in the fruit fly *Drosophila melanogaster*. Cell Metab 11: 35–46

Böhni R, Riesgo-Escovar J, Oldham S, Brogiolo W, Stocker H, Andruss BF, Beckingham K, Hafen E (1999) Autonomous control of cell and organ size by CHICO, a *Drosophila* homolog of vertebrate IRS1-4. Cell 97:865–875

Bokov A, Chaudhuri A, Richardson A (2004) The role of oxidative damage and stress inaging. Mech Ageing Dev 125:811–826

Boutros M, Agaisse H, Perrimon N (2002) Sequential activation of signaling pathways during innate immune responses in *Drosophila*. Dev Cell 3:711–722

Brand AH, Perrimon N (1993) Targeted gene expression as a means of altering cell fates and generating dominant phenotypes. Development 118:401–415

Branson K, Robie AA, Bender J, Perona P, Dickinson MH (2009) High-throughput ethomics in large groups of *Drosophila*. Nat Methods 6:451–457

Broughton SJ, Piper MD, Ikeya T, Bass TM, Jacobson J, Driege Y, Martinez P, Hafen E et al (2005) Longer lifespan, altered metabolism, and stress resistance in *Drosophila* from ablation of cells making insulin-like ligands. Proc Nat Acad Sci USA 102:3105–3110

Butler AA, Le Roith D (2001) Control of growth by the somatropic axis: growth hormone and the insulin-like growth factors have related and independent roles. Annu Rev Physiol 63:141–164

Campisi J, Vijg J (2009) Does damage to DNA and other macromolecules play a role in aging? If so, how? J Gerontol A Biol Sci Med Sci 64:175–178

Chen B, Retzlaff M, Roos T, Frydman J (2011) Cellular strategies of protein quality control. Cold Spring Harb Perspect Biol 3:a004374

Chen W, White MA, Cobb MH (2002) Stimulus-specific requirements for MAP3 kinases in activating the JNK pathway. J Biol Chem 277:49105–49110

Chistiakov DA, Sobenin IA, Revin VV, Orekhov AN, Bobryshev YV (2014) Mitochondrial aging and age-related dysfunction of mitochondria. Biomed. Res. Int. 2014 238463

Cho J, Hur JH, Walker DW (2011) The role of mitochondria in *Drosophila* aging. Exp Gerontol 46:331–334

Clancy DJ, Gems D, Harshman LG, Oldham S, Stocker H, Hafen E, Leevers SJ, Partridge L (2001) Extension of life-span by loss of CHICO, a *Drosophila* insulin receptor substrate protein. Science 292:104–106

Cohen E (2012) Ageing, protein aggregation, chaperones, and neurodegenerative disorders: mechanisms of coupling and therapeutic opportunities. Rambam Maimonides Med J 3:e0021

Cummings CJ, Sun Y, Opal P, Antalffy B, Mestril R, Orr HT, Dillmann WH, Zoghbi HY (2001) Over-expression of inducible HSP70 chaperone suppresses neuropathology and improves motor function in SCA1 mice. Hum Mol Genet 10:1511–1518

Curtis C, Landis GN, Folk D, Wehr NB, Hoe N, Waskar M, Abdueva D, Skvortsov D et al (2007) Transcriptional profiling of MnSOD-mediated lifespan extension in *Drosophila* reveals a species-general network of aging and metabolic genes. Genome Biol 8:R262

Dali-Youcef N, Lagouge M, Froelich S, Koehl C, Schoonjans K, Auwerx J (2007) Sirtuins: the 'magnificent seven', function, metabolism and longevity. Ann Med 39:335–345

Dietzl G, Chen D, Schnorrer F, Su KC, Barinova Y, Fellner M, Gasser B, Kinsey K et al (2007) A genome-wide transgenic RNAi library for conditional gene inactivation in *Drosophila*. Nature 448:151–156

Dudas SP, Arking R (1995) A coordinate upregulation of antioxidant gene activities is associated with the delayed onset of senescence in a long-lived strain of *Drosophila*. J Gerontol A Biol Sci Med Sci 50:B117–B127

Estevez M, Attisano L, Wrana JL, Albert PS, Massagué J, Riddle DL (1993) The *daf-4* gene encodes a bone morphogenetic protein receptor controlling *C. elegans*dauer larva development. Nature 365:644–649

Ferguson M, Mockett RJ, Shen Y, Orr WC, Sohal RS (2005) Age-associated decline in mitochondrial respiration and electron transport in *Drosophila melanogaster*. Biochem J 390:501–511

Finkel T, Holbrook NJ (2000) Oxidants, oxidative stress and the biology of ageing. Nature 408:239–247

Fleming JE, Walton JK, Dubitsky R, Bensch KG (1988) Aging results in an unusual expression of *Drosophila* heat shock proteins. Proc Nat Acad Sci USA 85:4099–4103

Gardner TS (1948) The use of *Drosophila melanogaster* as a screening agent for longevity factors; the effects of biotin, pyridoxine, sodium yeast nucleate, and pantothenic acid on the life span of the fruit fly. J Gerontol 3:9–13

Geuking P, Narasimamurthy R, Lemaitre B, Basler K, Leulier F (2009) A nonredundant role for *Drosophila* Mkk4 and hemipterous/Mkk7 in TAK1-mediated activation of JNK. PLoS ONE 4: e7709

Giannakou ME, Partridge L (2007) Role of insulin-like signalling in *Drosophila* lifespan. Trends Biochem Sci 32:180–188

Giannakou ME, Goss M, Partridge L (2008) Role of dFOXO in lifespan extension by dietary restriction in *Drosophila melanogaster*: not required, but its activity modulates the response. Aging Cell 7:187–198

Giannakou ME, Goss M, Junger MA, Hafen E, Leevers SJ, Partridge L (2004) Long-lived *Drosophila* with overexpressed dFOXO in adult fat body. Science 305:361

Gong WJ, Golic KG (2006) Loss of Hsp70 in *Drosophila* is pleiotropic, with effects on thermos tolerance, recovery from heat shock and neurodegeneration. Genetics 172:275–286

Grönke S, Clarke DF, Broughton S, Andrews TD, Partridge L (2010) Molecular evolution and functional characterization of *Drosophila* insulin-like peptides. PLoS Genet 6:e1000857

Guillozet AL, Weintraub S, Mash DC, Mesulam MM (2003) Neurofibrillary tangles, amyloid, and memory in aging and mild cognitive impairment. Arch Neurol 60:729–736

Hands S, Sinadinos C, Wyttenbach A (2008) Polyglutamine gene function and dysfunction in the ageing brain. Biochim Biophys Acta 1779:507–521

Harman D (1956) Aging: a theory based on free radical and radiation chemistry. J Gerontol 11:298–300

Harman D (1981) The aging process. Proc Natl Acad Sci USA 78:7124–7128

Harman D (1992) Free radical theory of aging. Mutat Res 275:257–266

Harshman LG, Haberer BA (2000) Oxidative stress resistance: a robust correlated response to selection in extended longevity lines of *Drosophila melanogaster*. J Gerontol A Biol Sci Med Sci 55:B415–B417

Hart FU, Bracher A, Hayer-Hartl M (2011) Molecular chaperones in protein folding and proteostasis. Nature 475:324–332

Helfand SL, Rogina B (2003a) From genes to aging in *Drosophila*. Adv Genet 49:67–109

Helfand SL, Rogina B (2003b) Genetics of aging in the fruit fly, *Drosophila melanogaster*. Annu Rev Genet 37:329–348

Herman MM, Miquel J, Johnson M (1971) Insect brain as a model for the study of aging. Age-related changes in *Drosophila melanogaster*. Acta Neuropathol 19:167–183

Hirth F (2010) *Drosophila melanogaster* in the study of human neurodegeneration. CNS Neruol Disord Drug Targets 9:504–523

Holzenberger M, Dupont J, Ducos B, Leneuve P, Géloën A, Even PC, Cervera P, Le Bouc Y (2003) IGF-1 receptor regulates lifespan and resistance to oxidative stress in mice. Nature 421:182–187

Hotamisligil GS (2006) Inflammation and metabolic disorders. Nature 444:860–867

Hwangbo DS, Gershman B, Tu MP, Palmer M, Tatar M (2004) *Drosophila* dFOXO controls lifespan and regulates insulin signalling in brain and fat body. Nature 429:562–566

Igaki T (2009) Correcting developmental errors by apoptosis: lessons from *Drosophila* JNK signaling. Apoptosis 14:1021–1028

Iliadi KG, Boulianne GL (2010) Age-related behavioral changes in *Drosophila*. Ann N Y Acad Sci 1197:9–18

Iliadi KG, Knight D, Boulianne GL (2012) Healthy aging—insights from *Drosophila*. Front Physiol 3:106

Jacinto E, Loewith R, Schmidt A, Lin S, Ruegg MA, Hall A, Hall MN (2004) Mammalian TOR complex 2 controls the actin cytoskeleton and is rapamycin insensitive. Nat Cell Biol 6: 1122–1128

Jazwinski SM (2000) Aging and longevity genes. Acta Biochim Pol 47:269–279

Johnson GL, Nakamura K (2007) The c-jun kinase/stress-activated pathway: regulation, function and role in human disease. Biochim Biophys Acta 1773:1341–1348

Jones MA, Grotewiel M (2011) Drosophila as a model for age-related impairment in locomotor and other behaviors. Exp Gerontol 46:320–325

Kabil H, Partridge L, Harshman LG (2007) Superoxide dismutase activities in long-lived *Drosophila melanogaster* females: chico1 genotypes and dietary dilution. Biogerontology 8:201–208

Kapahi P, Zid BM, Harper T, Koslover D, Sapin V, Benzer S (2004) Regulation of Lifespan in *Drosophila* by Modulation of Genes in the TOR Signaling Pathway. Curr Biol 14:885–890

Karpac J, Jasper H (2009) Insulin and JNK: optimizing metabolic homeostasis andlifespan. Trends Endocrinol Metab 20:100–106

Kenyon C (2005) The plasticity of aging: insights from long-lived mutants. Cell 120:449–460

Kimura KD, Tissenbaum HA, Liu Y, Ruvkun G (1997) *daf-2*, an insulin receptor-like gene that regulates longevity and diapause in *Caenorhabditis elegans*. Science 277:942–946

Kirby K, Hu J, Hilliker AJ, Phillips JP (2002) RNA interference-mediated silencing of Sod2 in *Drosophila* leads to early adult-onset mortality and elevated endogenous oxidative stress. Proc Natl Acad Sci USA 99:16162–16167

Kurapati R, Passananti HB, Rose MR, Tower J (2000) Increased hsp22 RNA levels in *Drosophila* lines genetically selected for increased longevity. J Gerontol A Biol Sci Med Sci 55: B552–B559

Lapointe J, Hekimi S (2010) When a theory of aging ages badly. Cell Mol Life Sci 67:1–8

Le Bourg E (2001) Oxidative stress, aging and longevity in *Drosophila melanogaster*. FEBS Lett 498:183–186

Leevers SJ, Weinkove D, MacDougall LK, Hafen E, Waterfield MD (1996) The *Drosophila* phosphoinositide 3-kinase Dp110 promotes cell growth. EMBO J 15:6584–6594

Liao PC, Lin HY, Yuh CH, Yu LK, Wang HD (2008) The effect of neuronal expression of heat shock proteins 26 and 27 on lifespan, neurodegeneration, and apoptosis in *Drosophila*. Biochem Biophys Res Commun 376:637–641

Lin YJ, Seroude L, Benzer S (1998) Extended life-span and stress resistance in the *Drosophila* mutant *methuselah*. Science 282:943–946

Loeb J, Northrop JH (1916) Is there a temperature coefficient for the duration of life? Proc Natl Acad Sci USA 2:456–457

Loeb J, Northrop JH (1917) On the influence of food and temperature upon the duration of life. J Biol Chem 32:103–121

Luckinbill L, Clare M (1985) Selection for life span in *Drosophila melanogaster*. Heredity 55:9–18

Luckinbill L, Arking R, Clare MJ, Cirocco WC, Buck S (1984) Selection for delayed senescence in *Drosophila melanogaster*. Evolution 38:996–1003

Luckinbill LS, Clare MJ (1987) Successful selection for increased longevity in *Drosophila*: analysis of the survival data and presentation of a hypothesis on the genetic regulation of longevity. Letter to the editor. Exp Gerontol 22:221–226

McCarroll SA, Murphy CT, Zou S, Pletcher SD, Chin CS, Jan YN, Kenyon C, Bargmann CI et al (2004) Comparing genomic expression patterns across species identifies shared transcriptional profile in aging. Nat Genet 36:197–204

McCord JM, Fridovich I (1969) Superoxide dismutase. An enzymatic function for erythrocuperin (hemocuperin). J Biol Chem 244:6049–6055

Min KJ, Yamamoto R, Buch S, Pankratz M, Tatar M (2008) *Drosophila* lifespan control by dietary restriction independent of insulin-like signaling. Aging Cell 7:199–206

Minois N (2000) Longevity and aging: beneficial effects of exposure to mild stress. Biogerontology 1:15–29

Minois N, Le Bourg E (1999) Resistance to stress as a function of age in *Drosophila melanogaster* living in hypergravity. Mech Ageing Dev 109:53–64

Minois N, Khazaeli AA, Curtsinger JW (2001) Locomotor activity as a function of age and life span in *Drosophila melanogaster* overexpressing hsp70. Exp Gerontol 36:1137–1153

Missirlis F, Phillips JP, Jackle H (2001) Cooperative action of antioxidant defense systems in *Drosophila.* Curr Biol 11:1272–1277

Mockett RJ, Sohal RS, Orr WC (1999) Overexpression of glutathione reductase extends survival in transgenic *Drosophila melanogaster* under hyperoxia but not normoxia. FASEB J 13:1733–1742

Morimoto RI (2008) Proteotoxic stress and inducible chaperone networks in neurodegenerative disease and aging. Genes Dev 22:1427–1438

Morimoto RI, Cuervo AM (2009) Protein homeostasis and aging: taking care of proteins from the cradle to the grave. J Gerontol A Biol Sci Med Sci 64:167–170

Morley JF, Brignull HR, Weyers JJ, Morimoto RI (2002) The threshold for polyglutamine-expansion protein aggregation and cellular toxicity is dynamic and influenced by aging in *Caenorhabditis elegans.* Proc Natl Acad Sci USA 99:10417–10422

Morrow G, Tanguay RM (2003) Heat shock proteins and aging in *Drosophila melanogaster.* Semin Cell Dev Biol 14:291–299

Morrow G, Battistini S, Zhang P, Tanguay RM (2004a) Decreased lifespan in the absence of expression of the mitochondrial small heat shock protein Hsp22 in *Drosophila.* J Biol Chem 279:43382–43385

Morrow G, Heikkila JJ, Tanguay RM (2006) Differences in the chaperone-like activities of the four main small heat shock proteins of *Drosophila melanogaster.* Cell Stress Chaperones 11:51–60

Morrow G, Samson M, Michaud S, Tanguay RM (2004b) Overexpression of the small mitochondrial Hsp22 extends *Drosophila* life span and increases resistance to oxidative stress. FASEB J 18:598–599

Muchowski PJ, Wacker JL (2005) Modulation of neurodegeneration by molecular chaperones. Nat Rev Neurosci 6:11–22

Muchowski PJ, Schaffar G, Sittler A, Wanker EE, Hayer-Hartl MK, Hartl FU (2000) Hsp70 and hsp40 chaperones can inhibit self-assembly of polyglutamine proteins into amyloid-like fibrils. Proc Natl Acad Sci USA 97:7841–7846

Netchine I, Azzi S, Le Bouc Y, Savage MO (2011) IGF1 molecular anomalies demonstrate its critical role in fetal, postnatal growth and brain development. Best Pract Res Clin Endocrinol Metab 25:181–190

Nichols CD, Becnel J, Pandey UB (2012) Methods to assay *Drosophila* behavior. J Vis Exp 7: pii:3795

Niedzwiecki A, Kongpachith AM, Fleming JE (1991) Aging affects expression of 70-kDa heat shock proteins in *Drosophila.* J Biol Chem 266:9332–9338

Nielsen MD, Luo X, Biteau B, Syverson K, Jasper H (2008) 14-3-3 Epsilon antagonizes FoxO to control growth, apoptosis and longevity in *Drosophila.* Aging Cell 7:688–699

Oldham S, Hafen E (2003) Insulin/IGF and target of rapamycin signaling: a TOR de force in growth control. Trends Cell Biol 13:79–85

Orr WC, Sohal RS (1993) Effects of Cu-Zn superoxide dismutase overexpression on life span and resistance to oxidative stress in transgenic *Drosophila melanogaster*. Arch Biochem Biophys 301:34–40

Orr WC, Sohal RS (1994) Extension of life-span by overexpression of superoxide dismutase and catalase in *Drosophila melanogaster*. Science 263:1128–1130

Pandey UB, Nichols CD (2011) Human disease models in *Drosophila melanogaster* and the role of the fly in therapeutic drug discovery. Pharmacol Rev 63:411–436

Parkes TL, Elia AJ, Dickinson D, Hilliker AJ, Phillips JP, Boulianne GL (1998) Extension of *Drosophila* lifespan by overexpression of human SOD1 in motor neurons. Nat Genet 19: 171–174

Partridge L, Gems D (2002) Mechanisms of ageing: public or private? Nat Rev Genet 3:165–175

Partridge L, Piper MD, Mair W (2005) Dietary restriction in *Drosophila*. Mech Ageing Dev 126:938–950

Pearl R, Parker SL (1921) Experimental studies on the duration of life I. Introductory discussion of the duration of life in *Drosophila*. Am Nat 60:481–509

Pearl R, Parker SL (1922) Experimental studies on the duration of life. II. Hereditary differences in duration of life in line-bread strains of *Drosophila*. Am Nat 56:174

Pérez VI, Bokov A, Van Remmen H, Mele J, Ran Q, Ikeno Y, Richardson A (2009) Is the oxidative stress theory of aging dead? Biochim Biophys Acta 1790:1005–1014

Pfeiffenberger C, Lear BC, Keegan KP, Allada R (2010) Locomotor activity level monitoring using the *Drosophila* Activity Monitoring (DAM) System. *Cold Spring Harb. Protoc.* 2010 pdb.prot5518

Phillips JP, Hilliker AJ (1990) Genetic analysis of oxygen defense mechanisms in *Drosophila melanogaster*. Adv Genet 28:43–71

Phillips JP, Campbell SD, Michaud D, Charbonneau M, Hilliker AJ (1989) Null mutation of copper/zinc superoxide dismutase in *Drosophila* confers hypersensitivity to paraquat and reduced longevity. Proc Natl Acad Sci USA 86:2761–2765

Piper MD, Partridge L (2007) Dietary restriction in *Drosophila*: delayed aging or experimental artefact? PLoS Genet 3:e57

Pletcher SD, Macdonald SJ, Marguerie R, Certa U, Stearns SC, Goldstein DB, Partridge L (2002) Genome-wide transcript profiles in aging and calorically restricted *Drosophila melanogaster*. Curr Biol 12:712–723

Raj K, Chanu SI, Sarkar S (2012) Decoding complexity of ageing. Cell Dev Biol 1:e117

Reiter LT, Potocki L, Chien S, Gribskov M, Bier E (2001) A systematic analysis of human disease-associated gene sequences in *Drosophila melanogaster*. Genome Res 11:1114–1125

Ritossa F (1962) A new puffing pattern induced by temperature shock and DNP in *Drosophila*. Experientia 18:571–573

Ritossa F (1996) Discovery of the heat shock response. Cell Stress Chaperones 1:97–98

Rogina B, Helfand SL (2004) Sir2 mediates longevity in the fly through a pathway related to calorie restriction. Proc Natl Acad Sci USA 101:15998–16003

Rogina B, Reenan RA, Nilsen SP, Helfand SL (2000) Extended life-span conferred by cotransporter gene mutations in *Drosophila*. Science 290:2137–2140

Rose M (1984) Laboratory evolution of postponed senescence in *Drosophila melanogaster*. Evolution 38:1004–1009

Rose M, Charlesworth B (1980) A test of evolutionary theories of senescence. Nature 287: 141–142

Rose MR, Charlesworth B (1981) Genetics of life history in *Drosophila melanogaster*. II. Exploratory selection experiments. Genetics 97:187–196

Ryder E, Ashburner M, Bautista-Llacer R, Drummond J, Webster J, Johnson G, Morley T, Chan YS et al (2007) The DrosDel deletion collection: a *Drosophila* genome wide chromosomal deficiency resource. Genetics 177:615–662

Salmon AB, Marx DB, Harshman LG (2001) A cost of reproduction in *Drosophila melanogaster*: stress susceptibility. Evolution 55:1600–1608

Saltiel AR, Kahn CR (2001) Insulin signalling and the regulation of glucose and lipid metabolism. Nature 13:799–806
Sarkar S, Singh MD, Yadav R, Arunkumar KP, Pitman GW (2011) Heat shock proteins: Molecules with assorted functions. Front Biol 6:312–327
Seto NO, Hayashi S, Tener GM (1990) Overexpression of Cu-Zn superoxide dismutase in *Drosophila* does not affect life-span. Proc Natl Acad Sci USA 87:4270–4274
Shaw P, Ocorr K, Bodmer R, Oldham S (2008) Drosophila aging 2006/2007. Exp Gerontol 43:5–10
Shaw PJ, Tononi G, Greenspan RJ, Robinson DF (2002) Stress response genes protect against lethal effects of sleep deprivation in *Drosophila*. Nature 417:287–291
Sies H, Cadenas E (1985) Oxidative stress: damage to intact cells and organs. *Philosophical.* Tran R Soc Lond Ser B, Biol Sci 311:617–631
Smith JM (1958) The effects of temperature and of egg laying on the longevity of *Drosophila subobscura*. J Exp Biol 35:832–842
Smith JM (1962) The causes of ageing. Proc. R. Soc. London Ser. B 157:115–127
Sohal RS (2002) Oxidative stress hypothesis of aging. Free Radic Biol Med 33:573–574
Sohal RS, Weindruch R (1996) Oxidative stress, caloric restriction, and aging. Science 273:59–63
Soti C, Csermely P (2003) Aging and molecular chaperones. Exp Gerontol 38:1037–1040
Stadtman ER (2006) Protein oxidation and aging. Free Radic Res 40:1250–1258
Sun J, Tower J (1999) FLP recombinase-mediated induction of Cu/Zn-superoxide dismutase transgene expression can extend the life span of adult *Drosophila melanogaster* flies. Mol Cell Biol 19:216–228
Sun Y, Yolitz J, Wang C, Spangler E, Zhan M, Zou S (2013) Aging studies in *Drosophila melanogaster*. Methods Mol Biol 1048:77–93
Tatar M (2010) Reproductive aging in invertebrate genetic models. Ann N Y Acad Sci 1204: 149–155
Tatar M, Bartke A, Antebi A (2003) The endocrine regulation of aging by insulin-like signals. Science 299:1346–1351
Tatar M, Khazaeli AA, Curtsinger JW (1997) Chaperoning extended life. Nature 390:30
Tatar M, Kopelaman A, Epstein D, Tu MP, Yin CM, Garofalo RS (2001) A mutant *Drosophila* insulin receptor homolog that extends life span and impairs neuroendocrine function. Science 292:107–110
Tatar M, Post S, Yu K (2014) Nutrient control of *Drosophila* longevity. Trends Endocrinol Metab 25:509–517
Teixeira-Castro A, Ailion M, Jalles A, Brignull HR, Vilaça JL, Dias N, Rodrigues P, Oliveira JF et al (2011) Neuron-specific proteotoxicity of mutant ataxin-3 in *C. elegans*: rescue by the DAF-16 and HSF-1 pathways. Hum Mol Genet 20:2996–3009
Theodosiou NA, Xu T (1998) Use of FLP/FRT system to study *Drosophila* development. Methods 14:355–365
Tower J (2011) Heat shock proteins and *Drosophila* aging. Exp Gerontol 46:355–362
Um SH, D'Alessio D, Thomas G (2006) Nutrient overload, insulin resistance, and ribosomal protein S6 kinase 1, S6K1. Cell Metab 3:393–402
Vermeulen CJ, Van De Zande L, Bijlsma R (2005) Resistance to oxidative stress induced by paraquat correlates well with both decreased and increased lifespan in *Drosophila melanogaster*. Biogerontology 6:387–395
Vijg J (2008) The role of DNA damage and repair in aging: new approaches to an old problem. Mech Ageing Dev 129:498–502
Voellmy R 2004 On mechanisms that control heat shock transcription factor activity in metazoan cells. Cell Stress Chaperones 9:122–133
Vowels JJ, Thomas JH (1992) Genetic analysis of chemosensory control of dauer formation in *Caenorhabditis elegans*. Genetics 130:105–123
Walker DW, Benzer S (2004) Mitochondrial "swirls" induced by oxygen stress and in the *Drosophila* mutant hyperswirl. Proc Natl Acad Sci USA 101:10290–10295

Wallace DC (2005) A mitochondrial paradigm of metabolic and degenerative diseases, aging, and cancer: a dawn for evolutionary medicine. Annu Rev Genet 39:359–407

Wang MC, Bohmann D, Jasper H (2003) JNK signaling confers tolerance to oxidative stress and extends lifespan in *Drosophila*. Dev Cell 5:811–816

Wang MC, Bohmann D, Jasper H (2005) JNK extends life span and limits growth by antagonizing cellular and organism-wide responses to insulin signaling. Cell 121:115–125

Wolf FW, Heberlein U (2003) Invertebrate models of drug abuse. J Neurobiol 54:161–178

Yadav R, Chanu SI, Raj K, Sarkar S (2013) Rise and Fall of Reactive Oxygen Species (ROS): implications in Aging and Neurodegenerative Disorders. Cell Dev. Biol. 1:e122

Yadav R, Kundu S, Sarkar S (2015) *Drosophila glob1* expresses dynamically and is required for development and oxidative stress response. Genesis. doi:10.1002/dvg.22902

Zeitlinger J, Bohmann D (1999) Thorax closure in *Drosophila*: involvement of Fos and the JNK pathway. Development 126:3947–3956

Zhao Y, Sun H, Lu J, Li X, Chen X, Tao D, Huang W, Huang B (2005) Lifespan extension and elevated hsp gene expression in *Drosophila* caused by histone deacetylase inhibitors. J Exp Biol 208:697–705

Zou S, Meadows S, Sharp L, Jan LY and Jan YN (2000) Genome-wide study of aging and oxidative stress response in *Drosophila melanogaster*. Proc Natl Acad Sci USA 97: 13726–13731

Dictyostelium discoideum: A Model System to Study Autophagy Mediated Life Extension

Punita Jain, Priyanka Sharma, Anju Shrivastava and Shweta Saran

Abstract Autophagy is a major catabolic process in eukaryotes that help degrade and recycle macromolecules and organelles. Recent evidences suggest autophagy to mediate cytoprotection and also help increase longevity. It acts as a central regulatory mechanism for aging/longevity in diverse eukaryotic species. In the present study we have exploited the lower eukaryotic model organism, *Dictyostelium discoideum* to study longevity, which could use autophagy as a major mechanism for increasing lifespan. This organism is unique as it follows a developmental programmed cell death by autophagy that is independent of caspases. It could be a beneficial organism to delineate the mechanism of increased longevity that follows autophagy.

1 Introduction

Aging is not a disease but an accumulation of damaged molecules over a lifetime, which leads to a decline in normal functions of the organism as well as lays a platform for many diseases. Thus, the central question that arises is "can we increase our lifespan without trading with its quality". Recent researches have emphasized on this aspect and today we have to some extent knowledge on the conserved genetic as well as biochemical pathways that are involved in longevity/aging. Today we know that tinkering with some signaling pathway or nutrition has led to an increase in longevity over a range of organisms.

Much of what we know about aging/longevity comes from the study of model organisms like yeast, *Drosophila, C. elegans* and mouse; even when we are sure that

Punita Jain and Priyanka Sharma have equally contributed for this chapter.

P. Jain · P. Sharma · S. Saran (✉)
School of Life Sciences, Jawaharlal Nehru University, New Delhi 110067, India
e-mail: ssaran@mail.jnu.ac.in; shweta_saran@hotmail.com

P. Jain · A. Shrivastava
Department of Zoology, University of Delhi, New Delhi 110007, India

© Springer Science+Business Media Singapore 2017

P.C. Rath et al. (eds.), *Topics in Biomedical Gerontology*,
DOI 10.1007/978-981-10-2155-8_2

they are not representatives of human aging. It has been proposed that similar mechanisms operate across species and possibly there could be an underlying conserved common mechanism of aging. In recent times there has been an explosion of research in the area of aging/longevity. The key pathways and molecules that have emerged out to playa role in life extensions, in most cases follow autophagy to increase lifespan. Some known conserved pathways that regulate lifespan in most organisms studied so far are target of rapamycin (TOR), AMPK, insulin/IGF signaling and dietary controls. Few small molecules like rapamycin, spermidine and resveratrol may directly or indirectly affect the above pathways to increase lifespan.

The terms "aging" and "longevity" is usually used interchangeably but there exists a clear demarcation (McDonald and Ruhe 2011). Aging deals with the functional decline of biological processes while longevity is a parameter to measure the length of time an individual remains alive in the absence of death due to extrinsic factors. It is for sure that aging is not a disease state. Recently Lepez-Otin et al. (2013) has laid emphasis on "nine tentative hallmarks" representing aging, which include genomic instability, telomere attrition, epigenetic alterations, loss of proteostasis, deregulated nutrient sensing, mitochondrial dysfunction, cellular senescence, stem cell exhaustion and altered intercellular communication. An understanding of all these hallmarks and their inter-relationship leading to aging still remains a challenge. Therefore, the need of the hour is to identify many more model systems for longevity studies, which could help identify and investigate the longevity genes as well as involvement of other signaling pathways. Model organisms having orthologs of many human genes could be best suited for such studies. Organisms having unique mode of life cycle could provide added advantage. Keeping this in view new model systems are coming up for researches in aging.

2 Model Systems to Study Longevity

The aging research community has been trying to measure the effects of potential treatments to carry out lifespan studies, which is next to impossible in humans due to varied known reasons. Studies in various model organisms suggest a surprising degree of similarity in the fundamental biology of cells and mechanism of aging. Thus, research in lower organisms remains relevant to mechanisms of human aging. It is still true that a mouse remains a mouse and yeast remains yeast and still maintains a difference from humans. Literature suggests that the translation of the outcomes from unicellular to multicellular, lower to higher organisms and ultimately to humans is quite different but the mechanisms remain similar. Since there are no available alternatives, it is useful to utilize as many different model systems as possible for understanding the mechanism(s) of aging. An added advantage in studying these organisms is their associated medical conditions, which can be created by genetically altering their lineages. Below, we discuss few of the model organisms that have provided immense knowledge regarding the mechanisms of aging/longevity.

2.1 *Saccharomyces Cerevisiae*

The advantage of using yeast as an aging model includes its fast growth and easy maintenance under laboratory conditions. Despite being unicellular organisms, it has been useful in identifying the conserved aging pathways that is shared amongst various organisms. In addition, easy genetic manipulations make yeast a powerful model system to study longevity. Two different types of aging are studied in yeast: replicative and chronological which allows the analysis of aging process in both dividing and non-dividing cells (Wasko and Kaeberlein 2014).

2.2 *Caenorabditis Elegans*

The analysis of the dauer larvae provided the first genetic evidence for longevity/aging. The involvement of insulin like growth factor 1 signaling in aging laid the foundation of research in modulating longevity. Also, the use of disease model and drug screens has helped in improving the quality of life and understanding of the molecular mechanisms of aging. This nematode is evolutionarily distant from human making it limiting as a model system (Bansal et al. 2015).

2.3 *Drosophila Melanogaster*

Fruit flies have many advantages as a model system for aging studies. Apart from having a short lifecycle it has more distinct tissues and organs. Infact, they also have proliferating stem cell populations in their gut, which allows analysis in proliferating populations. Flies share about 60 % of disease-related genes with humans, which makes them a desirable model. The only drawback is the maintenance of transgenic flies (Matthews et al. 2005).

2.4 *Danio Rerio*

Zebra fish is being successfully used as a model system for understanding human aging as it draws lot of parallels at the internal organization in terms of organs as compared to the whole organism level. It lives for about 2–3 years, which is not particularly beneficial, because its lifespan is similar to rodents, but it is more evolutionary distant from humans. The advantage of this model organism is its ability to regenerate its tissues, which help elucidate the mechanism of tissue regeneration and longevity (Lepilina et al. 2006).

2.5 Rodents

Mice are invaluable in aging research as they show approximately 99 % of human orthologs, which is significantly advantageous than the invertebrate model. Usually inbred mice are used for aging research, which shows high genetic relatedness, a fact that is significantly different from the human population. Thus, the results obtained may or may not hold true for humans.

Naked mole rats can provide clues to mechanisms of longevity and potential therapies in humans and hence are an extremely valuable model animal. These rats live for nearly 30 years and show age related pathologies similar to humans. There are several disadvantages of using them as laboratory animals including specific housing conditions like low light levels, high temperature and humidity. Very long lifespan also poses an obvious limitation on the variety of experiments that could be performed (Phelan 1992).

2.6 Primates

Rhesus macaque is a good model system to study longevity but unfortunately we do not have many age related studies with them. One of the main reasons is its housing under laboratory conditions and its long lifespan. The biggest advantage of using non-human primates for studying age related pathological conditions that cannot be recapitulated in mouse models (Mitchell et al. 2015).

Usage of various model organisms for aging studies has generated enormous useful information regarding the mechanism of aging/longevity. Observations from the animal models were helpful in the identification of the conserved pathways that regulate human aging/longevity. Model organisms are fundamental for aging research as there are limitations in using humans as subjects. Therefore, every model system does answer some questions in human aging/longevity. Thus, results obtained from many model systems would help understand few aspects in this area of research.

3 Autophagy Promotes Longevity

Aging is a multifactorial process and many mechanisms contribute to its decline, which lead to accumulation of damaged proteins and other cellular constituents, which is removed by autophagy. Autophagy, (a cellular self eating) lysosome mediated catabolic process to digest their own cellular content to the smallest components to be reutilized for survival. Many reports show that autophagy is required for life extension in various model organisms (Rubinsztein et al. 2011) especially those which are exposed to prolonged starvation. Literature suggests that lifespan extensions can be achieved by genetic manipulations of cellular processes like insulin/Insulin like growth factor 1 pathway, histone acetylations, p. 53 pathways etc. Other factors responsible are calorie restriction and target of rapamycin (TOR) pathway inhibition by drugs like rapamycin, resveratrol and spermidine. All these pathways lead to autophagy either directly or indirectly to increase the lifespan.

3.1 Genetics Manipulations Leading to Life Extension via Autophagy

Sirtuin 1, a class III histone deacetylase that require NAD^+ for its activity has been identified as a major gene responsible for longevity. Overexpression of *sirtuin 1* or its orthologs increases lifespan in flies, yeast and nematodes (Burnett et al. 2011). Under in vitro mammalian cells induce autophagy when *sirtuins* are overexpressed. Sirtuins mediate deacetylation of histones and several other cytosolic proteins like AMP-dependent kinase, AMPK and autophagy related genes like Atg8, Atg5 etc. Inhibition of *sirtuin 1* prevents autophagy. Resveratrol probably acts via the Sirtuin *1* and is its positive regulator.

A knockdown of p.53 in *C. elegans* showed induction of lifespan extensions and autophagy (Tavernarakis et al. 2008). TOR inhibition induces autophagy to increase lifespan and the same has been observed in many organisms (Madeo et al. 2010). TOR is an intracellular signal transduction pathway, which is conserved from lower to higher organisms that integrates various signals namely starvation, nutrients, growth factors, stress and hormones (Inoki et al. 2005). It increases protein synthesis, inhibits autophagy thus promoting growth.

3.2 Pharmacological Manipulations Affecting Longevity via Autophagy

Many anti-aging drugs are known today which increase lifespan by promoting autophagy. One such molecule, rapamycin, an inhibitor of TOR signaling, has already been shown to increase lifespan in a variety of organisms, including most recently, mice (Lamming et al. 2013). Rapamycin is known to have profound effects on health and longevity in various model organisms (Swer et al. 2014) and are being studied in humans as therapies for treating age-related diseases. Rapamycin (sirolimus) is a natural product initially identified as a fungicide (Sehgal 2003)and later found to be a potent immunosuppressant (Muthukkumar et al. 1995). It has a wide usage but its effects on other pathways or molecules are not understood. In recent years, signaling through the target of rapamycin (TOR) kinase has emerged as a key pathway involved in lifespan extension. Reduction in TOR signaling is sufficient to increase lifespan in yeast (Powers et al. 2006), *C. elegans* (Vellai et al. 2003), *Drosophila* (Bjedov et al. 2010) and mice (Stanfel et al. 2009) probably by modulating the mRNA translation and enhanced degradation of damaged macromolecules. Mice with reduced TOR signaling in adipose tissue showed resistance to diet induced obesity and metabolic diseases. Rapamycin is known to induce autophagy, which is independent of Sirtuin *1*, in other words they promote autophagy by non-overlapping mechanisms (Madeo et al. 2010). Rapamycin is now known to confer protection in model animals and cell based models of neurodegenerative diseases, cardiovascular disease and against a variety of cancers (Johnson et al. 2013a, b).

Resveratrol, a polyphenol present in red wine, prolongs lifespan via autophagy by inducing the *sirtuin 1* expression only if animals are fed with a high-calorie diet leading to obesity (Baur and Sinclair 2006). Studies show the potential of resveratrol in inducing both canonical and non-canonical pathways (Scarlatti et al. 2008). In *C. elegans* or in animals that express a functional *sirtuin 1* gene show lifespan extension mediated by resveratrol, whereas resveratrol abrogated the favourable effects on longevity when *Beclin 1/Atg 6* are knocked down (Morselli et al. 2011). Assembling the two different studies confirms that resveratrol elicits lifespan extension in *C. elegans* in *sirtuin 1*-dependent manner.

The polyamine, spermidine has been recently shown to increase lifespan in various model organisms by inducing autophagy (Eisenberg et al. 2009). Exogenous addition of spermidine also shows the same effect in various organisms tested. It significantly reduces the age-related oxidative protein accumulation in mice (Minois et al. 2011) and thus has been proposed as an anti-aging drug. Inhibition of autophagy by knockout of autophagy related genes abolish the lifespan promoting effect of spermidine in yeast (Eisenberg et al. 2009). Medical application of spermidine as a natural autophagy inducer is of most interest for future research and applications.

Many questions still need to be answered. It is true that autophagy related proteins decline with age and so does the efficiency of autophagy process. It suggests the possibilities of genetic, pharmacological and nutritional strategies that work via this pathway to serve as good targets for longevity. One that still remains to be answered is whether all longevity-promoting manipulations require autophagy? This may be true in lower organisms but does it still hold true for humans? Results obtained with neurodegenerative diseases also raise the question whether autophagy induction works better than interventions in the central nervous system (Kesidou et al. 2013). Much more rigorous researches in different model systems are required to answer these questions.

4 *Dictyostelium discoideum*: A Good Model System to Study Autophagy

Dictyostelium discoideum is a simple eukaryote that lives in the soil. It feeds on bacteria by phagocytosis and divides mitotically till there is sufficient food (Raper 1935). Upon depletion of food, the amoebae chemotax to common collecting points in response to the chemoattractant cAMP and initiate multicellular development (Konijn and Raper 1961). The unicellular amoebae, under nutrient starvation conditions enter multicellularity, which is not by repeated cell division but by coming together of spatially segregated cell types. Terminal differentiation in this organism leads to the formation of two cell types namely, the stalk and the spore cells. The spores are viable and each of them can germinate into an amoeba under favorable conditions whereas the stalk is made up of dead vacuolated cells, which fail to regrow under similar conditions. Multicellularity in *D. discoideum* is different from the other metazoans and genes involved during both the phases of life cycle are different and mutations in the developmentally regulated genes do not affect the viability of the organism (Gross 1994; Weeks 2000; Kessin 2001). Thus, it displays distinct characteristics of true cellular specialization, coherent cell movement, cell death and altruism.

The terminally differentiated stalk and spore cells are present in its precursor kinds as prestalk and prespore cells respectively during earlier developmental stages. These cells are specified and not terminally differentiated and also has the potential to show transdifferentiation (Browne and Williams 1993). At any given time during development, 15–20 % of the total cell population is destined to die by becoming stalk cells. A clear anterior-posterior pattern formation is visualized in the slugs where the anterior 1/5 is composed of prestalk cells while the rest is composed of prespore cells. Interspersed in the prespore region are the anterior-like cells (ALC) which biochemically is similar to the prestalk cells. The fruiting body formed includes a stalk that is made of vacuolized cellulose walled dead cells (Williams et al. 1987; Cornillon et al. 1994; Giusti et al. 2009) (Fig. 1).

D. discoideum occupies a strategic phylogenetic position which emerged after the divergence of the plants but before the animals and fungi. *Dictyostelium* represents one of the distinct successful transitions in evolution from unicellular eukaryotes to multicellularity. Since the fruiting body formed shows death with respect to stalk cells is thus considered as one of the most ancient developmental programmed cell death in eukaryotes. This death is neither by apoptosis nor necrotic but is vacuolar. Induction of this developmental cell death requires two signals: one being starvation that induces the complicated cAMP signal transduction pathway and second being the polyketide Differentiation inducing factor (DIF), a prestalk pathway promoter (Luciani et al. 2011). This also requires inositol 1, 4, 5-triphosphate receptor (IP3R), which governs the calcium fluxes from the endoplasmic reticulum into the cytosol. The prestalk cells show higher levels of intracellular free calcium (Saran et al. 1994) as compared to prespore cells; also high

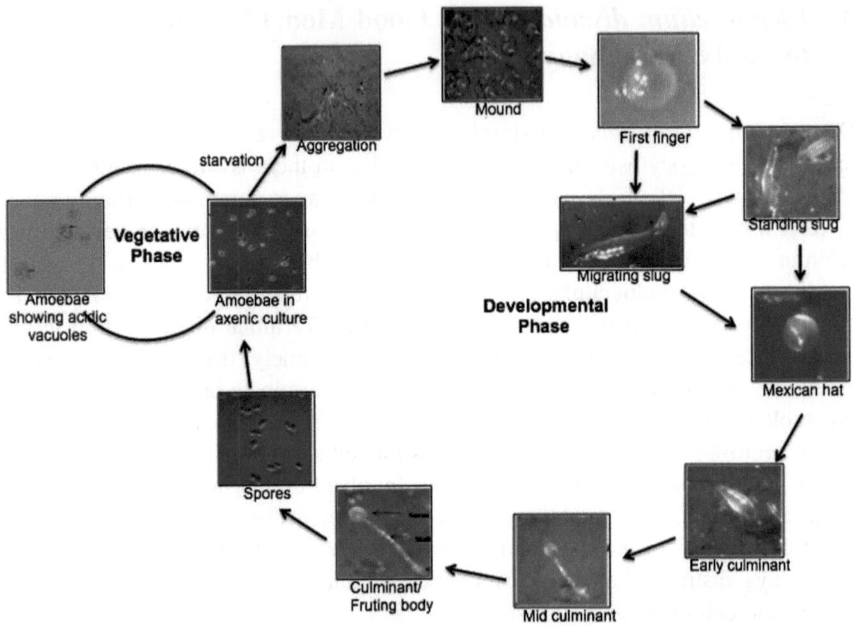

Fig. 1 Life cycle of *Dictyostelium discoideum.* Amoebae remain in vegetative cycle till the food is abundant. Starvation triggers the developmental cycle which undergoes various morphogenetic movements to ultimately form a fruiting body comprising two terminally differentiated cell types: stalk and spore. Here the vegetative cells were stained with the vital dye, neutral *red* that stains acidic vacuoles and appear *red* in colour under visible light. The acidic vacuoles are present in the prestalk/stalk cells during development

calcium levels can bias the cells towards the prestalk pathway (Abe and Maeda 1989).

The *Dictyostelium* genome has been fully sequenced and annotated (http://dictybase.org/). It is a 34 Mb genome having many genes that are homologous to the higher eukaryotes but are missing in yeast. Many genes are homologous to human genes making them a very useful model organism. Also approximately 33 genes are homologous to genes involved in human disease (Eichinger et al. 2005). Because of its similarities, it is now used to understand complex human diseases (Williams et al. 2006) and test anticancer drugs, immune cell diseases and bacterial intracellular pathogenesis. Behaviour of individual cells accounts for many phases of health and diseases, for example, cytokinesis for tissue maintenance and cell proliferation for cancer. Chemotaxis deals with inflammation, lymphocyte trafficking and axon guidance (Bozzaro 2013). Phagocytosis can be analyzed to understand antigen presentation. General basic features of embryogenesis like pattern formation, cell fate determination and differentiation can be always analyzed.

The ability to synchronize morphogenesis and a well-developed molecular genetics allows it to become a good model system for the analysis of various cell signaling as well as for the understanding of cellular and biological processes involved in multicellular development. Myosin mutations that cause cardiac myopathies (Fujita et al. 1997) and mechanism of action of lithium have recently been analyzed (Williams et al. 2002).

Multicellular development in case of *Dictyostelium* occurs in the absence of any external food source. The energy required for aggregation and morphogenesis comes from glycogenolysis and autophagy (Rabinowitz and White 2010). In case of *D. discoideum*, the stalk cell formation is due to autophagic cell death (ACD). With respect to programmed cell death, *Dictyostelium* offers advantages over other systems as it shows ACD in the absence of apoptosis, which is caspase dependent. There are no caspase or metacaspase genes identified in the *Dictyostelium* genome. Inactivation of the single paracaspase gene (Roisin-Bouffay et al. 2004) did not show any alteration in the cell death program but showed some degree of multicellular developmental defects. No ortholog of any known caspases or its family members were found in *Dictyostelium*, thus it offers a good model system to study non-apoptotic cell death in the absence of apoptosis.ACD is described as cell death with an autophagic component. The autophagic components are quite well known in *D. discoideum* but the mechanism of cell death (ACD) is still poorly defined. Out of the two signals, the first signal of starvation/cAMP initiates autophagy while the second signal, DIF-1 is required for the transition from autophagy to ACD. Autophagy is insufficient to trigger ACD as it requires an intact *Atg1*. Inactivation of *Atg1* blocks vacuolization as well as autophagy but not cell death (Giusti et al. 2009). All characteristics of necrotic cell death like generation of reactive oxygen species, depletion of ATP, rupture of plasma membrane are observed.

Autophagy induction and regulation is regulated by the energy and nutritional status of the cell. The nutritional sensor is the target of rapamycin (TOR), which is a serine threonine kinase and receives signals like nutrients, energy, calcium, amino acids, growth factors etc. There is one *Tor* gene in *D. discoideum,* which forms both the C1 and C2 complexes of which TORC1 is primarily involved in autophagy (Soulard et al. 2009). The autophagosomes originate from multiple sites in the cytoplasm and appear as punctate pattern in the cytoplasm, which can be quantitated. Several autophagy genes have been isolated in *Dictyostelium*, which can be grouped according to their functions. Some of the genes required for the origin, elongation, completion and degradation of the autophagosomes are known but their regulation is still not understood (Levine et al. 2011). Atg1 and the lipid kinase Vps34 (Maiuri et al. 2007) are required for the initiation of nucleation of the autophagosomes. Atg8 and Atg12 are required for the vesicle expansion and completion while Atg2, Atg9 and Atg18 play a role in the biogenesis of autophagy (Yorimitsu and Klionsky 2005). TORC1 functions upstream of Atg1 complex and is involved in cell growth and metabolism (Wullschleger et al. 2006).

5 *Dictyostelium discoideum*: As a Model to Study Longevity

Nearly all metazoans undergo the process of aging. Many model organisms have been used to elucidate the mechanism of aging especially to understand human aging. It is true that observations from all model organisms have added on to the understanding of the process of aging but as yet it has to be seen if they hold true for humans. One thing, which has emerged is that considerable number of pathways involved in longevity are conserved throughout evolution. Invertebrate model systems have given maximum answers in the field of aging research but the degree to which it is conserved between organisms is yet to be answered. Comparative functional genomics approach between yeast and *C. elegans* show conserved pathways involved in longevity suggesting that they may play similar roles in human aging. Also, there exists orthologs of aging related genes in mammals.

Organisms from various habitats across eukaryotic species are influenced by different extrinsic factors that may be responsible for age-associated mortality. However, many studies have shown the existence of few similar patterns of aging across species. Thus, the question remains is why evolution selected these processes in organisms, which are very divergent. Additionally, to what extent is the mechanism of longevity/aging conserved over evolution.

As discussed earlier many more model system for analyzing aging/longevity would prove beneficial. In fact the mechanism(s) involved in autophagy by the anti-aging drugs and their interactions still need to be elucidated. In the present study, we intend to present *D. discoideum* as a new model system to analyze longevity. There are many reasons contributing towards the process of aging. Some of the significant observations, point to dietary restrictions, as a measure for longevity but the mechanistic understanding of how it works is not yet understood. Using this model system we could understand the correlation between autophagy, TOR signaling, sirtuins, anti-aging drugs and their role in longevity. *Dictyostelium* provides a very good model system to delineate the mechanistic of these processes, as its genome is known to be close to the human genome. It follows a caspase independent cell death mechanism, which is ACD. Nutrient plays a major role in the transition from uni-to multicellularity.

By all means, *Saccharomyces cerevisiae* has identified the maximum number of genes that are involved in the control of aging/longevity. Aging in yeast is assayed primarily by measurement of replicative or chronological life span (Longo and Fabrizio 2012; Fabrizio et al. 2005). These are well-established methods to measure longevity. Yeast replicative lifespan is defined as number of times a single cell divides before its senescence. This method is useful to study aging process of mitotically active cells in multicellular organisms (Kaeberlein. 2010). Chronological life span is a suitable model for post-mitotic cells in multicellular organisms as it determines the length of time a cell can survive under non-proliferative conditions. i.e. number of days cells remain viable in stationary phase culture (Kaeberlein et al. 2007). Chronological aging helps investigate the different activities required for somatic maintenance, which is well conserved across species. The most important of them all is stress protection by anti-oxidants. In most cases mitochondrial dysfunction is associated with aging (Morino et al. 2006), which contributes to reactive oxygen species (ROS) production leading to damage of proteins, lipids and DNA (Bhat et al. 2015). This study is now being taken to other organisms. Even in yeast, the results obtained is limiting as we still do not have a clear picture of how the genetic factors controlling aging/longevity works.

Similar to yeast chronological aging/longevity model, chronological life span determinations was performed in *Dictyostelium discoideum*, wild type (A × 2) cells grown in axenic media containing standard amount of glucose. Chronological aging experiment was conducted by inoculating fresh A × 2(−80 °C) spores in axenic medium at 22 °C and allowed to grow till stationary phase was reached. The midpoint of stationary culture was considered as the beginning of the chronological ageing or t_0 h. Subsequently viability of non-proliferative cells from stationary culture was determined at different time points 12, 24, 36, 48 h to measure viability of the cells. This was carried out by three different methods (Fig. 2): MTT reduction assay (3-[4, 5-Dimethylthiazol-2-yl]-2, 5-diphenyltetrazolium bromide) (Swer et al. 2014), propidium iodide (PI) staining followed by flow cytometry analysis (FCM) (Ocampo and Barrientos 2011) and regrowth assay (Giusti et al. 2009). The methodologies followed are described in brief below.

- MTT assay: Cell viability was monitored at various time points by a slightly modified method of Mosmann (1983). Absorbance of the reduced dye was measured at 570 nm and the background reading at 630 nm was taken.
- Regrowth assay: A serial dilution of the stationary culture cells was taken for plating with bacteria on non-nutrient agar and subsequently allowed to incubate at 22 °C for 72 h. The clearing zone or plaques formed were counted for the analysis.
- Propidium iodide staining and flow cytometry analysis: Cell viability was analyzed using propidium iodide (PI), a fluorescent dye extensively used from bacteria to mammalian cells to discriminate between live (PI^-) and dead (PI^+) cells based on membrane permeability. Ocampo and Barrientos (2011) used it to determine the viability of stationary phase cells thus making it suitable for studying chronological aging. Approximately, a density of 1×10^6 cells/ml was incubated for 30 min at 22 °C in the presence of 1 μM PI. Cells were then subjected to flow cytometry. Analysis of cells at different time points ranging from 12–60 h was carried. The data obtained was utilized for the construction of a survival curve by plotting the percentage of PI^- cells versus time.

Our data shows that *Dictyostelium* could be used for the measurement of longevity using chronological lifespan extension model similar to yeast. The major genes involved in chronological aging pathways in almost all organisms studied so far are the TOR/S6 K pathway (Fabrizio et al. 2001) and the Ras/adenylate cyclase/PKA pathway (Longo et al. 1999; Fabrizio and Longo 2003). The TOR/S6 K pathway is activated by nutrition like amino acids etc. while the

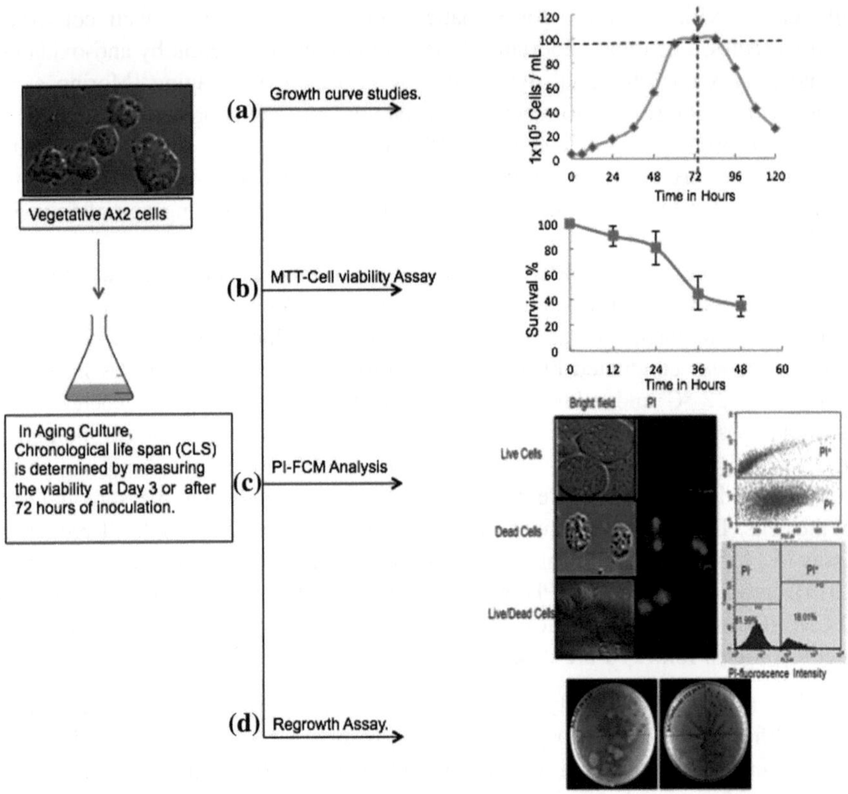

Fig. 2 Chronological longevity assay in *Dictyostelium discoideum*. The assay was based on that used for yeast longevity test. **a** Growth kinetics: A × 2 cells were grown in HL-5 medium till it reached the stationary phase ($\sim 1 \times 10^7$ cells/ml) after, which the rate of cell multiplication decreases and approaches zero due to lysing of cells. Here, the dashed lines represent extrapolations of stationary phase. Intersection of this plot represents the onset of stationary phase. This time point is considered as t_0 h to determine cell viability at different time points by MTT, PI-FCM and regrowth methods. **b** Cell viability assay using MTT: Cell viability of stationary phase cells is considered as initial survival (100 %) and was monitored by MTT-reduction assay, at different time points 12, 24, 36, 48 h. Here, data represents percentage of survival relative to t_0 as function of time in stationary phase. **c** Propidium iodide staining and Flow cytometer analysis: Both live (PI^-) dead (PI^+) cells were visualized in a given population. The stationary culture, which had a mixture of live and dead cells were visualized and further taken for FACS analysis. The percentage of cells in each population was accounted for. **d** Regrowth assay: Chronological survival of aged cells on *solid* medium was determined by counting the number of plaques or clearing of food seen

AC/PKA pathway is modulated largely by glucose etc. These signaling pathways interact with each other and regulate metabolism and stress. In case of *D. discoideum*, starvation activates the nutrient signal transduction pathway by inhibiting the TOR to induce autophagy (Swer et al. 2015). Also the energy sensor, AMPK is activated upon starvation, which decreases the level of ATP to ADP. The mechanistic of this is still not understood. NAD^+, an energy donor is utilized by Sirtuin for its activity, gets affected by nutrient depletion but how it regulates the Sirtuin 1 and AMPK is not yet known.

6 Sir2D of *D. discoideum* Promotes Lifespan Extension by Inducing Autophagy

As discussed earlier Sirtuin 1 promotes longevity in many organisms. There are five Sirtuins present in this organism, which may have a role to play in longevity. The Sirtuin family of proteins is both functionally and structurally conserved and is found in organisms ranging like eubacteria, archaea, eukaryotes and viruses (Raffaelli et al. 1999; Martin et al. 2001; Miller et al. 2003). Sirtuins are involved in both metabolic and chromatin regulations throughout evolution. The Sir2 was first discovered in *Saccharomyces cerevisiae* and was named after its ability to relieve gene silencing (Dali-Youcef et al. 2007). Phylogenetic analysis of sirtuins has divided the family into five different classes. Four of the five sirtuins from fall into class I and one in class III. We found that the sir2A and sir2C fall in one subclass while sir2B and sir2D fall in another subclass. sir2E could be seen as a separate clade from the rest of the sirtuins of *D. discoideum*. The *sirtuin* genes encode an important and complex family of proteins that participate in a wide array of physiological processes (North and Verdin 2004). In several species, caloric restriction has been shown to increase lifespan and decrease spontaneous rates of illness, such as insulin resistance, neurodegenerative disease, cancer and oxidative stress (Merksamer et al. 2013). Caloric restriction generally activates specific cellular signaling networks and therefore is thought that all chemicals that can induce sirtuin activity could possibly prove beneficial for therapeutics (Imai and Guarenete 2010; Johnson et al. 2013a, b). Sir2 is known to prolong the lifespan in budding yeast (Kaeberlein et al. 1999; Imai et al. 2000), flies and worms (Tissenbaum and Guarente 2001; Rogina and Helfand 2004). Humans have seven sirtuins namely SIRT1-7. SIRT1 of human plays a very important role in longevity. It is also reported that during fasting SIRT1 modulates gluconeogenesis in liver by deacetylating few important factors like PGC-1α, FOXO1, etc. (Brunet et al. 2004; Rodgers et al. 2005; Liu et al. 2008). SIRT1 is also found to function during inflammation, neurodegenerative diseases, cardiovascular disease and cancer (Luo et al. 2001; Yeung et al. 2004). SIRT2 (also called α-tubulin deacetylase) knockout delay entry into S-phase of the cell cycle (Li et al. 2007). It has been reported that SIRT2 specific inhibitor can ameliorate α-synuclein—mediated toxicity in Parkinson's disease (Outeiro et al. 2007). SIRT3, 4, and 5 are mitochondrial sirtuins as they have mitochondrial targeting sequence in their N-termini (Haigis et al. 2006). SIRT3 increases the acetyl-CoA by interacting with acetyl-CoA synthetase (Hallows et al. 2006). SIRT4 is known to control insulin secretion in pancreatic β-cells (Haigis et al. 2006) while SIRT5 is known to regulate urea cycle (Nakagawa et al. 2009). SIRT6 is nuclear localized and is known to be involved in DNA double strand break repair by regulating C-terminal binding protein (CtBP) and DNA protein kinase (McCord et al. 2009). SIRT7 is the least known sirtuin and possibly involved in regulating RNA polymerase I transcription (Ford et al. 2006).Thus, it seems reasonable to propose that the sirtuin protein family members could prove to represent novel molecular targets for the treatment of a variety of human diseases having a strong association with increasing age.

Sir2D, an ortholog of human SIRT1 was used to test this model of longevity in *D. discoideum* (Lohia 2015). We created an overexpressor and a knockout of *sir2D* by gene disruption followed by homologous recombination (Sutoh 1993) and tested for longevity (Fig. 3). Chronological ageing mechanism was studied in overexpressing cell line of *Ddsir2D.* As done with wild type cell population cell viability studies were conducted at different time points and as expected, at consecutive time intervals the percentage of PI¯ cells increased in comparison to wild type cells. This was further confirmed by the regrowth assay, which showed that the *sir2D* overexpressing cells survived 24 h more than the wild type cells confirming that in *D. discoideum* also Sirtuin 1 homolog increases longevity.

Longevity extension by overexpression of *sir2D* in *D. discoideum* is by autophagy induction (Fig. 3). Autophagic flux in cells can be monitored by microscopic visualization of the RFP-GFP-Atg8 tandem probe (Calvo-Garrido et al. 2011). Atg8 marks the autophagosomes and is taken as a marker for autophagy studies. RFP fluorescence is resistant to acidic pH of the lysosome whereas, GFP fluorescence gets quenched rapidly is a well-known fact. Thus, the red-green puncta

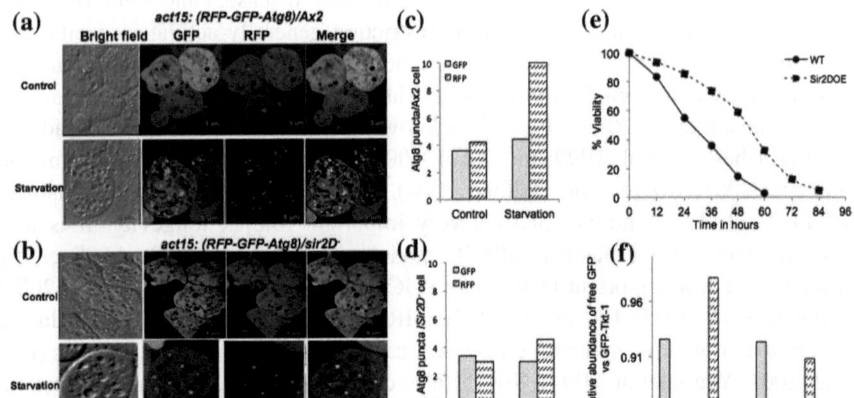

Fig. 3 Sir2D of *Dictyostelium discoideum* increases lifespan by inducing autophagy. Both wild type (**a**) and *sir2D*¯ (**b**) cells were transformed with tandemly tagged RFP-GFP-Atg8 (a marker for studying autophagic flux. Atg8 puncta/cell was counted using NIS elements software. The wild type **c** showed higher autophagy flux in comparison to the knockout cells (**d**). Scale bar: 5 μm. **e** Sir2D overexpressing cells show increased chronological lifespan as compared to wild type. **f** Autophagic flux by Western hybridization show lower ratio between free GFP and GFP–Tkt-1 in si2D knockout cells

marks the early autophagosomes and the red puncta that lack green fluorescence marks the fusion of autophagosomes with lysosomes. The *sir2D*$^{-}$ cells showed lower red puncta after starvation (Fig. 3) as compared to the control wild type. The percentage decrease in the red puncta count suggests a role for Sir2D in the process of autophagy. We found an increase in the red puncta in the wild type upon starvation as compared to the *sir2D*$^{-}$ cells. The *sir2D*$^{-}$ cells failed to activate autophagy in response to nutrient starvation as compared to wild type. Further, autophagic flux was also performed based on proteolytic cleavage (Calvo-Garrido et al. 2011). A significant decrease in free GFP levels in the *sir2D*$^{-}$ cells as compared to wild-type confirmed the decrease in autophagic flux (Fig. 3).

We also checked lifespan extensions by treating these cells with anti-aging drugs like spermidine, resveratrol and rapamycin. All these treatments induce autophagy and extend the chronological lifespan. Rapamycin treatment in *Dictyostelium* induces autophagy by the accumulation of ROS and intracellular free calcium (Swer et al. 2014) and vice versa. Figure 4 show the longevity extension in this organism in a dose dependent manner after spermidine treatment. We find that low

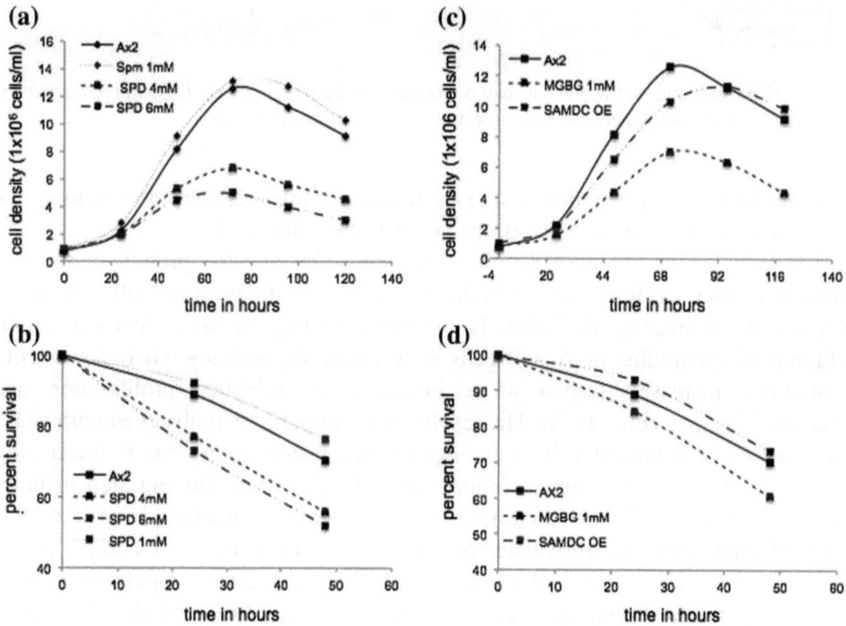

Fig. 4 Spermidine increases lifespan extension in *Dictyostelium discoideum.* **a** Shows the cell proliferation after addition of exogenous spermidine. Lower concentrations improve the proliferation over the control wild type cells. Higher concentrations inhibited cell proliferation. **b** Chronological life extensions were better in lower concentration of spermidine treatment. **c** Cell proliferation of overexpressing SAMDC that increases the spermidine levels in cells and treatment with MGBG that decreases the spermidine levels in comparison to wild type cells. **d** Chronological life extensions were better in cells having slightly higher net spermidine levels that are those which were treated with MGBG

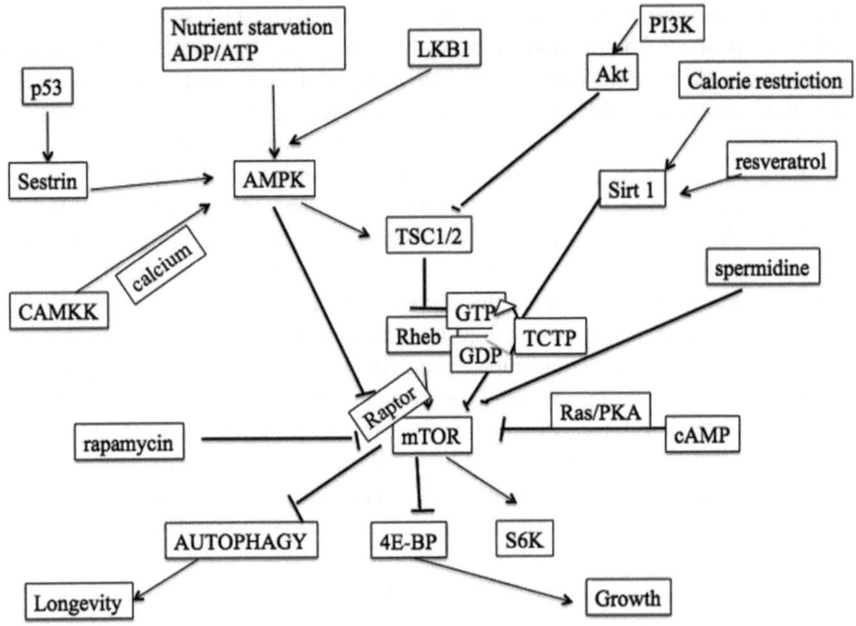

Fig. 5 Molecules and pathways that may control longevity in case of *Dictyostelium discoideum*. All possible interactions between the molecules are shown in the figure

concentrations of spermidine increases lifespan while higher doses induce cell death. The mechanism of these drugs is yet to be delineated.

We have earlier reported that both the transcript and the protein of the rate limiting enzyme ornithine decarboxylase were high in the prestalk/stalk cells in *D. discoideum* (Kumar et al. 2014). In the present study we show that exogenous addition of spermidine to A × 2 cells shows dose dependency where lower concentrations increase lifespan while higher doses inhibited proliferation and decreased lifespan (Fig. 4a, b). The results were similar to that observed with yeast, flies, worms and human cells. Our unpublished results show that at lower concentration spermidine treatment induces autophagy. A rate-limiting step in polyamine biosynthesis is the decarboxylation of *S*-adenosylmethionine (SAM) by SAMDC that yields decarboxylated SAM, which, in turn, donates its propyl amine moiety to form spermidine and spermine by two specific aminopropyl transferases, spermidine synthase and spermine synthase (Gilmour, 2007). Methylglyoxalbis-guanylhydrazone (MGBG) is a potent inhibitor of SAMDC (Regenass et al. 1992) and an antitumor drug with extremelytoxic side effects. Apart from addition of spermidine exogenously we created SAMDC overexpressing cells, which results in higher intracellular spermidine levels ultimately higher longevity (Fig. 4c, d). MGBG, which decreases net spermidine levels results in decreased lifespan.

Some of the signaling pathways and other molecules that may control longevity in *D. discoideum* by bringing about autophagy is shown in Fig. 5. Our preliminary unpublished data on the genes and other small molecules have confirmed their involvement in autophagy.

7 Conclusions

A major challenge is to find the interaction of these signaling pathways in inducing autophagy to increase the lifespan of an organism. Many intracellular signals recognized in *Dictyostelium* development are shown to be involved during vertebrate embryogenesis, synaptic transmission, or cellular interactions in plants or bacteria. We thus expect that they may show many parallel between the mechanisms of longevity with humans. We consider *D. discoideum* as one efficient model system to delineate these pathways and find the mechanism of action of these drugs. The beauty of this organism is to study autophagy in the absence of apoptosis as there are no known caspases identified in this organism. Since this organism shares high homology with humans, the result obtained here may hold true to certain extent in humans. Nutrient signaling plays an important role in the uni-to multicellularity in this organism the TOR pathway could be delineated well. The efforts made with this model organism would help understand the field of age-related research. Studies that could utilize genetic, nutritional and pharmacological manipulations to address the longevity issues could use *D. discoideum* as a model system to expand our knowledge.

Acknowledgments SS would like to thank Indian Council of Medical Research, India (54/41/CFP/GER/2011-NCD-II) and University Grants Commission, India (42-181/2013-SR) for the research grants. PJ thanks DST for the INSPIRE fellowship and PS thanks UGC for JRF.

References

Abe T, Maeda Y (1989) The prestalk/prespore differentiation and polarized cell movement in *Dictyostelium discoideum* slugs: a possible involvement of the intracellular Ca^{2+} concentration. Protoplasma 151:175–178

Bansal A, Zhu LJ, Yen K, Tissenbaum HA (2015) Uncoupling lifespan and healthspan in *Caenorhabditis elegans* longevity mutants. Proc Natl Acad Sci 112:277–286

Baur JA, Sinclair DA (2006) Therapeutic potential of resveratrol: the in vivo evidence. Nat Rev Drug Discov 5:493–506

Bhat AH, Dar KB, Anees S, Zargar MA, Masood A, Sofi MA et al (2015) Oxidative stress, mitochondrial dysfunction and neurodegenerative diseases; a mechanistic insight. Biomed Pharmacother 74:101–110

Bjedov I, Toivonen JM, Kerr F, Slack C, Jacobson J, Foley A et al (2010) Mechanisms of Life Span Extension by Rapamycin in the Fruit Fly *Drosophila melanogaster*. Cell Metab 11:35–46

Bozzaro S (2013) The model organism *Dictyostelium discoideum*. Methods Mol Biol 983:17–37

Browne LH, Williams KL (1993) Pure populations of *Dictyostelium discoideum* prespore and prestalk cells obtained by flow cytometry have different redevelopment characteristics at their cell surfaces. Cytometry 14:660–667

Brunet A, Sweeney LB, Sturgill JF, Chua KF, Greer PL, Lin Y, Tran H, Ross SE, Mostoslavsky R and Cohen HY (2004) Stress-dependent regulation of FOXO transcription factors by the SIRT1 deacetylase. Science 303:2011–2015

Burnett C, Valentini S, Cabreiro F, Goss M, Somogyvári M, Piper MD et al (2011) Absence of effects of Sir2 overexpression on lifespan in *C. elegans* and Drosophila. Nature 477:482–485

Calvo-Garrido J, Carilla-Latorre S, Mesquita A, Escalante R (2011) A proteolytic cleavage assay to monitor autophagy in *Dictyostelium discoideum.* Autophagy 7:1063–1068

Cornillon S, Foa C, Davoust J, Buonavista N, Gross JD, Golstein P (1994) Programmed cell death in *Dictyostelium.* J Cell Sci 107:2691–2704

Dali-Youcef N, Marie L, Sebastian F, Christian K, Kristina S and Johan A (2007) Sirtuins: the magnificent seven, function metabolism and longevity. Ann Med 39:335–345

Eichinger L, Pachebat J, Glöckner G, Rajendream MA, Sucgang R, Berriman M et al (2005) The genome of the social amoeba *Dictyostelium discoideum.* Nature 435:43–57

Eisenberg T, Knauer H, Schauer A, Büttner S, Ruckenstuhl C, Carmona-Gutierrez D et al (2009) Induction of autophagy by spermidine promotes longevity. Nat Cell Biol 11:1305–1314

Fabrizio P, Pozza F, Pletcher SD, Gendron CM, Longo VD (2001) Regulation of longevity and stress resistance by Sch9 in yeast. Science 292:288–290

Fabrizio P, Longo VD (2003) The chronological life span of *Saccharomyces cerevisiae.* Aging Cell 2:73–81

Fabrizio P, Gattazzo C, Battistella L, Wei M, Cheng C, McGrew K et al (2005) Sir2 blocks extreme life-span extension. Cell 123:655–667

Ford E, Voit R, Liszt G, Magin C, Grummt I, Guarente L (2006) Mammalian Sir2 homolog SIRT7 is an activator of RNA polymerase I transcription. Genes Dev 20:1075–1080

Fujita H, Sugiura S, Momomura S, Omata M, Sugi H, Sutoh K (1997) Characterization of mutant myosins of *Dictyostelium discoideum* equivalent to human familial hypertrophic cardiomyopathy mutants. Molecular force level of mutant myosins may have a prognostic implication. J Clin Invest 99:1010–1015

Gilmour SK (2007) Polyamines and non-melanoma skin cancer. Toxicol Appl Pharmacol 224:249–256

Giusti C, Tresse E, Luciani MF, Golstein P (2009) Autophagic cell death: analysis in *Dictyostelium.* Biochimica Biophysica Acta 1793:1422–1431

Gross JD (1994) Developmental decisions in *Dictyostelium discoideum.* Microbiol Rev 58:330–351

Haigis MC, Mostoslavsky R, Haigis KM, Fahie K, Christodoulou DC et al (2006) SIRT4 inhibits glutamate dehydrogenase and opposes the effects of calorie restriction in pancreatic beta cells. Cell 126:941 –954

Hallows WC, Lee S, Denu JM (2006) Sirtuins deacetylate and activate mammalian acetyl-CoA synthetases. Proc. Natl. Acad. Sci. 103 10230–10235

Imai S, Armstrong CM, Kaeberlein M, Guarente L (2000) Transcriptional silencing and longevity protein Sir2 is an NAD-dependent histone deacetylase. Nature 403:795–800

Imai S, Guarenete L (2010) Ten years of NAD-dependent SIR2 family deacetylases: implications for metabolic diseases. Cell 31:212–220

Inoki K, Ouyang H, Li Y, Guan KL (2005) Signaling by target of rapamycin proteins in cell growth control. Microbiol Mol Biol Rev 69:79–100

Johnson SC, Rabinovitch PS, Kaeberlein M (2013a) mTOR is a key modulator of ageing and age-related disease. Nature 493:338–345

Johnson LM, Robinson MM, Nair KS (2013b) Skeletal muscle aging and the mitochondrion. Trends Endocrinol Metab 24:247–256

Kaeberlein M, McVey M, Guarente L (1999) The SIR2/3/4 complex and SIR2 alone promote longevity in *Saccharomyces cerevisiae* by two different mechanisms. Genes Dev 13:2570–2580

Kaeberlein M, Burtner CR, Kennedy BK (2007) Recent developments in yeast aging PLoS genet 3:e84

Kaeberlein M (2010) Lessons on longevity from budding yeast. Nature 464:513–519

Kesidou E, Lagoudaki R, Touloumi O, Poulatsidou KN, Simeonidou C (2013) Autophagy and neurodegenerative disorders. Neural Regen Res 8:2275–2283

Kessin RH (2001) *Dictyostelium*—Evolution, cell biology, and the development of multicellularity. Cambridge Univ Press

Konijn TM, Raper KB (1961) Cell aggregation in *Dictyostelium discoideum*. Dev Biol 3:725–756

Kumar R, Rafia S, Saran S (2014) Cloning, expression and characterization of the ornithine decarboxylase gene from *Dictyostelium disocideum*. Int J Dev Biol 58:669–676

Lamming DW, Ye L, Sabatini DM, Baur JA (2013) Rapalogs and mTOR inhibitors as anti-aging therapeutics. J Clin Invest 123:980–989

Lepez-Otin C, Blasco MA, Partridge L, Serrano M, Kroemer G (2013) The hallmarks of aging. Cell 153:1194–1217

Lepilina A, Coon AN, Kikuchi K, Holdway JE, Roberts RW, Burns CG, Poss KD (2006) A dynamic epicardial injury response supports progenitor cell activity during zebrafish heart regeneration. Cell 127:607–619

Levine B, Mizushima N, Virgin HW (2011) Autophagy in immunity and inflammation. Nature 469:323–335

Li W, Zhang B, Tang J, Cao Q, Wu Y, Wu C, Guo, J, Ling EA, Liang F (2007) Sirtuin 2, a mammalian homolog of yeast silent information regulator-2 longevity regulator is an oligodendroglial protein that decelerates cell differentiation through deacetylating alpha-tubulin. J Neurosci 27 2606–2616

Liu Y, Dentin R, Chen D, Hedrick S, Ravnskjaer K, Schenk S et al (2008) A fasting inducible switch modulates gluconeogenesis via activator/coactivator exchange. Nature 456:269–273

Lohia R (2015) Role of HDACs during growth development and differentiation of *Dictyostelium discoideum*, Ph.D thesis, University of Delhi, New Delhi

Longo VD, Liou LL, Valentine JS, Gralla EB (1999) Mitochondrial superoxide decreases yeast survival in stationary phase. Arch Biochem Biophys 365:131–142

Longo VD, Fabrizio P (2012) Chronological aging in *Saccharomyces cerevisiae*. Subcell Biochem 57:101–121

Luciani MF, Giusti C, Harms B, Oshima Y, Kikuchi H, Kubohara Y, Golstein P (2011) Atg1 allows second-signaled autophagic cell death in *Dictyostelium*. Autophagy 7:501–508

Luo J, Nikolaev a Y, Imai S, Chen D, Su F, Shiloh A et al (2001) Negative control of p 53 by Sir2alpha promotes cell survival under stress. Cell 107:137–148

Madeo F, Tavernarakis N, Kroemer G (2010) Can autophagy promote longevity? Nat Cell Biol 12:842–846

Maiuri MC, Zalckvar E, Kimchi A, Kroemer G (2007) Self-eating and self-killing: crosstalk between autophagy and apoptosis. Nat Rev Mol Cell Biol 8:741–752

Martin PR, Shea RJ, Mulks MH (2001) Identification of a plasmid-encoded gene from *Haemophilus ducreyi* which confers NAD independence. J Bacteriol 183:1168–1174

Matthews KA, Kaufman TC, Gelbart WM (2005) Research resources for Drosophila: then expanding universe. Nat Rev Genet 6:179–193

McCord RA, Michishita E, Hong T, Berber E, Boxer LD, Kusumoto R et al (2009) SIRT6 stabilizes DNA-dependent protein kinase at chromatin for DNA double strand break repair. Aging 1:109–121

McDonald RB, Ruhe RC (2011) Aging and Longevity: Why Knowing the Difference Is Important to Nutrition Research. Nutrients 3:274–282

Merksamer PI, Liu Y, He W, Hirschey MD, Chen D and Verdin E (2013) The sirtuins, oxidative stress and aging: an emerging link. Aging 5:1–7, 135

Miller ES, Heidelberg JF, Eisen JA, Nelson WC, Durkin AS, Ciecko A et al (2003) Complete genome sequence of the broad host-range vibrio phage KVP40: comparative genomics of a T4-related bacteriophage. J Bacteriol 185:5220–5233

Minois N, Carmona-Gutierrez D, Madeo F (2011) Polyamines in aging and disease. Aging 3:716–732

Mitchell SJ, Scheibye-Knudsen M, Longo DL, de Cabo R (2015) Animal models of aging research: implications for human aging and age-related diseases. Ann Rev Ani Biosci 3:283–303

Morino K, Petersen KF, Shulman GI (2006) Molecular Mechanisms of Insulin Resistance in Humans and Their Potential Links With Mitochondrial Dysfunction. Diabetes 55(Supplement 2):S9–15

Morselli E, Mariño G, Bennetzen MV, Eisenberg T, Megalou E, Schroeder S et al (2011) Spermidine and resveratrol induce autophagy by distinct pathways converging on the acetylproteome. J Cell Biol 192:615–629

Mosmann T (1983) Rapid colorimetric assay for cellular growth and survival: application to proliferation and cytotoxicity assays. J. Immunological Methods 65:55–63

Muthukkumar S, Ramesh TM, Bondada S (1995) Rapamycin, a potent immunosuppressive drug, causes programmed cell death in B lymphoma cells. Transplantation 60:264–270

Nakagawa T, Lomb DJ, Haigis MC, Guarente L (2009) SIRT5 deacetylates carbamoyl phosphate synthetase 1 and regulates the urea cycle. Cell 137:60–70

North BJ, Verdin E (2004) Sirtuins: Sir2-related NAD-dependent protein deacetylases. Genome Biol 5:224

Ocampo A, Barrientos (2011) A Quick and reliable assessment of chronological life span in yeast cell populations by flow cytometry. Mech Ageing Dev 132:315–323

Outeiro TF, Kontopoulos E, Altmann SM, Kufareva I, Strathearn KE, Amore AM et al (2007) Sirtuin 2 inhibitors rescue alpha-synuclein mediated toxicity in models of Parkinson‘s disease. Science 317:516–519

Phelan JP (1992) Genetic variability and rodent models of human aging. Exp Gerontol 27:147–159

Powers RW, Kaeberlein M, Caldwell SD, Kennedy BK, Fields S (2006) Extension of chronological life span in yeast by decreased TOR pathway signaling. Genes Dev 20:174–184

Rabinowitz JD, White E (2010) Autophagy and metabolism. Science 330:1344–1348

Raffaelli N, Lorenzi T, Mariani PL, Emanuelli M, Amici A, Ruggieri S, Magni G (1999) The *Escherichia coli* NadR regulator is endowed with nicotinamide mononucleotide adenylyl transferase activity. J Bacteriol 181:5509–5511

Raper KB (1935) *Dictvostelium discoideum*: a new species of slime mold from decaying forest leaves. J Agric Res 50:135–147

Regenass U, Caravatti G, Mett H, Stanek J, Schneider P, Muller M, Matter A, Vertino P, Porter CW (1992) New S-adenosylmethionine decarboxylase inhibitors with potent anti tumor activity. Cancer Res 52:4712–4718

Rodgers JT, Lerin C, Haas W, Gygi SP, Spiegelman BM, Puigserver P (2005) Nutrient control of glucose homeostasis through a complex of PGC-1alpha and SIRT1. Nature 434:113–118

Rogina B, Helfand SL (2004) Sir2 mediates longevity in the fly through a pathway related to calorie restriction. Proc Natl Acad Sci 101:15998–16003

Roisin-Bouffay C, Luciani MF, Klein G, Levraud JP, Adam M, Golstein P (2004) Developmental cell death in *Dictyostelium* does not require paracaspase. J Biol Chem 279:11489–11494

Rubinsztein DC, Mariño G, Kroemer G (2011) Autophagy and aging. Cell 146:682–695

Saran S, Nakao H, Tasaka M, Iida H, Tsuji FI, Nanjundiah V, Takeuchi I (1994) Intracellular free calcium level and its response to cAMP stimulation in developing *Dictyostelium* cells transformed with jellyfish apoaequorin cDNA. FEBS Lett 337:43–47

Scarlatti F, Maffei R, Beau I, Codogno P, Ghidoni R (2008) Role of non-canonical Beclin 1-independent autophagy in cell death induced by resveratrol in human breast cancer cells. Cell Death Differ 15:1318–1329

Sehgal SN (2003) Sirolimus: its discovery, biological properties, and mechanism of action. Transplant Proc 35:7S–14S

Soulard A, Cohen A, Hall MN (2009) TOR signaling in invertebrates. Curr Opin Cell Biol 21:825–836

Stanfel MN, Shamieh LS, Kaeberlein M, Kennedy BK (2009) The TOR pathway comes of age. Biochim Biophys Acta 1790:1067–1074

Sutoh K (1993) A transformation vector for *Dictyostelium discoideum* with a new selectable marker *Bsr*. Plasmid 30:150–154

Swer PB, Lohia R, Saran S (2014) Analysis of rapamycin induced autophagy in *Dictyostelium discoideum*. Indian J Exp Biol 52:295–304

Swer PB, Mishra H, Lohia R, Saran S (2015) Overexpression of TOR (target of rapamycin) inhibits cell proliferation in *Dictyosteilum discoideum*. J Basic Micriobiol (In press)

Tavernarakis N, Pasparaki A, Tasdemir E, Maiuri MC, Kroemer G (2008) The effects of p53 on whole organism longevity are mediated by autophagy. Autophagy 4:870–873

Tissenbaum HA, Guarente L (2001) Increased dosage of a *sir-2* gene extends lifespan in *Caenorhabditis elegans*. Nature 410:227–230

Vellai T, Takacs-Vellai K, Zhang Y, Kovacs AL, Orosz L, Müller F (2003) Genetics: influence of TOR kinase on lifespan in *C. elegans*. Nature 426:620

Wasko BM, Kaeberlein M (2014) Yeast replicative aging: a paradigm for defining conserved longevity interventions. FEMS Yeast Res 14:148–159

Weeks G (2000) Signaling molecules involved in cellular differentiation during *Dictyostelium* morphogenesis. Curr Opin Microbiol 3:625–630

Williams JG, Ceccarelli A, McRobbie S, Mahbubani H, Kay RR, Early A, Berks M, Jermyn KA (1987) Direct induction of *Dictyostelium* prestalk gene expression by DIF provides evidence that DIF is a morphogen. Cell 49:185–192

Williams RS, Cheng L, Mudge AW, Harwood AJ (2002) A common mechanism of action for three mood-stabilizing drugs. Nature 417:292–295

Williams RS, Boeckeler K, Graf R, Muller-Taubenberger A, Li Z, Isberg RR et al (2006) Towards a molecular understanding of human diseases using *Dictyostelium discoideum*. Trends Mol Med 12:415–424

Wullschleger S, Loewith R, Hall MN (2006) TOR signaling in growth and metabolism. Cell 124:471–484

Yeung F, Hoberg JE, Ramsey CS, Keller MD, Jones DR, Frye RA, Mayo MW (2004) Modulation of NF-kappa B-dependent transcription and cell survival by the SIRT1 deacetylase. EMBO J 23:2369–2380

Yorimitsu T, Klionsky DJ (2005) Autophagy: molecular machinery for self-eating. Cell Death 12 (Suppl 2):542–552

Down Regulation of Sirtuins in the Serum of Frail Elderly as Compared to Non-frail Elderly: Promising Diagnostic Protein Marker for Frailty

Sharmistha Dey

Abstract Ageing, the accumulation of damage to molecules, cells and tissues over a lifetime, often leads to frailty and malfunction. The functional decline, disability and frailty are the common conditions of the geriatric syndromes. The syndrome of frailty is complex and still lack of a standard definition. As the exact cause of frailty of older people is not known, it will be a great challenge to develop biomarker that characterize the frailty syndrome and help in clinical diagnosis. This will provide a new ways of defining disability outcomes in epidemiological studies of the elderly. Based on above evidences, this study attempts to devise a new tool and develop sirtuins as a biological marker for frailty in old age. Most recent study has observed the low expression of sirtuin in old age. This study evaluated the sirtuin in frail person and compared with non-frail elderly person. The concentration of SIRT1, SIRT2 and SIRT3 were evaluated by Surface Plasmon resonance technology. In frail elderly the declining level of SIRT1, SIRT2 and SIRT3 were observed as compared to non-frail elderly. The pronounced declining level of sirtuin in frail elderly can provides an opportunity to develop this protein as a predictive marker in early stages of frailty with suitable cut off values.

Keywords Frailty · Serum · Sirtuin · Protein marker

1 Introduction

Ageing or senescence is a biological phenomenon operates at various levels of living the process. Starting at the molecular level of proteins and nucleic acids, senescence can affect cells, tissues and organs of the body of all living beings. There are two categories in modern biological theories of aging: programmed and damage theories. In programmed theory, the aging follows the biological system which associated

S. Dey (✉)
Department of Biophysics, All India Institute of Medical Sciences, Ansari Nagar, New Delhi 110029, India
e-mail: sharmistha_d@hotmail.com

© Springer Science+Business Media Singapore 2017

P.C. Rath et al. (eds.), *Topics in Biomedical Gerontology*,
DOI 10.1007/978-981-10-2155-8_3

with changes in gene expression, hormones which regulate insulin/IGF-1 signaling (IIS) pathway of aging (Van Heemst 2010) and immune system (Cornelius 1972). The damage theories emphasize on environmental stress that accumulates damages at various level cause aging. This damage include damages of cells and tissues (Gerschman et al. 1954), increase of oxygen based metabolism (Brys et al. 2007), cross linking of proteins damages the cells and tissues, accumulation of free radicals destroy the cells and tissues and finally the organs (Denham Harman 1956; Gerschman et al. 1954). This theory also involve the somatic DNA damage continuously which causes genetic mutation and accumulate with increasing age causing cells to deteriorate and lead to mitochondrial dysfunction. By Hayflick theory, the human cells can divide approximately 50 times and telomeres have been shown to shorten with each successive cell division (Campisi 2000).

According to United Nations Population Division, the aging population (above 60 years) in India is predicted to increase significantly from 8 % in 2010 to 19 % in 2050. This age group population is expected to become 323 millions which is greater than the total US population in 2012. Unfortunately, the care of the elderly has not been as advanced as the aging population is increasing. As the elderly population continues to grow, the impact of frailty is felt throughout families and pervading our economic, health care, and social systems and provision of effective in patient care for frail older people is becoming a key issue. Sirtuins levels were analyzed to be significantly lower among frail patients compared to the non-frail.

The term "frailty" is in use in clinical practice nearly for three decades and its understanding has improved as the elderly population has increased in most societies. It is commonly used to indicate older persons, at increased risk for morbidity and mortality. The syndrome of frailty is a multi-organ dysfunction in sub-cellular level, which affects older people and progresses irreversibly if not intervened at very early stage. It is noticeable by decreased of strength, increased weakness to stressors, as well as high risk for serious adverse health outcomes including disability, dependency, and mortality. It increases with ageing, but more so with biological ageing. In its early stage its clinical findings are very subtle, so diagnosis from clinical assessment is very difficult.

There is no standard definition for "frailty" in Geriatric Medicine literature. It is a phenotypical expression of a collapsing system which manifests as lack of muscle strength, exhaustion, unintentional weight loss, lack of desire to function, low physical activity, slow speed of movement and being bed bound. The "syndrome of frailty" is a concept rather than a disease and objective diagnostic parameters are still evolving.

There are several diagnostic criteria related to physical performance tests, but the criteria proposed by Fried et al. (2001) is most used in clinical practice and research. These definition suits North American and European populations with well established health care system. However, for Indian population there are several practical difficulties in using these tools because of social cultural reasons and lack of information on health status in childhood and adulthood. To obviate these difficulties there is a need for developing a new tool, which uses the principles of the above criteria but obviates the disadvantages, useful for older Indians.

Frailty may develop in the setting of chronic inflammation, dementia, heart failure and other chronic debilitating diseases. However in many older individuals, no obvious ill health may be present. As the exact pathway that leads to the state of frailty is still not clear, there is no diagnostic biological test available till date. Frailty is a common term used to denote older people at increased risk of mortality and morbidity. Frailty is analogous of "failure to thrive" in children. Syndrome of multi organ dysfunction at sub cellular level, which affects elderly community and progress irreversibly, if not intervened at an early stage.

Ageing or senescence is a strong risk factor for developing frailty with accumulation of senescent cells in tissues and organs over the lifetime (Heemels 2010). The accumulation of reactive oxygen species (ROS), increases with ageing due to the insufficient anti–oxidant molecules which causes disability and frailty in the elder people. The prolonged oxidative stress develop low muscle mass, alteration of metabolic system, inflammation, thus causing cellular damage and develop frailty. CR (Calorie Restriction) can moderate the oxidative damage and increase the life span of numerous organisms yeast, mice etc. Considering the strong association of frailty with senescence, it is essential to explore molecules linked to the process of senescence. Recent studies have identified Sirtuin protein molecules with a role in senescence which has been emerged as a critical epigenetic regulator of ageing.

Sirtuins (silent information regulator) are a conserved family of NAD-dependent proteins found in all domains of life. They regulate ribosomal DNA recombination, gene silencing, DNA repair, chromosomal stability and longevity. It also affects the calorie restriction, the only regiment that slow down ageing and extend lifespan. Calorie restriction is a term used to refer to the finding that life span of a creature can be increased by 50 % by feeding it a diet that is nutritious but 30 % less in calories.

In experimental small organisms, such as yeast, worms, and flies, it has been demonstrated that SIRT2 family proteins take part a vital role in the regulation of aging and longevity (Astrom et al. 2003; Howitz et al. 2003). Increasing the activity of SIRT2 and its orthologs extends the life spans of those organisms, while deactivation or mutations of the SIRT2 genes shorten their life spans (Kaeberlein et al. 1999; Rogina and Helfand 2004; Tissenbaum and Guarente 2001; Wood et al. 2004; Lin et al. 2000). In certain genetic backgrounds, SIRT2 family proteins are also required for the CR-mediated life span elongation in those organisms (Lin et al. 2002; Wang and Tissenbaum 2006; Dali-Youcef et al. 2007).

In mammals, seven sirtuin isoforms (SIRT 1-7) are known and so far SIRT1 is well studied (Schwer and Verdin 2008; Shin-ichiro 2009). Although it is still unclear whether SIRT1 regulates aging and longevity in mammals, several studies have verified that SIRT1 regulates critical metabolic responses to nutritional signal, specifically to low nutritional input, in several tissues, as well as cell survival in response to stress and damage, both features common to the response to CR in mammals (Harikumar and Aggarwal 2008).

Sirtuins have been termed as the "guardian of cell" for their anti ageing property. Researchers believe that drugs like resveratrol which boost sirtuin production can help fight diseases of ageing (Chen et al. 2005). The appeal of sirtuins as anti-ageing drug targets was elevated significantly by two related developments. First, studies carried out in yeast led to the proposal that SIRT1 acts as a key downstream mediator of life-span extension from dietary constraint. Dietary constraint lacking mal-nutrition is known to increase life span in multiple organisms and to hold-up the appearing of variety of age-associated diseases. Second, the small molecule resveratrol was identified as an activator of SIRT1 and was observed to increase duration of life in lower organism like yeast, worms, flies etc. More recently, resveratrol has also been found to protect mice against deterioration of health consequences of a high-fat diet, an effect attributed to activation of SIRT1. It has been found that SIRT1 activation is a prerequisite for both resveratrol and calorie restriction to expand the longevity of life in multi-cellular organisms, together with evidence that calorie restriction has beneficial effects on age-associated morbidity in primates suggests that sirtuin activity is of significance to human aging (Bordone et al. 2007; Barger et al. 2008; Ferrara et al. 2008).

The potential marker for early detection of frailty remains elusive till date. Considering the strong association of frailty with senescence, it is necessary to search molecules linked to the process of senescence. The study of the proteins in human body fluids (e.g. blood, urine, saliva etc.) is very good approach to identify marker for disease. As a result of secretion and leakage many proteins originated from different tissues of human come into the serum by circulation (Taylor 1969). The concentration of these proteins reflects human physiological or pathological state as suggested by several earlier reports (Anderson and Anderson et al. 2002; Thadikkaran et al. 2005). As frailty is not any tissue or specific organ related condition, hence serum samples considered to be ideal source for biomarkers. In this cross sectional study, evaluation of serum sirtuin concentration was assessed in frail and non-frail older subjects with an objective of examining it as a marker of frailty in old age. As a result of secretion and leakage many proteins originated from different tissues of human come into the serum by circulation.

2 Methods

2.1 Study Group

In this study group, 200 elderly (≥ 60 years of age) patients were included in from Geriatric Medicine Outpatient Department, AIIMS. The study was approved by the Ethics Committee (IESC/T-270/01.07.2011) of AIIMS and all participants provided written informed consent. After the assessment for frailty status for frailty by Fried's criteria it was found that 119 (59.5 %) were non-frail and 81 (40.5 %) as frail. According to demographic characteristics of the cases frail were found to be in higher age group compare to non-frail. Diabetes mellitus (34 %) and hypertension (32 %) were common co-morbidities and higher among frail subjects. No significant difference was observed for COAD (chronic obstructive airways disease)

Fig. 1 Gender based percentage of frail and non-frail in study group

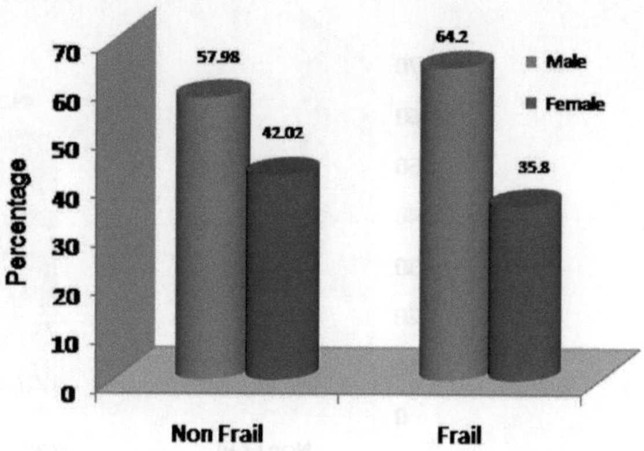

prevalence between Frail and Non-frail. Proportion of female was lower for both Frail and Non-frail (Fig. 1). Prevalence of Diabetes is more prominent in Frail (Fig. 2). No difference in MMSE between Frail (25.94 ± 2.25) and Non-Frail (24.95 ± 2.87). No significant in mean BMI (Body Mass Index) difference between Frail and Non-Frail.

2.1.1 Evaluation of Serum SIRT1 and SIRT2 by Surface Plasmon Resonance (SPR)

The concentration of sirtuins (1, 2, 3) were determined by label free SPR by immobilizing antibody of sirtuin on different flow cells on CM5 sensor chips by amine coupling methods and then serum of patients and control were passed over it to get the corresponding Resonance unit (RU). The concentration of the sirtuins (1, 2, 3) present in each samples were evaluated from standard curve. The standard curve were plotted with different concentrations of recombinant sirtuins (1, 2, 3) proteins passed over antibody on sensor chips and corresponding RU's obtained.

3 Results and Discussion

The SIRT 1, 2, 3 were found to be decreased significantly in case of frail compare to non-frail subjects (Fig. 3), still after adjustment for age, sex, diabetes mellitus, hypertension, cognitive status (MMSE scores) and number of co morbidities (Table 1). SIRT1 (Table 2) and SIRT3 (Table 3) levels were lower in case of frail subject irrespective of age (>60), sex, diabetes mellitus, hypertension and number of co-morbidities; however SIRT2 was not significantly lower in compare to non-frail.

Receiver Operating Characteristic (ROC) curves were created for each protein. Based on the SPR data, the area under curve was calculated from ROC to measure

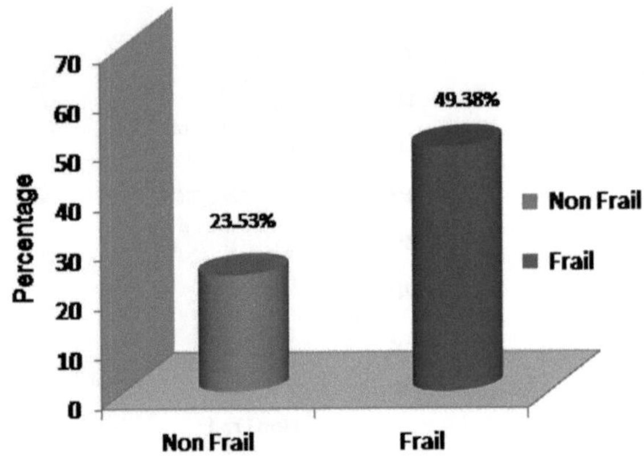

Fig. 2 Prevalence of diabetes among frail and non-frail

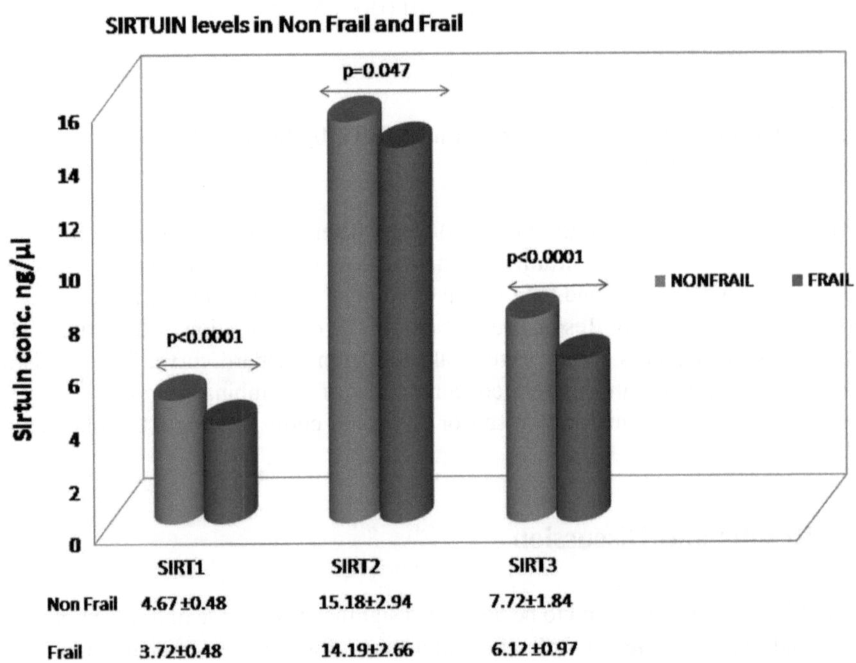

Fig. 3 Sirtuin levels are significantly lower in case of frail as compared to non-frail

the utility of each protein as potential marker for frailty. In this study lower level was associated with frail disease condition; hence the ROC curves were constructed to detect non-frails. The area under curve for SIRT1 (0.9037) and SIRT3 (0.7988) claimed that both can be used to detect non-frail (Fig. 2). SIRT1 at a cutoff value of <4.29 ng/µl can detect frailty with sensitivity of 81.48 % and specificity of

Table 1 Concentration of SIRT1, SIRT2 and SIRT3 in serum of frail and non-frail elderly (ng/μl) represented as mean ± SE

Protein	Unadjusted/Adjusted	Non-frail	Frail	*p*-value
SIRT1	Unadjusted	4.67 ± 0.04	3.72 ± 0.63	<0.0001
	Adjusted	4.68 ± 0.05	3.71 ± 0.07	<0.0001
SIRT2	Unadjusted	15.19 ± 0.27	14.19 ± 0.29	<0.01
	Adjusted	15.20 ± 0.28	14.17 ± 0.36	0.047
SIRT3	Unadjusted	7.71 ± 0.17	6.12 ± 0.11	<0.0001
	Adjusted	7.7 ± 0.15	6.14 ± 0.19	<0.0001

Table 2 Level of serum SIRT1 (ng/μl) with respect to different characteristics of frail and non-frail elderly

	Non-frail	Frail	*p*-value
Age			
60–69	4.69 ± 0.5	3.70 ± 0.61	<0.0001
70–79	4.60 ± 0.43	3.69 ± 0.53	<0.0001
≥80	5.12 ± 0.11	3.84 ± 0.67	0.02
Gender			
Male	4.64 ± 0.40	3.77 ± 0.59	<0.0001
Female	4.70 ± 0.56	3.64 ± 0.53	<0.0001
Diabetes			
Yes	4.65 ± 0.58	3.69 ± 0.49	<0.0001
No	4.67 ± 0.44	3.76 ± 0.64	<0.0001
Hypertension			
Yes	4.73 ± 0.53	3.68 ± 0.57	<0.0001
No	4.64 ± 0.46	3.76 ± 0.58	<0.0001
No. of comorbidities			
0 or 1	4.69 ± 0.45	3.77 ± 0.67	<0.001
2	4.56 ± 0.60	3.77 ± 0.51	<0.001
≥3	4.77 ± 0.52	3.51 ± 0.46	0.001

79.83 %. Similarly, at cut off value <6.61 ng/μL SIRT3 has a sensitivity of 70.37 % and specificity of 70.59 % to detect frailty. Cutoff value for SIRT2 was not calculated since the area under curve was very low (Fig. 4). However, the diagnostic efficiency of SIRT1 to distinguish frail from non-frail was more accurate as compared to SIRT3.

As the exact cause of frailty of older people is not known, it will be a great challenge to develop biomarker that characterize the frailty syndrome and help for clinical diagnosis. Though the lifespan has increases in our population by the success of modern medicine, but still most of the elderly end up in frail. Some biological parameter have been used for frailty as a non-specific biomarkers like: inflammation (Leng et al. 2002, 2004; Leng et al. 2007, 2009), vitamin-E (Ble et al. 2006), central adiposity, serum albumin, oxidative stress (Wu et al. 2009), and 25 hydrooxy vitamin-D level (Hirani et al. 2013). This study will provide a new ways

Table 3 Level of serum SIRT3 (ng/μl) with respect to different characteristics of frail and non-frail elderly

	Non-frail	Frail	*P*-value
Age			
60–69	7.67 ± 1.96	5.93 ± 0.96	0.001
70–79	7.76 ± 1.6 0	6.11 ± 1.04	<0.0001
≥ 80	8.46 ± 2.72	6.30 ± 0.79	0.01
Gender			
Male	7.72 ± 1.84	6.18 ± 1.01	<0.0001
Female	7.70 ± 1.86	6.0 ± 0.91	<0.0001
Diabetes			
Yes	7.50 ± 1.48	5.90 ± 0.97	<0.0001
No	7.78 ± 1.94	6.53 ± 0.93	<0.0001
Hypertension			
Yes	7.37 ± 1.5 0	6.09 ± 0.86	<0.0001
No	6.13 ± 1.06	7.81 ± 1.92	<0.0001
No. of comorbidities			
0 or 1	7.81 ± 1.91	6.24 ± 1.24	<0.0001
2	7.25 ± 1.50	6.09 ± 0.83	0.001
≥3	7.74 ± 1.77	5.93 ± 0.63	0.00

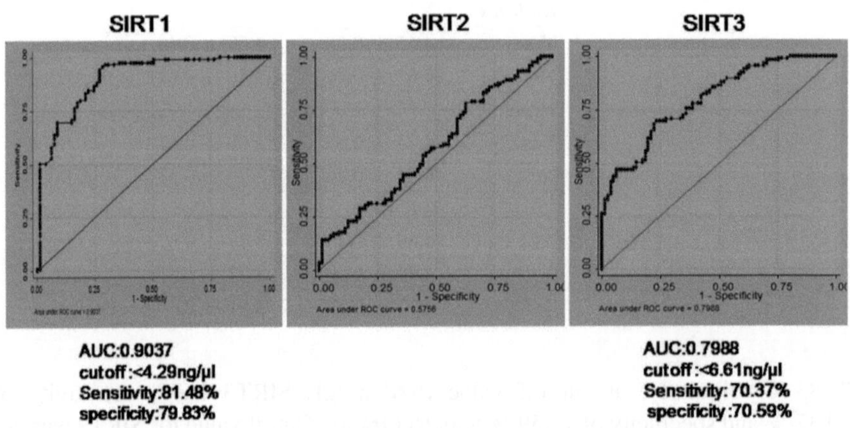

Fig. 4 ROC analysis showing the area under *curve* for SIRT1, SIRT2 and SIRT3 to distinguish frail from non-frail subjects

of defining disability outcomes in epidemiological studies of the elderly. This study is the first attempts to develop sirtuins as a protein marker which can be use as a new clinical tool for detection for frailty in old age. The parameters of the proteins obtain can differentiate frail from non-frail at early onset of the syndrome.

Acknowledgements Authors acknowledge research section of AIIMS for proving financial support for purchasing chemicals.

References

Anderson NL, Anderson NG (2002) The human plasma proteome: history, character, and diagnostic prospects. Mol Cell Proteomic 1:845–867

Astrom SU, Cline TW, Rine J (2003) The Drosophila melanogaster sir2+ gene is nonessential and has only minor effects on position-effect variegation. Genetics 163:931–937

Bordone L, Cohen D, Robinson A, Motta MC, van Veen E, Czopik A, Steele AD, Crowe H, Marmor S, Luo J, Gu W, Guarente L (2007) SIRT1 transgenic mice show phenotypes resembling calorie restriction. Aging Cell 6:759–767

Barger JL, Kayo T, Vann JM, Arias EB, Wang J, Hacker TA, Wang Y, Raederstorff D, Morrow JD, Leeuwenburgh C, Allison DB, Saupe KW, Cartee GD, Weindruch R, Prolla TA (2008) A low dose of dietary resveratrol partially mimics caloric restriction and retards aging parameters in mice. PLoS ONE e 3:2264

Brys K, Vanfleteren JR, Braeckman BP (2007) Testing the rate-of-living/oxidative damage theory of aging in the nematode model Caenorhabditis elegans. Exp Gerontol 42:845–851

Ble A, Cherubini A, Volpato S, Bartali B, Walston JD, Windham BG, Bandinelli S, Lauretani F, Guralnik JM, Ferrucci L (2006) Lower plasma vitamin E levels are associated with the frailty syndrome: the In CHIANTI study. JGerontol A BiolSci Med Sci 61:278–283

Cornelius E (1972) Increased incidence of lymphomas in thymectomized mice–evidence for an immunological theory of aging. Experientia 28:459

Chen D, Steele AD, Lindquist S, Guarente L (2005) Increase in activity during calorie restriction requires SIRT1. Science 310:1641

Campisi J (2000) Cancer, aging and cellular senescence. In Vivo 14:183–188

Dali-Youcef N, Lagouge M, Froelich S, Koehl C, Schoonjans K, Auwerx J (2007) Sirtuins: the ‘magnificent seven’, function, metabolism and longevity. Ann Med 39:335–345

Ferrara N, Rinaldi B, Corbi G (2008) Exercise training promotes SIRT1 activity in aged rats. Rejuvenation Res 11:139–150

Fried LP, Tangen CM, Walston J, Newman AB, Hirsch C, Gottdiener J, Seeman T, Tracy R, Kop WJ, Burke G, McBurnie MA (2001) Frailty in older adults: evidence for a phenotype. J Gerontol A Biol Sci Med Sci 56:146–156

Gerschman R, Gilbert DL, Nye SW, Dwyer P, Fenn WO (1954) Oxygen poisoning and xirradiation: a mechanism in common. Science 119:623–626

Heemels M-T (2010) Ageing. Nat Insight 464:503–544

Howitz KT, Bitterman KJ, Cohen HY, Lamming DW, Lavu S, Wood JG, Zipkin RE, Chung P, Kisielewski A, Zhang LL, Scherer B, Sinclair DA (2003) Small molecule activators of sirtuins extend Saccharomyces cerevisiae lifespan. Nature 425:191–196

Harman D (1956) Aging: a theory based on free radical and radiation chemistry. J Gerontol 11 (3):298–300

Harikumar KB, Aggarwal BB (2008) Resveratrol: a multitargeted agent for age-associated chronic diseases. Cell Cycle 7:1020–1035

Hirani V, Naganathan V, Cumming RG, Blyth F, Le Couteur DG, Handelsman DJ, Waite LM, Seibel MJ (2013) Associations Between Frailty and Serum 25-Hydroxyvitamin D and 1,25-Dihydroxyvitamin D Concentrations in Older Australian Men: The Concord Health and Ageing in Men Project. J Gerontol A BiolSci Med Sci 68:1112–1121

Kaeberlein M, McVey M, Guarente L (1999) The SIR2/3/4 complex and SIR2 alone promote longevity in Saccharomyces cerevisiae by two different mechanisms. Genes Dev 13:2570–2580

Lin SJ, Defossez PA, Guarente L (2000) Life span extension by calorie restriction in S. cerevisiae requires NAD and SIR2. Science 289:2126–2128

Lin SJ, Kaeberlein M, Andalis AA, Sturtz LA, Defossez PA, Culotta VC, Fink GR, Guarente L (2002) Calorie restriction extends *Saccharomyces cerevisiae* lifespan by increasing respiration. Nature 418:344–348

Leng S, Chaves P, Koenig K, Walston J (2002) Serum interleukin-6 and hemoglobin as physiological correlates in the geriatric syndrome of frailty: a pilot study. J Am Geriatr Soc 50:1268–1271

Leng SX, Cappola AR, Andersen RE, Blackman MR, Koenig K, Blair M, Walston JD (2004) Serum levels of insulin-like growth factor-I (IGF-I) and dehydroepiandrosterone sulfate (DHEA-S), and their relationships with serum interleukin-6, in the geriatric syndrome of frailty. Aging Clin Exp Res 16:153–157

Leng SX, Xue QL, Tian J, Walston JD, Fried LP (2007) Inflammation and frailty in older women. J Am Geriatr Soc 55:864–871

Leng S, Xue Q, Tian J (2009) Association of neutrophil and monocyte counts with frailty in community-dwelling older women. Exp Gerontol 44:511–516

Rogina B, Helfand SL (2004) Sir2 mediates longevity in the fly through a pathway related to calorie restriction. Proc Natl Acad Sci USA 101:15998–16003

Schwer B, Verdin E (2008) Conserved metabolic regulatory functions of sirtuins. Cell Metab 7:104–112

Shin-ichiro (2009) SIRT1 and Caloric Restriction: An Insight Into Possible Trade-Offs Between Robustness and Frailty. Imai Curr Opin Clin Nutr Metab Care 12:350–356

Taylor WH (1969) Serum enzymes in the diagnosis of disease. Br J Anaesth 41:227–234

Thadikkaran L, Siegenthaler MA, Crettaz D, Queloz PA, Schneider P, Tissot JD (2005) Recent advances in blood-related proteomics. Proteomics 5:3019–3034

Tissenbaum HA, Guarente L (2001) Increased dosage of a sir-2 gene extends lifespan in Caenorhabditis elegans. Nature 410:227–230

Van Heemst D (2010) Insulin, IGF-1 and longevity. Aging Disease 1:147–157

Wood JG, Rogina B, Lavu S, Howitz K, Helfand SL, Tatar M, Sinclair D (2004) Sirtuin activators mimic caloric restriction and delay ageing in metazoans. Nature 430:686–689

Wang Y, Tissenbaum HA (2006) Overlapping and distinct functions for a Caenorhabditis elegans SIR2 and DAF-16/FOXO. Mech Ageing Dev 127:48–56

Wu IC, Shiesh SC, Kuo PH, Lin XZ (2009) High oxidative stress is correlated with frailty in elderly chinese. J Am Geriatr Soc 57:1666–1671

Morphology of the Human Pancreas During Development and Aging

Renu Gupta, Shubhi Saini, Saroj Sharma, Tony G. Jacob and T.S. Roy

Abstract Pancreas gets affected by fibrosis associated with aging. This study analyzed the age-related fibrotic changes in the ductular system of the pancreas. After obtaining necessary ethical clearances, twelve human fetal and thirty post-natal and adult pancreas were collected and processed to obtain resin-embedded sections for transmission electron microscopy and paraffin-embedded sections for H&E staining and light microscopy. The sections were analyzed qualitatively and quantitatively. The human pancreas had mature zymogen granules in the exocrine part and secretory granules in the endocrine part by 20th week of gestation. The amount of connective tissue and acini increased with age. After the 3rd decade, there was increased fibrosis. This began around small and medium sized ducts. Changes in the epithelium of ducts were seen in later decades. There was a direct correlation between area of the ducts and increasing age. There was increased fibrosis in and around the islets of Langerhans. The number of fibroblasts and stellate cells increased with age. The

Electronic supplementary material The online version of this article (doi:10.1007/978-981-10-2155-8_4) contains supplementary material, which is available to authorized users.

R. Gupta
Department of Anatomy, All India Institute of Medical Sciences, Jodhpur
Rajasthan, India
e-mail: drrenu.gupta79@gmail.com

S. Saini · T.G. Jacob
Department of Anatomy, All India Institute of Medical Sciences,
New Delhi 110029, India
e-mail: shubh2685@gmail.com

S. Sharma
Dr. B.S. Ambedkar Medical College and Hospital, Rohini, Delhi 110085, India
e-mail: drsaroj2001@gmail.com

T.S. Roy (✉)
Department of Anatomy, All India Institute of Medical Sciences,
Room no. 1018, Ansari Nagar, New Delhi 110029, India
e-mail: tsroyaiims@gmail.com; tsroy@aiims.ac.in

© Springer Science+Business Media Singapore 2017

P.C. Rath et al. (eds.), *Topics in Biomedical Gerontology*,
DOI 10.1007/978-981-10-2155-8_4

increased fibrosis with increasing age that first appears around the small and medium sized ducts may be due to increased number of pancreatic stellate cells.

Keywords Stellate cells · Zymogen granules · Islets of langerhans · Fibrosis · Nucleator · Transmission electron microscopy

1 Introduction

Pancreas is the largest digestive gland that has both exocrine and endocrine functions. The exocrine component consists of serous acini packed into small lobules that secrete a range of enzymes involved in the digestion of lipids, proteins and carbohydrates. The endocrine component forms 1–2 % of the volume of the pancreas and is scattered throughout the substance of the gland and controls glucose metabolism and upper gastrointestinal motility (Junqueira and Carneiro 1995). Both the exocrine and endocrine components are involved in various disease processes during one's lifetime. Acute pancreatitis (AP), an acute inflammatory condition affecting the exocrine pancreas, annually affects between 5 and 80 people per 100,000 (Whitcomb 2006). In chronic pancreatitis (CP) there are chronic inflammatory cells within the pancreas, progressive fibrosis, sclerosis and parenchymal atrophy. Malfunction of the endocrine component in CP leads to diabetes mellitus (DM) (Kloppel and Maillet 1993). Beyond developing on a background of CP, pancreatic cancer arises usually from the ductal elements of the exocrine gland and rarely from the acinar cells themselves (Klöppel et al. 1980). It is known now that the adult pancreas can regenerate in a suitable environment even after severe injury related to AP. Further, transplanted adult pancreatic tissue in both experimental animals and diabetic patients can differentiate into islet cells and produce insulin (Lohr et al. 1989; Bonner Weir et al. 1993). Hence, it becomes important to know various aspects related to development and aging with respect to acini, ducts and islets.

In the present study we investigated the detailed histology of pancreatic acinar and ductular system during development and aging. Since the ductular cells regulate the differentiation of the acinus and the islets, we also studied the exocrine, endocrine and some of the interstitial components.

2 Materials and Methods

2.1 Specimen Collection

2.1.1 Collection of Fetuses

Twelve fetuses were collected from the Department of Obstetrics and Gynecology, All India Institute of Medical Sciences, New Delhi in accordance with the protocol approved by the Institutional Ethics Committee. Due consent was taken from the parents of the fetuses for the use of the material for research purposes. The fetuses were measured for crown-rump length, foot length (Mandarim-de-Lacuda 1990; Sailaja et al. 1996) and biparietal diameter to correlate with fetal age and then fixed in 4 % buffered paraformaldehyde and stored 4 °C to minimize the postmortem changes. The pancreas of the fetuses was dissected out after initial fixation and a portion of the pancreas was also immersed in modified Karnovsky's fixative (4 % paraformaldehyde, 1 % glutaraldehyde in 0.1 M phosphate buffer, pH 7.4).

2.1.2 Collection of Adult Pancreata

Thirty human pancreata from persons aged between the 40th post-natal day and 62 years were collected from the mortuary of the Department of Forensic Medicine, All India Institute of Medical Sciences; New Delhi in accordance with the protocol approved by the Institutional Ethics Committee. (Consent was deemed unnecessary by the Ethics Committee for the use of post-mortem samples). The pancreas was immersed in 4 % paraformaldehyde and preserved at 4 °C to minimize postmortem autolytic changes. The middle part of the pancreas was removed after fixation and a sagittal slice was processed for the study. In addition, a small portion of the pancreas was also immersed in modified Karnovsky's fixative as above.

2.2 Tissue Preparation

After fixation of the whole pancreas, it was dissected from the surrounding connective tissue and a small sagittal slice from middle part near the neck of the organ was placed in fresh fixative at 4 °C. After proper fixation the slice was washed under tap water for 5–6 h for removing the extra fixatives. The tissue slice was then processed for paraffin embedding hematoxylin and eosin staining and microscopy.

2.3 *Tissue Processing*

The pancreatic tissue fixed in buffered paraformaldehyde was washed in running tap water and processed for paraffin embedding; whereas the pancreatic tissue fixed in modified Karnovsky's fixative was washed in phosphate buffer (pH 7.4) and processed for resin embedding.

2.3.1 Hematoxylin and Eosin Staining and Measurements

The paraffin blocks were sectioned at 5 μm, taken on glass slides, stained with hematoxylin and eosin and mounted with DPX. The stained sections of fetal and adult pancreas were examined under Zeiss Axiophot Research microscope (Oberkochen, Germany). The sections were assessed qualitatively and quantitatively. The areas of the pancreatic ducts were estimated from the captured images using a video camera attached to the microscope. The nucleator probe of the StereoInvestigator software (MicroBrightfield Inc., USA) was used to measure the area of pancreatic ducts.

2.3.2 Microtomy and Staining for Light and Transmission Electron Microscopy

Semithin sections (1 μm) were cut with a glass knife on Reichert Ultramicrotome and stained with Toluidine blue (1 % solution in 1 % borax). The stained sections were utilized not only for morphological assessment but also to determine the area of interest for ultramicrotomy. Trimmed resin blocks of pancreas were further sectioned using glass knives to obtain ultrathin sections (50–60 nm thick, silver colored). These were picked up on 300-mesh copper grids. These ultrathin sections of the pancreas were stained with uranyl acetate and lead citrate and observed under the transmission electron microscope (Philips Morgagni 268D TEM, Field Emission Inc., Eindhoven, the Netherlands). Photomicrographs were obtained by a CCD camera, stored and analyzed further.

2.4 *Statistical Analysis*

The data were expressed as mean ± standard deviation (S.D.). Ducts were divided into three types, according to the luminal size: the large duct (had luminal area more than 500 μ^2), the medium sized duct (had luminal area between 150–499 μ^2), and the smaller ducts, (the area was less than 150 μ^2) (Table 1). The pancreata were divided into various groups according to the age of the deceased at the time of death. Group 1-before birth; Group 2-immediately after birth up to 10 years; Group

Table 1 Area of pancreatic ducts in fetuses (different gestational age) and adults (in different decades)

	Areas of small ducts (<150 mm^2) Mean (SD)			Ares of medium sized ducts (150–499 mm^2) Mean (SD)			Areas of large sized ducts (>500 mm^2) Mean (SD)		
Ages	A	B	C	A	B	C	A	B	C
Prenatal (n = 5)	113.72 (39.6)	516.89 (143.79)	1676.35 (452.60)*	302.89 (22.47)	1402.24 (625.61)	2715.62 (960.68)	1159.53 (346.96)	2673.402 (536.66)	4219.064 (706.53)**
Postnatal (n = 3)	94.37 (35.79)	426.79 (150.80)	1082.15 (239.68)*	247.23 (22.89)	974.91 (146.08)	2370.86 (941.03)	963.20 (382.39)	2455.15 (1024.93)	4550.60 (1709.95)**
2nd Decade to 4th decade (n = 13)	88.07 (27.90)	545.99 (223.22)	1336.04 (430.90)*	271.55 (70.11)	1085.12 (364.88)	2817.70 (929.69)	2558.52 (419.45)	6821.94 (1309.72)	19291.84 (3286.26)**
>5th decade (n = 10)	82.18 (23.00)	488.25 (223.93)	1898.24 (696.72)*	318.05 (77.34)	1264.39 (474.67)	4094.27 (2240.94)	1724.57 (982.98)	4154.52 (2398.28)	14161.13 (9480.46)**

A Luminal area of duct, *B* Area of duct along with epithelium, *C* Area of duct along with connective tissue around it. H and p value for fibrous wall around the small ducts were respectively 7.75, 0.051. H and p value for fibrous wall around the medium sized ducts were respectively 2.45, 0.485. H and p value for fibrous wall around the large ducts were respectively 10.38, 0.016

*$p < 0.05$

**$p < 0.01$

3-up to 35 years of age; Group 4-above 35 years. These data were analyzed using Kruskal-Wallis non-parametric test followed by post hoc test (Mann-Whitney-U test). To test the increase the amount the connective around the duct after birth and up to 65 years, Pearson's correlation test was applied. Pearson's correlation test was also used for comparison of data for area of lumen of duct and total area of the duct, including its wall. The software SPSS (v12, IBM, USA) was utilized for the statistical analysis.

3 Observations

3.1 Macroscopic Anatomy of Developing and Aging Pancreas

The pancreas in the fetus was fleshy and multilobulated. The neonatal pancreas had all of the features of an adult pancreas and its various subdivisions could be recognized. Its head was proportionately large in newborn and it was continuous with the body and tail. The inferior border of the head of pancreas was in contact with the 'C' shaped duodenum. The body and tail passed upwards and to the left, to reach the hilum of the spleen. The fibrous connective tissue capsule was thin in the fetal and neonatal period; extending up to 3rd decade. Thereafter, the thickness of the connective tissue gradually increased up to the 7th decade. There were no other visible changes in the gross appearance in the adult and aged population.

3.2 Microscopic Anatomy

3.2.1 Fetal Pancreas

At the twelfth gestational week (GW) the pancreas contained numerous ducts and few acini. There was abundant connective tissue containing blood vessels between the ducts and acini. Simple columnar or cuboidal cells lined the ducts. Acini had small lumina and were lined by columnar to cuboidal cells. Extra cellular matrix was loose and irregular and contained some undifferentiated dark cells. Blood vessels contained RBCs and were lined by single layer of flattened epithelium.

At the fourteenth GW, the acini were small but their lumina were clearly visible. Ducts were lined by simple cuboidal to columnar epithelium. Connective tissue surrounded the ducts and acini. The extra-cellular matrix was disorganized. Ultrastructurally, the acini of the pancreas at 14 gestational weeks had elongated cells with little cytoplasm and large, open-phase nuclei having a single nucleolus. The apical parts of these cells were in close apposition and showed small vacuoles. There were small spaces between clusters of these cells. There were many ducts and

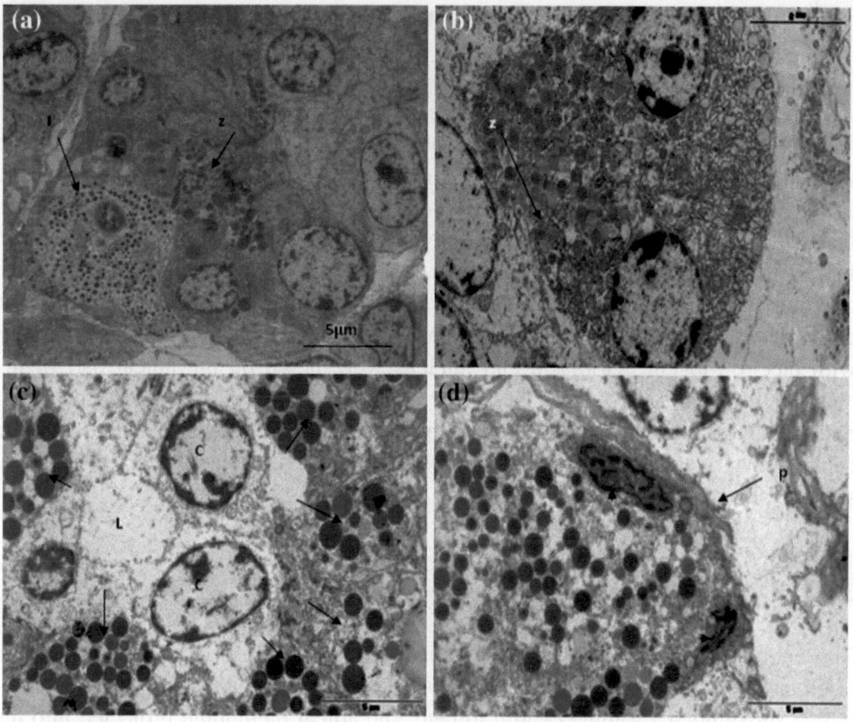

Fig. 1 Electronmicrographs of the pancreas showing **a** an acinus in a 20 week old fetus and its clear lumen (L) with few cells showing zymogen granules (z). In close proximity to the acini, the mature form of the islets (I) can be seen. **b** A single secretory acinar cell in a 40 post natal day infant, having two nuclei, evidenced by absence of cell membrane between the nuclei. Note the zymogen granules (z) in the apical part of the cell. **c** Centroacinar cells (C) covering the apical part of the acinar cells (arrows) in an 18 year old. Note the lumen (L) of the acinus. The centroacinar cells had few organelles and diffusely distributed polyribosomes. They form the beginning of the smallest ducts. **d** A pancreas from a 25 years old individual shows an acinus with undifferentiated cell (arrowhead), along the acinar process (p) of a stellate cell

in some areas the ductular cells contained endocrine cells among the ductal cell. These cells contained multiple small electron-dense granules in the cytoplasm (Fig. 1).

At the twentieth GW, large acini with clearly visible lumina were visible in the parenchyma of the pancreas. Ducts were lined by simple columnar epithelium. Some of the larger ducts had compound epithelium. Connective tissue around the ducts was well organized. Extra cellular matrix (ECM) was organized and less compact. Under the electron microscope, the cells of the acini showed some zymogen granules towards their apical part (Fig. 1). Undifferentiated cells were scattered in the ECM especially outside the basement membrane of the acini and the duct (Fig. 1). Numerous mature endocrine cells of the islets of Langerhans lay adjacent to the acini and ducts (Fig. 1).

3.2.2 Pancreas of Early Postnatal, Infantile Period and First Decade of Life

The pancreata of this period of life had similar appearance. They contained differentiated acini. Many acini showed zymogen granules near their apices. Ducts were lined by simple columnar or cuboidal epithelium and had well-organized connective tissue around them. There were a few undifferentiated dark cells seen scattered in extra cellular matrix even up to 5 months after the birth. Ultrastructurally, the majority of the cells of the acini showed well-formed zymogen granules with a diameter of 500–1000 nm and they were mostly found towards the apical region, establishing an apical to basal polarity to the cells. Centroacinar cells with electron-lucent cytoplasm and nucleus were noted on the apical part of the acinar cells (Fig. 1). Some of the acinar cells had two nuclei (Fig. 1). The larger ducts showed columnar epithelium with tight junctional complexes between the cells. Some of the ductal cells were observed to have monocilia on their apical surface (Suppl. Fig. 1).

3.2.3 Second Decade

The exocrine component of pancreas consisted of closely packed secretory acini. Each acinus was made up of irregular cluster of pyramid shaped cells, the apices of which surrounded a minute central lumen. The apical membrane projected into the lumen as microvilli. The acinar cells were typical protein secreting cells. The nuclei were basally located and surrounded by basophilic cytoplasm with rough endoplasmic reticulum; the apices of the cells were packed with eosinophilic secretory granules (called zymogen granules). Some acinar cells contained greater amount of zymogen granules and some had lesser in amount at same time. The center of the acini frequently contained one or more pale nuclei of centroacinar cells with sparse pale-stained cytoplasm; these represented the terminal lining cells of minute ducts. Adjacent acini were separated by inconspicuous supporting tissue containing numerous capillaries.

The larger ducts were mainly confined towards the central part of the pancreas, whereas the medium sized ducts were present in the intermediate area (between the periphery and the central part of the gland in a sagittal section of the gland). The smaller ducts were present in the periphery of the gland. In our qualitative study it was noted that major changes in the ductular system occurred in the fourth decade, therefore the quantitative data were analyzed and expressed in the following manner. Small ducts were seen to be lined by a single layer of cuboidal epithelium with centrally placed round or oval nucleus. Medium sized ducts were lined by low columnar to cuboidal cells. Both type of ducts had many dark cells in their basal regions. Large ducts were lined by a single layer of tall columnar cells, which had brush border on their luminal side. The large ducts had few dark cells in their basal regions. Some of these dark cells extended from basement membrane to lumen. Some of the ductal cells had dark granules in their apical cytoplasm (Fig. 2).

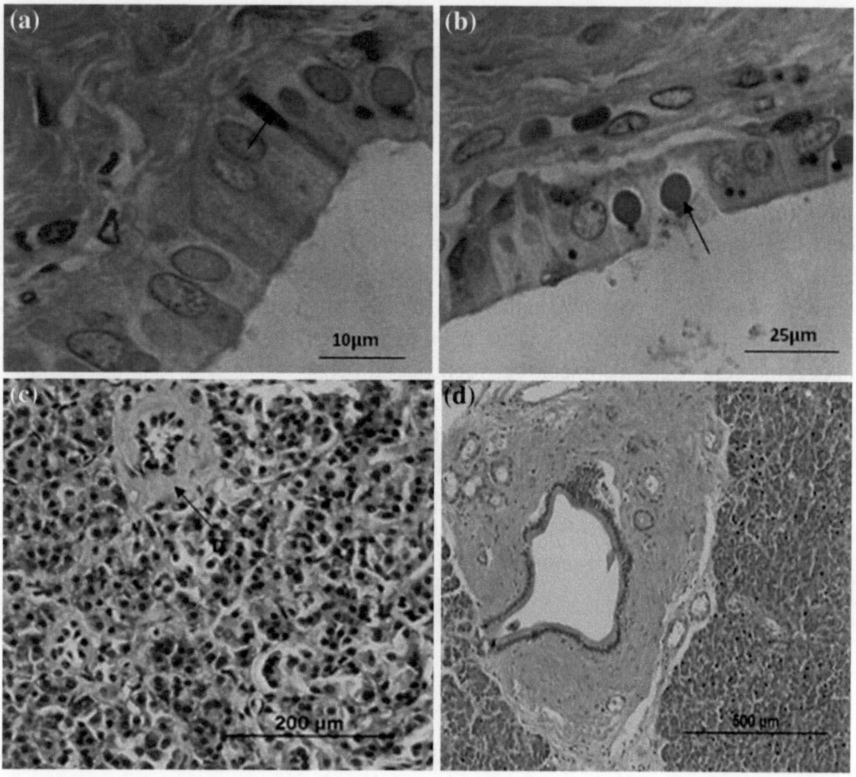

Fig. 2 **a** Photomicrograph of 27 years pancreas showing a large duct lined by single layer of tall columnar cells. These ducts show *dark* cells in the basal region. Some of these dark cells extend from basement membrane to lumen (*arrow*). Toluidine *blue* stain. **b** Photomicrograph of 27 years pancreas showing large duct epithelium in which some of the cells had *dark granules* in the apical cytoplasm. Toluidine blue stain. **c** Photomicrograph of 52 years pancreas showing small duct fibrosis (D). H & E stain. **d** Photomicrograph of 60 years pancreas showing distortion of duct lumen (*arrow*). H & E stain

Ultrastructurally, it was noted that ductal cells possessed round or oval nuclei, rough endoplasmic reticulum, mitochondria and a well-developed Golgi apparatus. The epithelium had a large number of mucus-secreting cells that had many mucin granules in their apical zone. There were also some endocrine cells that could be seen near the basal region of the epithelium.

The centroacinar cells were seen close to the luminal aspect of the acinar cells (Fig. 1). Centroacinar cells closely resembled ductal cells. Ultrastructurally, the centroacinar cells had fewer organelles and diffusely distributed polyribosomes. They had small mitochondria that were located towards the luminal surface, where the Golgi apparatus was also seen. The nucleus was round or oval with marginal indentation, usually having at least one nucleolus. Some centroacinar cells had complexly arranged, long cytoplasmic processes extending between the acinar cells.

The endocrine part of pancreas (islets of Langerhans) were scattered among the exocrine glandular tissue. They were spheroidal in shape, measuring 52–210 μm in diameter and the cells in the islets were organized as irregular cords. They were composed of groups of secretory cells, supported by a fine network of connective tissue, containing numerous fenestrated capillaries. The same connective tissue formed a delicate capsule that surrounded each islet. The endocrine cells were small with pale staining granular cytoplasm and these were in contrast to the large acinar cells of surrounding exocrine pancreas that stained strongly with H&E. Ultrastructurally, these endocrine cells were filled with electron dense secretory vesicles that had a maximum diameter of 300 nm with no polarity to the cells. Apart from the clustered endocrine cells in the islet of Langerhans, some unitary endocrine cells were also seen between ductal cells.

The extra cellular matrix in the pancreas was distributed throughout the gland, but was more prominent around the large ducts and blood vessels. Small number of macrophages, neutrophils, mast cells, plasma cells, lymphocytes and fat cells were seen scattered in the ECM.

Stellate cells were also seen in the ECM around the acini. These cells were spindle-shaped that also contained lipid vesicles and intermediate filaments in their cytosol.

3.2.4 Third Decade

The pancreas in third decade of life did not show any major differences from the histological features observed in the second decade.

3.2.5 Fourth Decade

In the pancreata of the fourth decade there was increased amount of ECM. Increased amount of fibrous tissue was noted in the ECM and this was confined to the periductal region of small and medium sized ducts. Fibroblasts were often seen in a perilobular position.

3.2.6 Fifth Decade

Pancreas of the fifth decade had fatty infiltration and fibrosis around the ducts, epithelial hyperplasia that appeared as papillary projections into the lumen of the duct (Fig. 2). Increased number of fibroblast cells was observed in the ECM around the ducts. There was disruption of the basement membrane of the acinar epithelium.

The islets were seen to be more dispersed and scattered in small fragments within the lobules of the pancreas.

Table 2 Pearsons correlation test (The fetal samples were excluded)

		AGECAT1	LOWLUMI	LOWEPI	LOWCONNE	MEDLUM	MEDEPI	MEDCONNE	HIGLUM	HIGEPI	HIGCONNE
AGECAT1	Pearson Correlation	1	−0.021	0.174	0.550[a]	0.390	0.367	0.424[b]	−0.077	−0.120	−0.077
	Sig. (2-tailed)		0.922	0.417	0.005	0.059	0.078	0.039	0.719	0.576	0.721
	N	24	24	24	24	24	24	24	24	24	24
LOWLUMI	Pearson Correlation	−0.021	1	0.415[b]	0.433[b]	−0.024	0.133	0.143	−0.481[b]	−0.512[b]	−0.470[b]
	Sig. (2-tailed)	0.922		0.044	0.034	0.913	0.536	0.506	0.017	0.011	0.020
	N	24	24	24	24	24	24	24	24	24	24
LOWEPI	Pearson Correlation	0.174	0.415[b]	1	0.593[a]	0.107	0.355	0.066	−0.212	−0.275	−0.258
	Sig. (2-tailed)	0.417	0.044		0.002	0.618	0.089	0.761	0.319	0.193	0.223
	N	24	24	24	24	24	24	24	24	24	24
LOWCONNE	Pearson Correlation	0.550[a]	0.433[b]	0.593[a]	1	0.365	0.441[b]	0.336	−0.325	−0.376	−0.320
	Sig. (2-tailed)	0.005	0.034	0.002		0.080	0.031	0.109	0.121	0.070	0.128
	N	24	24	24	24	24	24	24	24	24	24
MEDLUM	Pearson Correlation	0.390	−0.024	0.107	0.365	1	0.791[a]	0.612[a]	−0.296	−0.250	−0.337
	Sig. (2-tailed)	0.059	0.913	0.618	0.080		0.000	0.001	0.160	0.239	0.107
	N	24	24	24	24	24	24	24	24	24	24
MEDEPI	Pearson Correlation	0.367	0.133	0.355	0.441[b]	0.791[a]	1	0.842[a]	−0.198	−0.157	−0.174
	Sig. (2-tailed)	0.078	0.536	0.089	0.031	0.000		0.000	0.353	0.463	0.417
	N	24	24	24	24	24	24	24	24	24	24
MEDCONNE	Pearson Correlation	0.424[b]	0.143	0.066	0.336	0.612[a]	0.842[a]	1	−0.245	−0.209	−0.169
	Sig. (2-tailed)	0.039	0.506	0.761	0.109	0.001	0.000		0.249	0.327	0.429
	N	24	24	24	24	24	24	24	24	24	24
HIGLUM	Pearson Correlation	−0.077	−0.481[b]	−0.212	−0.325	−0.296	−0.198	−0.245	1	0.948[a]	0.980[a]
	Sig. (2-tailed)	0.719	0.017	0.319	0.121	0.160	0.353	0.249		0.000	0.000
	N	24	24	24	24	24	24	24	24	24	24

(continued)

Table 2 (continued)

		AGECAT1	LOWLUMI	LOWEPI	LOWCONNE	MEDLUM	MEDEPI	MEDCONNE	HIGLUM	HIGEPI	HIGCONNE
HIGEPI	Pearson Correlation	−0.120	−0.512[b]	−0.275	−0.376	−0.250	−0.157	−0.209	0.948[a]	1	0.941[a]
	Sig. (2-tailed)	0.576	0.011	0.193	0.070	0.239	0.463	0.327	0.000		0.000
	N	24	24	24	24	24	24	24	24	24	24
HIGCONNE	Pearson Correlation	−0.077	−0.470[b]	−0.258	−0.320	−0.337	−0.174	−0.169	0.980[a]	0.941[a]	1
	Sig. (2-tailed)	0.721	0.020	0.223	0.128	0.107	0.417	0.429	0.000	0.000	
	N	24	24	24	24	24	24	24	24	24	24

LOWLUMI Luminal area of small ducts, *LOWEPI* Area of small duct along with lining epithelium, *LOWCONNE* Area of small duct along with connective tissue, *MEDLUM* Luminal area of medium ducts, *MEDEPI* Area of medium sized duct along with lining epithelium, *MEDCONNE* Area of medium sized duct along with connective tissue, *HIGLUM* Luminal area of large ducts, *HIGEPI* Area of large sized duct along with lining epithelium, *HIGCONNE* Area of large sized duct along with connective tissue

[a]Correlation is significant at the 0.01 level (2-tailed).

[b]Correlation is significant at the 0.05 level (2-tailed)

3.2.7 Sixth Decade Onwards

In this age group, stellate cells were more commonly seen around the basal parts of the acinus. Small processes of the cytoplasm of the centroacinar cells covered a variably small to large portion of the luminal aspect of the neighboring acinar cells. Large ducts showed distortions of their lumen and their epithelial cells had accumulation of lipid droplets in their apical region and mucinous granules in their cytoplasm (Fig. 2). The amount of fibrous tissue increased in and around the islets.

3.3 Quantitative Observations-Stereological Estimates

The nucleator probe was used for the estimation of areas of pancreatic ducts. The measurements of the ducts were taken with respect to their lumen (A); lumen including epithelium (B) and duct lumen with the fibrous wall of the duct (C).

The estimations of area have been summarized in Table 1. There was significant increase of connective tissue around the large sized ducts ($H = 10.38$, $p = 0.016$) and area of connective tissue around the small ducts ($H = 7.75$, $p = 0.051$) with aging, at all age groups studied. There was no statistically significant increase in area of the connective tissue around the medium sized ducts with aging ($H = 2.44$, $p = 0.485$). A maximum increase in the connective tissue was noted in 3rd age category, up to 35 years old and onwards.

For Pearsons correlations test, data of prenatal and early postnatal groups were excluded from the sample. The Pearsons Correlations test showed significantly increase in the connective tissue organization around the small ducts (55 %, $p = 0.005$) and connective tissue around the medium sized ducts (42.4 %, $p = 0.039$) between 5 months pancreas to 65 years pancreas. But no significant changes were observed in the connective tissue around the large sized ducts (77 %, $p = 0.72$) (Table 2, Fig. 3). Maximum increase in fibrosis occurred in the fourth decade.

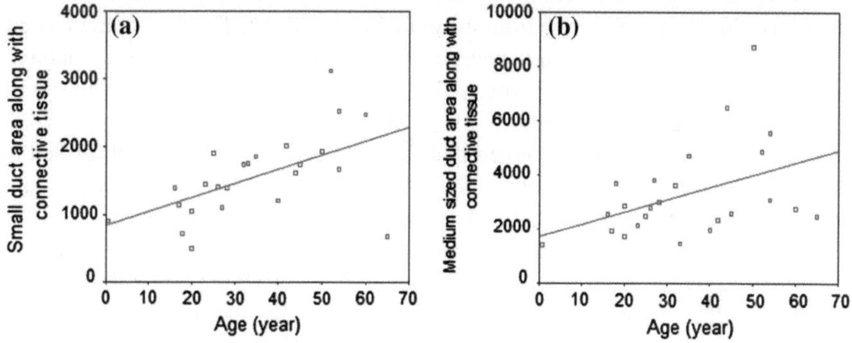

Fig. 3 **a** Correlation of the area of small ducts of the pancreas (including the connective tissue surrounding them) with age. **b** Correlation of the area of medium sized ducts of the pancreas (including the connective tissue surrounding them with age

4 Discussion

This study provides for the first time, the detailed light and electron microscopic morphology of the developing and age related changes in the exocrine and endocrine part of the human pancreas in Indian population. In this study we also observed similarities and differences between fetal and adult pancreas. We have used the unbiased stereological technique to measure the various ducts, its epithelium and the connective tissue organization around the ducts, which has not been reported earlier.

The pancreas was small during fetal period and grew gradually to the adult size during the 2nd decade. Connective tissue around the whole gland was very less up to the 3rd decade, after which it increased dramatically up to the 7th decade, the maximum age group available in the present study.

4.1 Pancreatic Development

The pancreatic exocrine cells differentiate from the endodermal ventral and dorsal pancreatic buds respectively. Initially, the buds are solid, surrounded by undifferentiated mesenchyme. Later these buds proliferate several times forming smaller terminal buds. Canaliculi appear in between the cells of the solid buds thereby forming the acini. The endocrine part of pancreas form in the similar manner from the solid buds. The terminal buds separate from the main bud to form isolated group of cells during 8–10 embryonic weeks (Hamilton et al. 1957). At first, the endocrine cells are found in the walls of ducts or within buds arising from them; later they aggregate as pancreatic islets. Later, in weeks 10–15, some of the primitive ducts differentiate into acinar cells. (Collins 1995). The origin of pancreatic endocrine cells in the islet however is controversial. It has been suggested that they arise from neural crest cells (Larson 2001). Evidence from cell tracing experiments some involving the quail chick chimera system and from transgenic animals however suggested that these cells arise from the gut tube endoderm. The mesenchymal cells form connective tissue around ducts, acini, lobules and lobes of the pancreas; thereby it forms a skeletal support of the gland. Some of the cells remain undifferentiated throughout the life.

In the present study, it was noted that the pancreatic acini were well formed by 12th week of gestation. Previous authors have shown that the zymogen granules and acinar cell markers could be detected at 12–16 gestational weeks (Collins 1995), confirming our observations. However, the proteolytic pancreatic enzymes are produced by the acini from the fifth month onwards (Hamilton et al. 1957). In our study, ultrastructural investigation revealed that the duct cells and endocrine cells were in close apposition, indicating that there may be similar endodermal origin of exocrine as well as endocrine part of pancreas. A close proximity of the acini and the isolated endocrine cells were observed during 14 gestational week suggested that the exocrine and endocrine part arise from a common source in the embryonic life. To confirm the theory of neural crest origin of these cells, further study is needed using cell markers like nestin, which is used for neuroendocrine cells (Zulewski et al. 2001). Rutter (1980) and Githens (1993) had shown that the ductal epithelial cells give rise to endocrine cells of the islets. This linkage remains maintained in the post-natal period too. In an earlier report, Bouwens et al. (1997) showed that there were insulin-positive cells found in pancreatic ducts that could be labeled with CK19. This suggested that endocrine cells of the pancreas were being generated from ductal stem cells in the human fetal pancreas. These cells were far fewer in the adult human pancreas (Bouwens et al. 1994), suggesting that there is little neogenesis of cells in adult humans. The presence of electron dense granules in the endocrine cells indicated that the endocrine component was functional during the 14th gestational week. Collins (1995) had shown that endocrine secretion starts 8–10 gestational weeks. Our study confirms observations of Collins (1995). However, it is reported that insulin secretion starts approximately in the fifth month of intrauterine life (Sadler 2006).

4.2 *Pancreatic Acini*

In the adult pancreas, the acinar cells were typical protein secreting cells consisting of greater amount of zymogen granules towards the apex of the cell, central nuclei surrounded by basophilic cytoplasm with rough endoplasmic reticulum. Differential distribution of zymogen granules in a single acinus indicated that all the cells of an acinus were not active at a particular time. Some cells are active while other cells are in resting phase. The resting cells contained fewer zymogen granules. This observation needs further morphological and correlated physiological study.

Occasional presence of double nucleus in cells has been reported in hepatocytes, which are highly metabolically active cells (Ross et al. 2003; Young et al. 2006). In the present study binucleated acinar cells were noted during early postnatal period and in adult pancreas. Cells with double nucleus are present during mitosis in physiological conditions. Double nucleus may be present in cancerous cells (Bojan and Brian 1980). This indicates that pancreas is also a highly active organ like the liver. From extensive literature survey, we conclude that this feature has not been reported earlier.

4.3 *Centroacinar Cells*

The center of the acini frequently contained one or more pale nuclei of centroacinar cells with sparse pale staining cytoplasm; these represented the terminal lining cells of minute ducts. Centroacinar cell regulates the secretion from acinar cells under normal physiological condition (Kern 1971). The tonofibrillary architecture of centroacinar cells helps them act as a valve that regulates the amount of zymogen granules that are released by the acinar cells. In the present study, we observed that centroacinar cells covered almost the entire luminal surface of acinar cells with long cytoplasmic processes. The long and wide processes virtually separate all the acinar cells all along their adjacent lateral surfaces. Occasionally, we observed that these cytoplasmic processes extended along the basal aspect of the acinar cells, separating the acinar cells from the subjacent basement membrane. The acinar cells that are undergoing degeneration and necrosis are covered by centroacinar cell processes along their basal regions. This may have a role in expelling the degenerating cell into the lumen of the acinus. Thereby the centroacinar cell has a key role to play in the normal functioning of the pancreas.

4.4 Ductular System

The three-dimensional structure of pancreatic ductular system is likened to the branching of a tree in which the interlobular and intralobular ducts, and the intercalated ducts represent the small branches; some of the intercalated ducts are seen to be anastomosing with each other. The intercellular secretory canaliculi extend from the lumen to the basement membrane of the acinus. The lumen of all pancreatic ducts (the interlobular, intralobular and intercalated) is round and has a smooth surface. The luminal aspect of the epithelial cell is covered by many microvilli and a monocilium. The length of the monocilium is short in comparison to the diameter of pancreatic duct. Each epithelial cell has a different density of microvilli. The intraductal cilium has a core of nine microtubules. This is different from the number of microtubules seen in cilia of the epithelial cells of either the uterine tube or bronchial epithelium. The number of microtubules in the cilia decreases distally along its length. The epithelial cells have numerous pits that have elongated cilia on them (Ashizawa et al. 1997).

Squamous, cuboidal and tall columnar epithelial cells line the smaller ductules, intercalated ducts and the major ducts of the pancreas. Epithelial cells are separated from each other by tight junctions and lie on a basement membrane. Integrity of the junctional complex and the basement membrane are important for development of acute or chronic pancreatitis. In the present study, microvilli and occasional cilia were noted in larger ducts, which regulate the flow of the pancreatic secretion, which are in agreement with studies conducted by Nagata and Monno 1984. Accumulation of lipid inside the epithelial cells of the larger ducts was noted frequently in the aged pancreas. Accumulation of membrane bound lipid inside the cell indicates fatty degeneration of these cells (Ghadially 1997; Cotran et al. 1999). A marked feature of the ducts in our study was periductal fibrosis of the larger ducts, which started during 4th decade and continued up to 8th decade. Ultrastructurally, it was noted that there were thickening of the basement membrane along with fibrosis. In addition to fibrosis there were epithelial hyperplasia in the large ducts and papillary projections observed inside the lumen of the smaller ducts.

The peripheral parts of the glands showed the formation of fibrotic foci that involved one or two lobules. Here, the acinar cells were supplanted by connective tissue. The pattern of fibrosis is mainly intralobular because the fibrotic changes look as if they start within the lobule and not in perilobular space. Epithelial hyperplasia, papillary projections and periductal fibrosis started very early in Indian populations whereas it started in 6th decade in German populations (Detlefsen et al. 2005). No clinical study is available for comparison. Therefore this observation needs an in depth study.

Mazhuga et al. (1981) described three different kinds of cells within the epithelium of the pancreatic ducts in patients with CP: (i) main ductal cells that have "light" and "dark" forms, which have differing levels of development of their biosynthetic apparatus and are in different states of functional activity; (ii) endocrine cells that have features of chromaffin-like and D-cells of the gut; and (iii) migrant cells like histocytes, granulocytes and lymphocytes. In the present study, dark cells were noted in the lining epithelium of all varieties of ducts (Fig. 2a). Presences of dark cells in tissues are a controversial subject in cell biology. There are reports that these cells may be a fixation artifact, shrunken version of light cells or at different physiological state, protein synthesizing cell, dying or dead cells (Ghadially 1997). The human pancreas in addition to cuboidal to columnar epithelium also contains small dark cells in the basal portion of the duct. These cells contain round hyper chromatic nuclei with clear cytoplasm. These darkly nucleated cells are not prominent in normal pancreas but are visible in hyperplastic ducts. Their clear cytoplasm probably contains sphingolipids and hence appears empty in paraffin embedded tissues. These cells also have been suggested to act as stem cells (Feyrter 2005). Equivalent dark cells were visible in the fetal pancreas also signifying their special role in pancreatic growth and development. The process of degeneration and regeneration of these cells in the large ducts accompany simultaneously, as evidenced by multiple dark cells or stem cells in these ducts. Similarity of fetal and adult dark cells has not been studied in detail.

In the present study, the pancreatic ducts were divided into three types on the basis of their area. We also quantified the area of the lining epithelium, and fibrous wall beyond the epithelium, to observe the changes with aging. Quantitative observations of pancreatic duct revealed a significant increase amount of the connective tissue around the small and medium sized duct with aging. Louis and Brenda (1973), observed a steady increase in duct caliber with age throughout the whole of the pancreas head, body, and tail and this increase was at the rate of 0–8 % per year. Due to the small sample size in our study, it is too early to predict a similar trend for Indian population. Increase in the connective tissue component of the small and medium sized ducts in the present study indicate the vulnerability to pancreatic diseases in the Indian population in early ages. This needs further confirmation using a larger sample size.

4.5 Islets of Langerhans

The Islets of Langerhans were observed scattered among the acini of the exocrine portion. They were surrounded by thin capsule and fine network of connective tissue along with capillaries. In present study, we observed increased amount of fibrous tissue within the islets as well as around the islets in aged pancreas. Not only was the fibrous tissue increased in islets but also they were scattered in small fragments within the lobules of the pancreas. Earlier literature showed that in rats there is an increase in size of pancreatic islets with advancing age. In aged rats, the population of small islets loses their ability to respond rapidly to glucose in vitro. This ability is retained in the large islets (Adelman 1989).

Movassat et al. (1995) observed that when there was a reduction in the total beta-cell mass in adult GK rats, it was associated with an obvious change in the architecture of some islets: only the larger islets showed signs of changes in the mantle-core relationship. This was because of the marked fibrosis, wherein clusters of beta cells were separated by connective tissue.

Recent studies of the human pancreas seem to indicate that the reduced beta cell mass seen in early stages of diabetes are caused by accelerated apoptosis of the beta cells. One of the reasons that may be causing the death of islet cells may be the increasing fibrosis within the islets and the inadequate capacity of beta-cells to proliferate in diabetic patients. Pancreatic stellate cells (PSCs) are involved in the progression of fibrosis within pancreatic islet in patients with type 2 diabetes (Hayden et al. 2007; Kim et al. 2008). The histo-morphology of islets from patients with type 2 diabetes showed the presence of an inflammatory process that was evidenced by the occurrence of cytokines, apoptotic cells, immune cell infiltration, amyloid deposits, and fibrosis (Donath et al. 2008). Although we have not estimated the inflammatory or apoptotic markers, the fibrotic changes in the islets of the aged pancreas alarms a need for in depth study of aging of islet cells in Indian population, which may contribute literature on the onset of type2 diabetes.

4.6 Pancreatic Fibrosis

Chronic pancreatitis (CP) is characterized mainly by fibrosis (Ammann et al. 1996; Klöppel et al. 2004), but this fibrosis may also be seen in a "normal pancreas," i.e., the pancreas of persons with no history of any pancreatic disease or any macroscopically visible pancreatic pathology. Pitchumoni et al. (1984) first suggested that there may be an association between fibrosis and aging in the pancreas. Shimizu et al. (1989) also described fibrotic changes in normal pancreatic specimens, but age-related changes were documented only in Stamm's (1984) study. Kloppel and Maillet (1993) observed that there were variations in the ducts that drained fibrotic lobules. These changes in the duct were defined as "ductal papillary hyperplasia" (Klöppel et al. 1996) and were recently recognized as a precursor lesion of pancreatic ductal adenocarcinoma (Hruban et al. 2001). Stamm's (1984) observed that the there was a patchy pattern of interlobular fibrosis in the normal pancreas that was clearly related to the age of the individual and that the narrowing of ducts was due to epithelial hyperplasia. The pattern of fibrosis was that of multifocal intralobular fibrosis because most pancreata contained numerous fibrotic foci. This pattern was called "patchy lobular fibrosis in the elderly," (PLFE). Shimizu et al. (1989) observed that the commonest changes seen in the ducts are epithelial hyperplasia (88 %) and periductal fibrosis (74 %). These authors observed that the ducts showed cystic widening (62 %) and intraluminal protein precipitates (40 %), within the parenchyma as an intralobular fibrosis or total fibrosis of lobules (88 %). These fibrotic changes increased with aging, both in the number of the foci and their extent. In the present study, we noted papillary projections and epithelial hyperplasia beginning at the age of 44 years and it increased gradually up to 65 years, the maximum age studied. This signifies that in Indian population pancreatic ductal changes starts earlier than in Europeans (Detlefsen et al. 2005). This needs further study including a larger population size.

4.7 *Pancreatic Stellate Cells*

The main source of extra cellular matrix (ECM) proteins in the diseased pancreas are activated pancreatic stellate cells (PSCs). These cells store vitamin A, and have characteristics of fibroblasts. Hence, they have morphological and biochemical similarities to similar cells in the liver, known as Itocells. Cytokine mediators that promote fibrogenesis activate PSCs. The PSCs then proliferate, show characteristics of myofibroblasts and increase the production of ECM proteins. In a normal pancreas, PSCs constitute approximately 4 % of all pancreatic cells and are found around acini. They have cytoplasmic lipid droplets that contain retinoids and cytoskeletal proteins such as desmin and glial fibrillary acidic protein (GFAP) (Apte et al. 1998). PSCs have an important part to play in pancreatic fibrogenesis.

The source of fibrosis in aged specimens of pancreas is probably pancreatic stellate cells, which become active during aging. When activated, PSCs display a decrease in the number of lipid droplets and begin accumulating smooth muscle actin-alpha. These features are similar to those of myofibroblasts. Most of the fibrotic foci that were found in "normal" pancreata showed cells that were positive immunohistochemically for α-SMA, and sometimes desmin. In the present study, the increased fibrosis noted around the pancreatic ducts during 7th decade may be dependent on activated stellate cells.

4.8 Age Changes in Pancreas

Aging is defined as the increasing accumulation of changes with time that are either linked to or are responsible for perpetually increasing vulnerability to disease and death, which is the final event of age. Textbooks of anatomy and histology mention that a thin, fibrous capsule surrounds the pancreas and that pancreatic ducts are bounded by delicate strands of connective tissue (Junqueira and Carneiro 1995; Klimstra 1997). However, these books do not mention about the increasing fibrosis in the pancreas as a function of age. This may be because to date there are very few studies that have described fibrosis in pancreas unaffected by any disease. Pitchumoni et al. (1984), Shimizu et al. (1989), and Stamm (1984) showed that the pancreas has fibrotic changes that increase with the age of the person but they did not grade the fibrosis nor discussed its pathogenesis. One study quantified fibrosis in the normal pancreas and compared it to that found in CP (Valderrama et al. 1991) and two other studies showed that there are ectasias in the pancreatic duct system and noted that in these the fibrosis found around the ducts increased with age (Schmitz-Moormann and Hein 1976; Allen-Mersh 1985). In another study, photometric evaluation was done to determine the collagen content in histological sections of the pancreas with CP; however, the authors did not provide any information regarding either the distribution of fibrotic foci or the association of fibrosis with age (Valderrama et al. 1991).

In our study we observed an overall increase amount of fibrous tissue around the ducts with aging; more around the small and medium sized ducts in comparison with large sized ducts. Results from earlier literature report that there is decrease in weight of the pancreas after the seventh decade of life in humans, and it becomes harder and more atrophic (Geokas et al. 1985; Laugier et al. 1991; Andrew 1994). In addition to histological changes such as ductal epithelial hyperplasia and intralobular fibrosis, acinar cell degranulation has also been described. It is difficult, often times, to discriminate between physiological age related changes and significant pathologic alterations in the morphology of the pancreatic ducts.

5 Summary and Conclusions

The fetal pancreas contained mainly ducts, few acini, numerous centroacinar cells, rare stellate cells and large amounts of undifferentiated mesenchymal tissue. Acini had a functional structure by 12th GW. A close association between acini and endocrine cells indicated a common germ cell origin. With aging, the cytoplasm of ductal epithelial cells showed numerous lipoid bodies. The connective tissue around ducts also increased. The number of stellate cells and centroacinar cells increased with aging. Pancreatic stellate cells are probably the principal mediators of fibrogenesis and its consequences that occur during aging. This study needs to be extended to a larger population and with more molecular techniques to draw conclusive inferences regarding certain pathophysiological processes that have been observed.

References

Adelman RC (1989) Changes in pancreatic hormones during aging. Prog Clin Biol 287:163–166

Allen-Mersh TG (1985) What is the significance of pancreatic ductal mucinous hyperplasia? Gut 26:825–833

Ammann RW, Heitz PU, Klöppel G (1996) Course of alcoholic chronic pancreatitis: a prospective clinicomorphological long-term study. Gastroenterology 111:224–231

Andrew W (1994) Senile changes in the pancreas of Wistar institute rats and of man with special regard to similarity of locale and cavity formation. Am J Anat 74:97–127

Apte MV, Haber PS, Applegate TL, Norton ID, McCaughan GW, Korsten MA, Pirola RC, Wilson JS (1998) Periacinar stellate shaped cells in rat pancreas: identification, isolation, and culture. Gut 43:128–133

Ashizawa N, Endoh H, Hidaka K, Watanabe M, Fukumoto S (1997) Three-dimensional structure of the rat pancreatic duct in normal and inflamed pancreas. Microsc Res Tech 37(5–6):543–556

Bojan F, Brian C (1980) Fine structure of rat liver during chronic intoxication with two heterocyclic N- nitrosamine: N- nitrosopiperidine and the non- carcinogen, 2, 2′, 6, 6′ – tetramethyl- N- nitroso – piperidine. Carcinogenesis 1(12):961–974

Bonner Weir S, Baxter LA, Schuppin GT, Smith FE (1993) A second pathway for regeneration of adult exocrine and endocrine pancreas: a possible recapitulation of embryonic development. Diabetes 42:1715–1720

Bouwens L, Lu WG, De Krijger (1997) Proliferation and differentiation in the human fetal endocrine pancreas. Diabetologia 40:398–404

Bouwens L, Wang RN, De Baly E, Pipeleers DG, Klöppel G (1994) Cytokeratins as markers of ductal cell differentiation and islet neogenesis in the neonatal rat pancreas. Diabetes 43:1279–1283

Collins P (1995) Embryology and Development. In: Williams PL (ed) Gray's Anatomy, 38th edi. Churchill Livingstone, New York, pp 186–187

Collins P (2002). Embryology of pancreas. In: Howard ER, Stringer MD, Columbani PM (eds) Surgery of liver, bile duct and pancreas in children, part 8. London, Arnold, pp 479–492

Cotran RS, Kumar V, Collins T (1999) In: Robbins Pathologic Basis of Disease. 6th ed. Harcourt Asia PTE Ltd, Singapore, pp 39–40

Detlefsen S, Bence S, Bernd F, Klöppel G (2005) Pancreatic fibrosis associated with age and ductal papillary hyperplasia. Virchows Arch 447:800–805

Donath MY, Schumann DM, Faulenbach M, Ellingsgaard H, Perren A, Ehses JA (2008) Islet inflammation in type 2 diabetes: from metabolic stress to therapy. Diab Care 31: Suppl 2: 161–164

Feyrter F (2005) Uber Die Pripheren Endokrinen Drusen des Menschen. Are there any stem cells in the pancreas? Pancreas 31(2):108–118

Geokas MC, Conteas CN, Majumdar PN (1985) The aging gastrointestinal tract, liver, and pancreas. Clin Geriatr Med 1:177–205

Ghadially FN (1997) In: Ultrastructural pathology of the cell and matrix. 4th ed, vol 2. Butterworth–Heinemann, Oxford, pp. 1014–1037

Githens S (1993) Differentiation and development of the pancreas in animals. In: Go VL, Dimagno EP, Gardner JD, Lebenthal E, Reber HA, Scheele GA (eds) The pancreas: biology, pathobiology, and disease. 2nd ed. Raven Press, New York, 21–55

Hamilton WJ, Boyd JD, Mossaman HW (1957) In Human Embryology, 2nd ed. W Heffer and Son's Ltd, Cambridge, pp 208–210

Hayden MR, Karuparthi PR, Habibi J, Wasekar C, Lastra G, Manrique C, Stas S, Sowers JR (2007) Ultrastructural islet study of early fibrosis in the Ren2 rat model of hypertension. Emerging role of the islet pancreatic pericyte-stellate cell. JOP 8(6):725–738

Hruban RH, Adsay NV, Albores-Saavedra J, Compton C, Garrett ES, Goodman SN, Kern SE, Klimstra DS, Klöppel G, Longnecker DS, Lüttges J (2001) Offerhaus GJA. Pancreatic intraepithelial neoplasia. A new nomenclature and classification system for pancreatic duct lesions. Am J Surg Pathol 25:579–586

Junqueira LC, Carneiro J (1995) Pancreas. In: LC Junqueira, J Carneiro (eds) Basic histology, 8th ed. Lange Medical Publications, Los Altos

Kern HF (1971) Die Feinstruktur des exokrinen Pankreasgewebe. Zeitshrift fur Zellforschung 113:322–343

Kim JW, Ko SH, Cho JH, Sun C, Hong OK, Lee SH, Kim JH, Lee KW, Kwon HS, Lee JM, Song KH, Son HY, Yoon KH (2008) Loss of beta-cells with fibrotic islet destruction in type 2 diabetes mellitus. Front Biosci 13:6022–6033

Klimstra DS (1997) Pancreas. In: Histology for pathologists. 2nd edn. Lippincott-Raven, Philadelphia, pp 613–647

Klöppel G, Bommer G, Heitz PU (1980) Immunocytochemistry of the endocrine pancreas in chronic pancreatitis and pancreatic carcinoma. In: Podolsky S, Vishwanathan M (eds) The spectrum of the diabetic syndromes, secondary diabetes. Raven Press, New York, pp 147–151

Klöppel G, Detlefsen S, Feyerabend B (2004) Fibrosis of the pancreas: the initial tissue damage and the resulting pattern. Virchows Arch 445:1–8

Kloppel G, Maillet B (1993) Pathology of acute and chronic pancreatitis. Pancreas 8:659–670

Klöppel G, Solcia E, Longnecker DS, Capella C, Sobin LH (1996) Histological typing of tumours of the exocrine pancreas. In: WHO International histological classification of tumours. 2nd ed. Springer, Berlin, Heidelberg, New York

Larson WJ (2001) In: Human embryology. 3rd ed. Churchill Livingstone, New York, pp 240–244

Laugier R, Bernard JP, Berthezene P, Dupuy P (1991) Changes in pancreatic exocrine secretion with age: pancreatic exocrine secretion does decrease in the elderly. Digestion 50:202–211

Lohr M, Lubbersmeyer J, Otremba B, Klapdor R, Grossner D, Klöppel G (1989) Increase of B-cells in the pancreatic remnant after partial pancreatectomy in pigs: an immunocytochemical and functional study. Virchows Arch B Cell Pathol Mol Pathol 56:277–286

Louis K, Brenda S (1973) Changes in pancreatic morphology associated with aging. Gut 14: 962–970

Mandarim-de-Lacuda CA (1990) Foot length growth related to crown- rump length, gestational age and weight in human staged fresh fetuses: An index for anatomical and medical use. Surg Radiol Anat 12:103–107

Mazhuga PM, Medvetskiĭ EB, Rodionova NV (1981) Ultrastructure of the ductal epithelium in the human pancreas in chronic pancreatitis. Tsitol Genet 15(5):15–24

Movassat J, Saulnier C, Portha B (1995) Beta-cell mass depletion precedes the onset of hyperglycaemia in the GK rat, a genetic model of non-insulin-dependent diabetes mellitus. Diab Metab 21(5):365–370

Nagata A, Monno S (1984) Ultrastructure of pancreatic duct and pancreatic ductal cell. Cell 16:397–402

Pitchumoni CS, Glasser M, Saran RM, Panchacharam P, Thelmo W (1984) Pancreatic fibrosis in chronic alcoholics and nonalcoholics without clinical pancreatitis. Am J Gastroenterol 79:382–388

Ross MH, Kaye GI, Pawlina W (2003) Histology a text and atlas, 4th ed. Lippincott Williams & Wilkins, Baltimore, pp 533–567

Rutter WJ (1980) The development of the endocrine and exocrine pancreas. Fitzgerald, PJ, Morrison AB (eds) In the pancreas. Williams and Wilkins, Baltimore, pp 30–38

Sadler TW (2006) Langman's medical embryology, 10th ed. Lippincott Williams & Wilkins, Baltimore, pp 215–216

Sailaja K, Ahuja RK, Gopinath G (1996) Biparital diameter: a useful measure for determining gestation age of human abort uses. Natl Med J India 9:165–167

Schmitz- Moormann P, Hein J (1976) Changes of the pancreatic duct system associated with aging: their relations to parenchyma. Virchows Arch A Pathol Anat Histol 371(2):145–152

Shimizu M, Hayashi T, Suitor Y, Itch H (1989) Interstitial fibrosis in the pancreas. Am J Clin Pathol 91:531–534

Stamm BH (1984) Incidence and diagnostic significance of minor pathologic changes in the adult pancreas at autopsy: a systematic study of 112 autopsies in patients without known pancreatic disease. Human Pathol 15:677–683

Valderrama R, Navarro S, Campo E, Camps J, Gimenez A, Pares A, Caballeria J (1991) Quantitative measurement of fibrosis in pancreatic tissue. Evaluation of a colorimetric method. Int J Pancreatol 10:23–29

Whitcomb DC (2006) Clinical practice: Acute pancreatitis. N Engl J Med 354:2142–2150

Young B, Lowe JS, Stevens A, Heath HW (2006) Wheater's functional histology—A text and colour atlas, 5th edi. Churchill Livingstone, New York, pp 288–302

Zulewski H, Abraham EJ, Gerlach MJ, Daniel PB, Moritz W, Muller B, Vallejo M (2001) Multipotential nestin-positive stem cells isolated from adult pancreatic islets differentiate ex vivo into pancreatic endocrine, exocrine, and hepatic phenotypes. Diabetes 50:521–533

Part II
Metabolic and Molecular Changes in Aging

Insulin-like Growth Factor-1 and Its Related Signalling During Aging: Modulation by Dietary Restriction

Ibanylla Kynjai Hynniewta Hadem and Ramesh Sharma

Abstract Aging is characterized as the changes of several biological processes in an organism as a function of time. The most important system engaged during the aging process is the endocrine system. Insulin-like growth factor-1(IGF-1), a single chain polypeptide synthesized under the action of growth hormone (GH), regulates a variety of developmental and metabolic processes of an organism. The primary source of serum IGF-1 is the liver; however, it is also synthesized in other tissues where it functions as autocrine and paracrine hormone. It acts by binding to its cell surface receptor and amplifies its action by regulating downstream molecules like PI3K, Akt, SIRT, FOXO, NF-κB, thereby, forming a network of signalling pathways. The link of IGF-1 and its signalling pathway to the process of aging is of significant interest because it is involved in controlling various cellular processes that determine the lifespan of an organism. Dietary restriction (DR), an intervention where the intake of food is lessened but without causing any malnourishment, has been found to be an effective modulator of the IGF-1 signalling pathway. An outline of IGF-1 and its signalling pathway in aging process and how DR acts as a mediator of various players of the pathway that may enhance longevity in various model organisms will be illustrated.

Keywords IGF-1 · Sirtuins · FOXO · NF-κB · Aging · DR

1 Introduction

Aging is defined as a process that is genetically determined and environmentally modulated resulting in a gradual decline of physiological functions ultimately leading to death. Many theories explaining the causes of aging have been put forward, which can be categorized as molecular, cellular and systemic theories

I.K.H. Hadem · R. Sharma (✉)
Biochemistry Department, North-Eastern Hill University,
Shillong 793022, India
e-mail: sharamesh@gmail.com

© Springer Science+Business Media Singapore 2017

P.C. Rath et al. (eds.), *Topics in Biomedical Gerontology*,
DOI 10.1007/978-981-10-2155-8_5

(Sharma and Dkhar 2014). The molecular and cellular changes observed during aging lead to accumulation of damaged proteins (Dkhar and Sharma 2014), modification of enzymes important for metabolism (Kharbhih and Sharma 2014; Majaw and Sharma 2015) and alteration of immune response (Adler et al. 2007; Montecino-Rodriguez et al. 2013) thus decreasing the functional capability of the organism.

Many hormones are shown to be involved with longevity processes and aging has also been associated with the dysregulation of the endocrine system. With aging, there is a decreased secretion of GH (Sonntag et al. 1980) and glucocorticoids (GC) receptors, a mediator of GC action, important for anti-inflammatory and stress responses (Sharma and Dutta 2006). Among others, IGF-1 has also been implicated to play crucial part in longevity processes in a wide spectrum of organisms (Kenyon 2011). In mammals, aging and its associated diseases have been linked to the altered regulation of the IGF-1/Akt signalling pathway (Jung and Suh 2015).

IGF-1 is composed of 70 amino acid residues and is a 7.5 kDa single chain polypeptide. Together with the receptor, IGF binding protein (IGFBP) and IGFBP proteases, the insulin-related peptides consisting of insulin, IGF-1 and IGF-2, form the IGF system (Humbel 1984). The isoforms of IGF-1 that are expressed either by liver or locally by other tissues have different functional characteristics as suggested by current studies. IGF-1 is ubiquitous and may act independently of GH action (Clemmons 2007; Musarò et al. 2007; Schug and Li 2010).

In human, the gene encoding the IGF-1 peptide is positioned at the long arm of the chromosome 12q23-23 and it has six exons including two leader exons. The multiple transcripts generated by alternative splicing of the exons code for a pre-pro IGF-1 protein with different signalling peptide leader sequences. An identical 70 amino acid mature peptide is generated following posttranslational processing of all transcript isoforms and it is conserved across species, having B, C, A and D domain (Brissenden et al. 1984; Shavlakadze et al. 2005; Oberbauer 2013). Depending on the type of leader sequence present, there are two biological isoforms of IGF-1 transcript viz., class 1 and class 2 (Yang et al. 1995). Class 1 transcript uses exon 1 and is preferred in the autocrine/paracrine form whereas the endocrine type that is released into circulation, utilizes exon 2, and is characterised as the class 2 transcript (Sussenbach et al. 1992). In order to promote an effective endocrine action class 2 has a usual signal peptide whereas class 1 has a longer signal peptide, which most likely inhibits its secretion. Hence, class 2 is found to be highly expressed in liver that is responsible for the main source of IGF-1 in serum while class 1 have diminished capacity to bind to binding proteins thus, rendering it highly mitogenic suggesting its autocrine/paracrine functions (O'Sullivan et al. 2002).

The 48 % homology between the B, C, A and D domain of the mature IGF-1 protein and insulin suggests that they evolve from a common precursor gene (Humbel 1984). Besides structural homology, IGF-1 shares several similar biological activities with insulin. There is differential tissue distribution between insulin and IGF-1 receptors (Froesch et al. 1985). Although GH mostly controls the synthesis of IGF-1 from liver, however IGF-1 also has feedback inhibition on GH production (Berelowitz et al. 1981). IGF-1 also suppresses insulin and glucagon secretion. IGF-1 treatment induces lipolysis and reduces proteolysis. IGF-1 increases insulin sensitivity by either decreasing GH synthesis or by acting directly on insulin target tissues or by reduction of insulin levels leading to increasing insulin receptors. IGF-1 treatment was shown to improve metabolic functions of diabetic patients (Hussain et al. 1995). One of the first signalling pathways distinctively shown to affect the aging process is the insulin/IGF-1 signalling cascade (Kenyon 2010). The function of insulin/IGF-1 signalling during aging was first implicated by the work carried on Daf-2 (orthologue of insulin/IGF-1 receptor) signalling pathway in *C. elegans*. Mutations of any of the genes in this pathway, which includes *daf-2, age-1*(PI3K), *akt-2*(Akt), *daf-16* (FOXO) and *daf-18*(PTEN), extend lifespan in *C. elegans* (Masse et al. 2005; Kenyon 2010). The GH pathway that signals through the IGF1/insulin network has subsequently been linked to mammalian aging. In rodents, null mutation of $igf1r^{-/-}$ gene has a detrimental effect; however, $igf1r^{-/-}$ mice have an extended lifespan than their wild type littermates(Holzenberger et al. 2003).

Additionally, as compared to the normal mice, several mutant mouse models e.g., Ames, Snell, and GHRKO mice having low level of IGF-1 were shown to have an increased lifespan (20–70 %) (Salminen and Kaarniranta 2010). Other studies have reported that transgenic mice with high expression of serum IGF-1 level and IGF-1 receptor, due to the presence of short p44 isoform of p53 that induces hyperactivation of p53 signalling, have a diminished lifespan and have age-associated diseases (Maier et al. 2004). The insulin/IGF-1 paradox is categorized under the theory of antagonistic pleiotropy, which reasons that this growth factor is necessary during early developmental phases but can be detrimental in later stages of an organism's lifespan as it can promote processes characteristic of aging. (Rincon et al. 2004).

2 IGF-1 and Its Related Signalling Cascades

Cell function depends on the transmission of signals by various signalling molecules like growth factors, hormones and neurotransmitters. A coordinated activation or inactivation of proteins is initiated through signal transduction pathways that may occur at the cell membrane, cytoplasm or nucleus. The initiation of a signalling pathway usually involves the activation of the receptor on binding of the ligand. The activated receptor can then activates downstream signalling proteins by phosphorylation or dephosphorylation at tyrosine, serine or threonine residues by suitable protein kinases/phosphatases. When growth factors bind to the tyrosine kinase receptor, it recruits and activates one of the first signalling intermediates the phosphoinositide 3-kinases (PI3K). The major downstream target of PI3K is Akt/PKB, which is of considerable interest due to its role in many cellular processes including proliferation, differentiation, cell survival, protein synthesis and gene expression regulation (Datta et al. 1999; Manning and Cantley 2007).

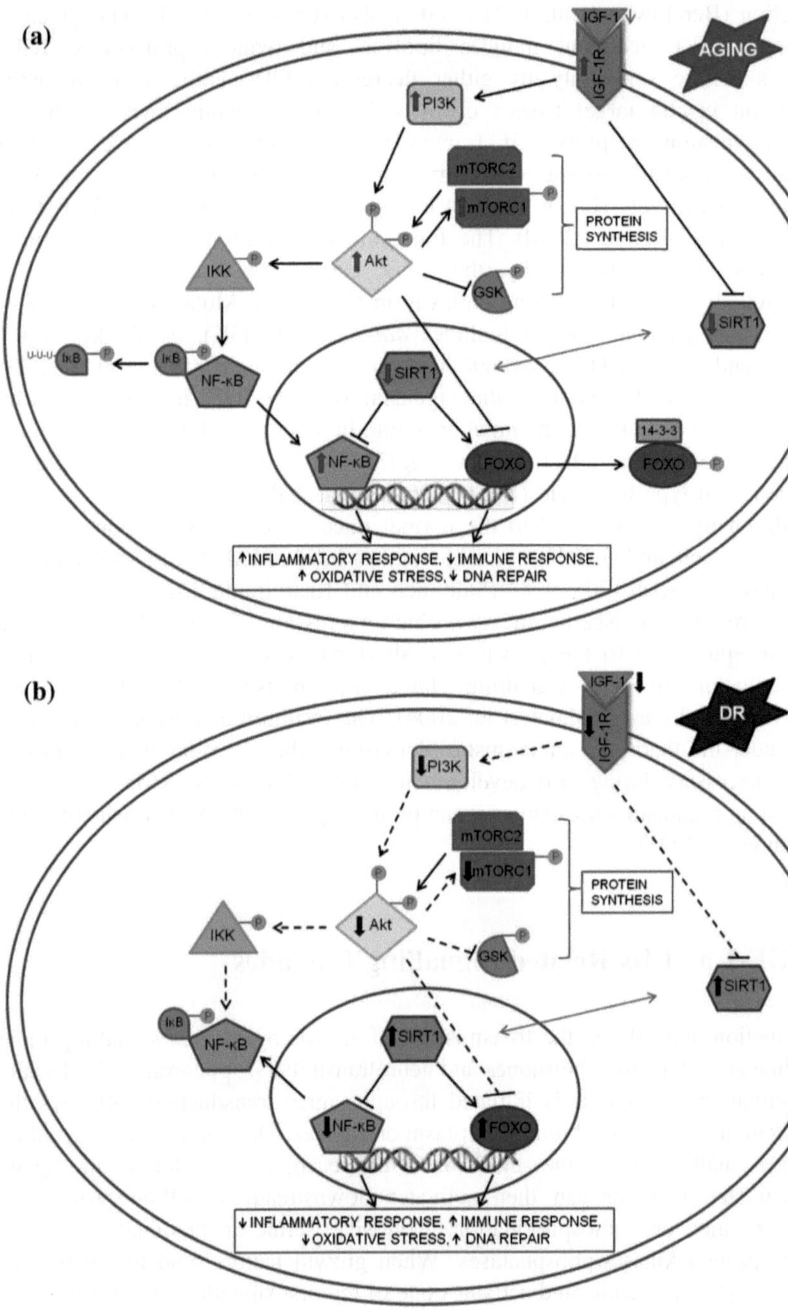
(a)
IGF-1
IGF-1R
AGING
PI3K
mTORC2
mTORC1
PROTEIN SYNTHESIS
Akt
IKK
GSK
SIRT1
U-U-U IκB
IκB
NF-κB
SIRT1
NF-κB
FOXO
14-3-3
FOXO
↑INFLAMMATORY RESPONSE, ↓IMMUNE RESPONSE, ↑OXIDATIVE STRESS, ↓DNA REPAIR
(b)
IGF-1
IGF-1R
DR
PI3K
mTORC2
mTORC1
PROTEIN SYNTHESIS
Akt
IKK
GSK
SIRT1
IκB
NF-κB
SIRT1
NF-κB
FOXO
↓INFLAMMATORY RESPONSE, ↑IMMUNE RESPONSE, ↓OXIDATIVE STRESS, ↑DNA REPAIR

◀ **Fig. 1** Diagrammatic illustration of the IGF-1 signalling pathway. IGF-1 binds to and activates its receptor, IGF-1R, that in turn activates PI3K leading to activation of Akt, a central molecule. mTORC2 further activates Akt by phosphorylation at Ser473 which then phosphorylates and activates mTORC1. Akt also phosphorylates and inactivates GSK. Activation of mTORC1 and inhibition of GSK regulates protein synthesis. Akt can activate NF-κB by phosphorylating IKK (inhibitor of IκB kinase) thereby inhibiting IκB. Akt, on the other hand, inhibits FOXO transcription factor by phosphorylation leading to its translocation to the cytoplasm wherein it is retained by interaction with 14-3-3 chaperone. IGF-1 also crosstalks with sirtuin (SIRT1) and inhibits its activity. **a** Aging upregulates IGF-1 signalling pathway thereby stimulating Akt activity, increasing mTOR and NF-kB signalling while reducing FOXO cascade. This results in increased inflammatory and oxidative responses and decreased immune responses and **b** DR reduces the IGF-1 signalling pathway that leads to activation of FOXO transcription factors and inactivation of NF-κB and mTOR. Reduced IGF-1 signalling by DR also leads to enhanced SIRT1 activity. SIRT1 deacetylates NF-κB and FOXO causing an inactivation and activation of the respective proteins, which results in decreased inflammatory and oxidative responses and increased DNA repair and immune competence giving rise to beneficial effects of DR on healthy lifespan. (↓—Decrease, ↑—Increase, ⩛—Reduced signalling, ⊥—Inhibition, bold arrows indicate increased or decreased expression)

The highly conserved IGF-1 signalling pathway regulates many of the cellular and tissue functions of an organism. By binding to IGF-1R (IGF-1 receptor), it initiates a signalling cascade through the B domain (Gauguin et al. 2008). It can also bind to the insulin receptor but with a 100-fold lower affinity (Dupont et al. 2001). When IGF-1 binds to its receptor, it activates several signalling cascades, which promotes cell survival, proliferation and differentiation (Datta et al. 1999). There are two major signalling cascades that IGF-1 activates, PI3 K/Akt pathway and the mitogen-activated protein (MAP) kinase pathway. The PI3 K/Akt pathway is the central pathway engaged by insulin/IGF-1 during aging and in improving the degenerative diseases associated with it (Martin et al. 2006). A principal constituent of this pathway is Akt, alternatively known as protein kinase B (PKB). Akt regulates many cellular processes like protein synthesis and degradation. The signalling through mammalian/mechanistic target of rapamycin (mTOR) and glycogen synthase kinase (GSK) is important for protein synthesis while the signalling through the transcription factors of FOXO family is necessary for protein degradation (Manning and Cantley 2007). Besides being a target of the IGF-1 signalling pathway, Akt is also a substrate for several other growth factors and cytokines (Datta et al. 1999).

When IGF-1 binds to IGF-1R, with an inherent tyrosine kinase, it leads to its activation and autophosphorylation. The activated receptor serves as a docking site for the insulin receptor substrate (IRS) protein, which becomes activated upon phosphorylation. The IRS protein then phosphorylates and activates PI3K by binding through its SH2 domain which in-turn phosphorylates the membrane bound phosphotidylinositol 4,5-bisphosphate (PIP_2) to generate the second messenger, phophotidylinositol 3,4,5-triphosphate (PIP_3). A vital target of PIP_3 is a protein serine/threonine kinase, Akt that is present in the cytoplasm. Both Akt and phosphoinositide dependent kinase (PDK1) are recruited by PIP_3 to the plasma membrane by binding through their PH domain. When bound to PIP_3, PDK1 phosphorylates Akt at Thr 308 thereby activating it. To be fully activated, Akt requires to be phosphorylated by another protein kinase, mammalian target of rapamycin complex 2 (mTORC-2) at the Ser473 residue (Alessi et al. 1996; Manning and Cantley 2007). The fully activated Akt can further activate several other target proteins that regulate cellular homeostasis. Some of the substrates of Akt include pro-apoptotic protein like BAD or anti-apoptotic protein like NF-kB, forkhead transcription factors of the FOXO family and GSK. Akt can also phosphorylate and inhibit tumour sclerosis complex (TSC) 1 and 2 which otherwise prevents Rheb, a GTPase acting protein. Rheb on activation phosphorylates and activates the protein mTORC-1, which further activates many other signalling cascades. IGF-1 signalling is regulated in a tissue-dependent manner (Huang and Tindall 2007; Salminen and Kaarniranta 2010; Zoncu et al. 2011). We have recently reported an age-dependent regulation of IGF-1/PI3K/Akt signalling in a tissue-specific manner in mice (Hadem and Sharma 2015).

2.1 IGF-1 and FOXO

The insulin/IGF-1/FOXO pathway was the first lifespan-related pathway discovered (Kenyon 2011). In the worm *C. elegans, daf-16* (orthologue of mammalian FOXOs) is likely the major regulator of longevity and its protein DAF-16 is negatively regulated by the insulin/IGF-1 signalling pathway (Barthel et al. 2005). Subsequently, studies indicate that Akt can phosphorylate and inhibit the function of mammalian FOXO proteins upon insulin/IGF-1 activation. The FOXO proteins belong to the class O of the forkhead family of transcription factors that have a conserved DNA-binding domain called the forkhead box, FOX. In invertebrates, only one FOXO gene is present, *daf-16* in *C. elegans* and *dFOXO* in *Drosophila* (Van Der Heide et al. 2004; Huang and Tindall 2007). However, mammals are shown to contain four FOXO genes, *FOXO1, FOXO3, FOXO4* and *FOXO6.* Among the four proteins, FOXO1 and FOXO3 are present in almost all tissues. FOXO4 expression is abundant in kidney, muscles and colorectal tissues and FOXO6 expression is primarily in the liver and brain. FOXOs have four highly conserved functional motifs viz., a forkhead, nuclear localization, nuclear export and transactivation domains (Obsil and Obsilova 2008).

Through posttranslational processing, the transcriptional activities of FOXO proteins are tightly regulated and these regulations, activation or inactivation, ensure that the precise target genes are transcribed in response to stimuli. One of the main regulators of FOXO proteins, mediated through Akt, is insulin/IGF-1 signalling. Akt phosphorylates FOXO at three conserved residues, which results in the export of the protein to the cytoplasm and thus prevents the transcription of FOXO dependent genes (Vogt et al. 2005; Obsil and Obsilova 2008). The FOXO protein is retained in the cytoplasm by binding of the chaperone protein 14-3-3 preventing the re-entry of FOXO proteins in to nucleus (Van Der Heide et al. 2004). Akt can phosphorylate FOXO protein at a second site leading to decreased binding affinity to DNA by incorporating a negative charge on a positively charged binding domain (Vogt et al. 2005). Acetylation and deacetylation processes that alter its DNA-binding affinity also regulate FOXO proteins by either enhancing or lowering its function. In response to stress, SIRT1 deacetylates FOXO and increases their DNA-binding actions, which leads to enhanced transcription of antioxidant enzyme (Brunet et al. 2004).

FOXOs are involved in many cellular processes that include metabolism, apoptosis, proliferation and stress resistance (Barthel et al. 2005; Huang and Tindall 2007). Mutation of the *daf-2* and *age-1* genes leads to lifespan extension by increasing resistance to oxidative stress, while mutation of *daf-16* reverses the extension of lifespan. *In D. Melanogaster,* mutation in *Inr (insulin/IGF-1receptor)* gene and the over-expression of *dFOXO* increases the lifespan of the fly (Libina et al. 2003; Kenyon 2010). Mice, whose insulin/IGF-1 receptor or signalling pathway efficiency is lowered, showed a 33 % increase in lifespan compared to their wild type (Holzenberger et al. 2003). Furthermore, the genes regulated by FOXO transcription factors responsible for stress resistance are conserved between invertebrates and higher organisms, thus, indicating that these proteins are evolutionarily conserved and are important intermediates of insulin/IGF-1 signalling, hence their activation can be beneficial during aging processes (van Heemst 2010).

2.2 *IGF-1 and NFκB*

Nuclear Factor κB (NF-κB) is a transcription factor that plays an important role during stress response. Studies using motif mapping suggest that NF-κB and the IKK/NF-κB signalling pathway is one of the key mediators of aging (Adler et al. 2007; Tilstra et al. 2011). NF-κB or Rel family of transcription factors have five subunits: p65/RelA, RelB, c-Rel, p105/p50, and p100/p52 (Sen and Baltimore 1986). An over-expression of p65/Rel A or c-Rel of the NF-κB subunit induced a senescent phenotype in cultured cells (Bernard et al. 2004). Tilstra et al. (2011) illustrated that natural aging and progeroid mouse model (ERCC1-deficient mice with defective DNA repair system) lead to an up-regulation of transcriptional activity of NF-κB in cells from various tissues, and genetic reduction of p65 subunit slowed down the onset of age-related pathology. As compared to younger individuals, studies of the skin-derived human fibroblast from aged individuals showed an increase in the activation level of NF-κB and increased in gene expression responsible for inflammation. The reduction in NF-κB gene expression results in a positive overturn in the age-related pathology and gene expression from skin of aged mouse demonstrates that blocking NF-κB gene expression is beneficial to reverse the age-related degeneration (Adler et al. 2007).

The NF-κB transcription factors may be activated by oxidative, genotoxic and inflammatory stresses. Many age-associated diseases viz., atherosclerosis, diabetes, sarcopenia are due to the abnormal signalling of NF-κB, which are well-documented (Tilstra et al. 2011). Since their discovery in the 1980s, NF-κB is implicated in the regulation of numerous cytokines and chemokines actions. Many different proteins can trigger the activation of NF-κB, which include the toll like receptors (TLR) ligands, tumour necrosis factors (TNF), interleukins and IGF-1 among various other growth factors (Madrid et al. 2000; Salminen and Kaarniranta 2010). There are two different pathways by which NF-κB can be activated, the canonical and non-canonical pathway and during the aging process, the most likely route by which NF-κB is stimulated is through the canonical pathway. In the canonical pathway, there is the phosphorylation of the IκB, an inhibitory protein with a nuclear export signal that binds to NF-κB thereby concealing the nuclear localization signal (NLS). The phosphorylation and degradation of IκB in the proteosomes lead to the exposure of the NLS present in the NF-κB subunits and cause the import of NF-κB into the nucleus which can trigger the expression of various genes important for inflammatory responses, cell growth and other regulated genes (Tilstra et al. 2011). The overlapping of functions relating to stress response, cell survival and proliferation observed between NF-κB and IGF-1 signalling pathway implies that there is a link between the two pathways. Studies by several investigators have established this link via the PI3K/Akt signalling (Madrid et al. 2000; Salminen and Kaarniranta 2010). The NF-κB system is activated by the PI3K/Akt pathway either by stimulating the transactivation potential of p65/RelA thereby suppressing apoptosis, or by the direct phosphorylation at Ser181of IKKβ by PDK1 which eventually induces the translocation of NF-κB complexes into the nucleus and the related genes are activated. Further reports showed that anti-apoptotic action mediated through NF-κB could be stimulated by GH, IGF-1 and insulin (Madrid et al. 2000; Mitsiades et al. 2002). During growth phase, the inhibition of apoptosis and autophagy are ideal, however during aging a slow senescent phenotype could be created due to the accumulation of damaged macromolecules within cells thereby provoking inflammatory responses (Salminen and Kaarniranta 2010).

2.3 IGF-1 and Sirtuins

Recent findings have focused on the role of sirtuins in aging. Regulation of skeletal muscles mass by sirtuins is potentially through the alteration of the IGF-1 and its associated signalling pathway (Tran et al. 2014). Sirtuins belong to a class III NAD^+-dependent deacetylases that are highly preserved from single cell organisms like bacteria to a multicellular organisms including human. Silent Information Regulator 2 (SIR2) was first discovered to extend lifespan in yeast. The indication that sirtuins are anti-aging genes comes from the fact that the over expression of the orthologue in the worm *C. elegans* and the fly *Drosiphila* causes a life-extension effect (Kaeberlein et al. 1999; Chen et al. 2008). Mammalian sirtuins are of seven types (SIRT1-7) differing in their N-terminal and C-terminal domains but having a core domain constituting the conserved NAD^+-binding and catalytic domain. They are localized in separate compartments of the cell and have different specific targets (diverse physiological functions) (Frye 2000). SIRT1 possesses the highest structural and functional homology to the yeast SIR2 and located in both nucleus and cytoplasm. Although the primary function of yeast SIR2 has been shown to deacetylate histones protein, however, mammalian sirtuins have other nonhistones nuclear protein targets, mostly transcription factors and cofactors associated with many physiological functions (Tanno et al. 2007). Sirtuins are shown to be involved in a number of cellular processes comprising regulatory effect on gene silencing, glucose metabolism, lipid homeostasis and cell proliferation (Nakagawa and Guarente 2011). SIRT1 has both nuclear and cytosolic substrates. Many nuclear proteins are deacetylated by SIRT1; these include p53, NF-κB, PGC-1α and forkhead family of proteins (Tanno et al. 2007). Proteins of the IGF-1 signalling cascades are shown to interact with the Sir2 in *C. elegans* (Berdichevsky et al. 2006). Locally produced IGF-1 in the heart has been shown to posses cardio-protective effect through crosstalk with SIRT1 via the PDK1 and SGK1 instead of the typical PI3K/Akt pathway (Schug and Li 2010). In skeletal muscle of mice, IGF-1 negatively regulates SIRT1 and SIRT6 and its downstream signalling. Overexpression of SIRT6 has been shown to reduce circulating IGF-1. In mouse neural cells, silencing or overexpression of SIRT1 leads to an increased or decreased IGF-1 and Akt signalling (Kanfi et al. 2012; Sansone et al. 2013).

3 Dietary Restriction Modulation

Besides the genetic makeup of an organism, environmental factors play a pivotal role during the course of aging and one such factor is nutrition. It has been shown that the metabolic syndrome such as obesity decreases the lifespan of mice (Fadini et al. 2011). One of the mechanisms that control lifespan extension as observed in various organisms, ranging from yeast to mammals, is dietary restriction (DR). DR is a limitation in food consumption without causing malnourishment. It has been suggested that both the increase in mean and maximum lifespan and the extent of dietary restriction form a linear relationship but the increase in years does not persist indefinitely with increased restriction. DR of about 55–60 % gives the maximum effect and restrictions greater that this has a negative impact on lifespan (Speakman and Mitchell 2011). There are many health benefits associated with DR in tested animal models. It has been demonstrated that DR reduces the susceptibility of rodents to the incidence of spontaneous cancer in several tissues. The protective effect of DR was seen by reduction in initiation, promotion and progression of tumours. Even if DR was started at mid-adulthood, it slowed cancer and aging (Longo and Fontana 2010).

DR increases insulin sensitivity and hence, alleviates the metabolic syndrome. The metabolic abnormalities were reversed by 30 % DR for 3 months (Bowman et al. 2010). The other health benefits of DR include the action on the immune system. Aging leads to a compromised immune system with a decline in T-cell function. Investigation by Walford et al. (1973) suggests that DR augments the humoral response in mice after one year of restriction by delaying the maturation of the immune system. It has been shown to increase the production and enhance activity of T lymphocytes and reverse the age-related reduction in interleukin 2 (IL2), which are contributory factors for cancer resistance and diminished autoimmune diseases in mice (Yang et al. 2009). DR has also been seen to improve immune response by suppression of NF-κB through IKK complex cascade by modulating ROS dependent mechanism (Kim et al. 2002).

The beneficial effect of DR is quite universal and there is a systemic rejuvenation of the organism as a whole. Investigations of the effect of DR on metabolic enzymes have shown that DR regulates both the activity and protein expression level of enzymes. DR upregulates liver type arginase, an enzyme important for nitrogenous waste disposal (Majaw and Sharma 2015). DR also regulates energy metabolism by upregulating the enzyme inorganic pyrophosphatase (Kharbhih and Sharma 2014). Acetylcholinesterase (AChE), an enzyme that catalyses the breakdown of acetylcholine, is important for preventing neuronal damage, essential for brain development and improving learning and memory. In aging brain, loss of cholinergic functions contributes to decline in cognitive functions. DR improves brain function by acting as a natural inhibitor of AChE and thereby maintaining the level of the already declining acetylcholine (Suchiang and Sharma 2011). Neurodegenerative diseases are due to the aging of the nervous system and DR can enhance the function of the nervous system by influencing various longevity signalling molecules like insulin/IGF-1, sirtuins and FOXO transcription factor that have been shown to regulate lifespan. These proteins help prevent neurodegenerative diseases mostly by enhancing stress resistance (Martin et al. 2006).

One of the consistent results seen with DR is its anti-oxidative action based on its ability to suppress oxidative-related alterations and age-associated diseases caused by oxidative stress. DR curbs the formation of reactive oxygen species (ROS) that can otherwise cause the oxidation of macromolecules (proteins, lipids and nucleic acid) leading to their loss of functions. DR enhances the activity of glutathione peroxidise, superoxide dismutase and catalase the major antioxidant enzymes important for resistance to oxidative stress. To improve the cellular reaction to stress, DR also induces heat shock proteins (HSP) (Aly et al. 1994; Dkhar and Sharma 2014).

Aging and dietary restriction interact through partially overlapping mechanisms in the activation of the conserved signalling pathways and one of that pathway is the insulin/IGF-1 signalling pathway. Moderate DR (30 %) reduces serum IGF-1 protein level by 30 % in old mice and the IGF-1/Akt signalling is regulated in an age- and tissue-dependent manner (Hadem and Sharma 2015). Such DR had lead to an elevated PI3K level while pAkt/Akt ratio is decreased in most of the tissues of aged mice. Other investigators have reported a reduction in both the gene and protein expression level of IGF-1 when animals are subjected to DR (Breese et al. 1991). Based on the observations in invertebrates, where mutants with a reduction in the insulin/IGF-1 pathway is associated with augmented lifespan, the reduction of the IGF-1 level with DR can be linked to its beneficial effect in lifespan extension (Libina et al. 2003; Kenyon 2011). Furthermore, a study by Al-Regaiey et al. (2005) indicates that when the GH receptor/GH-binding protein gene was disrupted it leads to a reduction of IGF-1 level; the DR effects on reducing IGF-1 may also act through overlapping but distinct mechanism. Studies in mutant mice like Snell dwarf and Ames dwarf indicate that they are resistant to cancer and have prolonged lifespan and are associated with reduced IGF-I level (Salminen and Kaarniranta 2010). One of the mechanisms by which IGF-1 mediates its effect on lifespan and health-span is by reducing oxidative stress. This is achieved by the over-expression of FOXO protein in response to diminished IGF-1 signalling. The reduction in the insulin/IGF-1 signalling, as illustrated by other researchers, resulted in increased FOXO activity that are essential in the regulation of several cellular processes like glucose metabolism, ROS detoxification, DNA repair and cell death (Libina et al. 2003; Barthel et al. 2005; Al-Regaiey et al. 2005). It has been demonstrated that the insulin/IGF-1 signalling central molecule, Akt can directly regulate the activity of mTOR. Recent studies proposed that one of the beneficial effects of DR and lowered IGF-1 level in longevity act through PI3K/Akt/mTOR signalling by reducing mTOR activity and significantly increasing lifespan. Besides, being activated by IGF-1, energy sensing pathway such as sirtuins and AMP-activated protein kinase (AMPK) can negatively regulate mTORC1 activity (Solon-Biet et al.2015).

The activity of SIR2 was increased upon DR and the expression of *SIR2* is crucial for extension of lifespan in many yeast strains studied. As the function of sirtuin requires NAD^+, this indicates that there is a connecting link between metabolism and aging, which is further regulated by diet (Haigis and Guarente 2006). SIRT1 expression in multiple tissues is augmented when animals are subjected to DR. Cohen et al. (2004) reported that in the liver of rat, SIRT1 is upregulated during DR and insulin/IGF-1 negatively regulates its expression. The increased expression of SIRT1 is mediated by the action of FOXO proteins (Nemoto et al. 2004). Additionally, other investigators have reported that SIRT1 can in turn deacetylates FOXO protein and stimulate the expression of antioxidant enzyme. SIRT1 is also involved in regulating immune response by deacetylating the RelA/p65 component of NF-κB at Lys310 residue and inhibiting its transactivation capacity (Yeung et al. 2004).

DR for short duration resulted in the reduction of the activity of NF-κB in the kidney of old mice (Jung et al. 2009). The positive effect of DR has been attributed to the inhibition of NF-κB and immune response. As demonstrated by Lin et al. (2004), FOXO3 protein in mice can inhibit NF-κB signalling, as mice deficient in FOXO3 induces different auto-inflammatory response causing elevated inflammation in several tissues. DR leads to the reduction of inflammatory responses by the repression of the insulin/IGF-1 signalling pathway and activation of FOXO proteins.

4 Conclusion

IGF-1 is one of the key mediators of aging and its downstream signalling pathway regulates many cellular processes like cell proliferation, differentiation, cell survival and apoptosis, which are all important for proper functioning of the organisms during aging. This signalling pathway, however, interacts with other longevity-associated pathways to potentiate the aging process. In this review, we observed that Akt is the central molecule that serves as a link between the different important pathways associated with lifespan regulation by phosphorylation and activation of many kinases and through interaction with the FOXO proteins. Dietary restriction, an important modulator of lifespan extension, acts by repressing the insulin/IGF-1 signalling pathway. This pathway is a beneficial regulator of the lifespan as its reduction leads to the activation of FOXO factors and sirtuins that enhances longevity of an organism. Various signalling pathways such as, the NF-κB signalling, the FOXO factors, the insulin/IGF-1 signalling and the longevity molecule SIRT are inter-connected and exhibit crosstalk that, may ultimately determine the lifespan of an organism. IGF-1 being the central regulatory factor is modulated by DR to enhance the longevity of an organism (Fig. 1).

Acknowledgments The authors thank UGC-DRS, DST-FIST, DBT infrastructure and Biochemistry Department of North-Eastern Hill University, Shillong for the research funds and facilities. IKH Hadem is thankful to CSIR (Council of Scientific and Industrial Research), New Delhi for providing financial assistant as junior/senior research fellowships (09/347(0189)/2010-EMR-1).

References

Adler AS, Sinha S, Kawahara TLA, Zhang JY, Segal E, Chang HY (2007) Motif module map reveals enforcement of aging by continual NF-kB activity. Genes Dev 21:3244–3257

Alessi DR, Andjelkovic M, Caudwell B, Cron P, Morrice N, Cohen P, Hemmings BA (1996) Mechanism of activation of protein kinase B by insulin and IGF-1. EMBO J 15:6541–6551

Al-Regaiey KA, Masternak MM, Bonkowski M, Sun L, Bartke A (2005) Long-lived growth hormone receptor knockout mice: interaction of reduced insulin-like growth factor I/insulin signaling and caloric restriction. Endocrinology 146:851–860

Aly KB, Pipkin JL, Hinson WG, Feuers RJ, Duffy PH, Lyncook L, Hart RW (1994) Chronic caloric restriction induces stress proteins in the hypothalamus of rats. Mech Ageing Dev 76:11–23

Barthel A, Schmoll D, Unterman TG (2005) FoxO proteins in insulin action and metabolism. Trends Endocrin. Met. 16:183–189

Berdichevsky A, Viswanathan M, Horvitz HR, Guarente L (2006) *C. elegans* SIR-2.1 interacts with 14-3-3 proteins to activate DAF-16 and extend life span. Cell 125:1165–1177

Berelowitz M, Szabo M, Frohman LA (1981) Somatomedin-C mediates growth hormone negative feedback by effects on both hypothalamus and the pituitary. Science 212:1279–1281

Bernard D, Gosselin K, Monte D, Vercamer C, Bouali F, Pourtier A, Vandenbunder B, Abbadie C (2004) Involvement of Rel/nuclear factor-kappaB transcription factors in keratinocyte senescence. Cancer Res 64:472–481

Bowman TA, Ramakrishnan SK, Kaw M, Lee SJ, Patel PR, Golla VK, Bourey RE, Haram PM, Koch LG, Britton SL, Wisloff U, Lee AD, Najjar SM (2010) Caloric restriction reverses hepatic insulin resistance and steatosis in rats with low aerobic capacity. Endocrinology 151:5157–5164

Breese CR, Ingram RL, Sonntag WE (1991) Influence of age and long-term dietary restriction on plasma insulin-like growth factor-I (IGF-1), IGF-1 gene expression, and IGF-1 binding-proteins. J Gerontol 46:B180–B187

Brissenden JE, Ullrich A, Francke U (1984) Human chromosomal mapping of genes for insulin-like growth factors I and II and epidermal growth factor. Nature 310:781–784

Brunet A, Sweeney LB, Sturgill JF, Chua KF, Greer PL, Lin Y, Tran H, Ross SE, Mostoslavsky R, Cohen HY (2004) Stress-dependent regulation of FOXO transcription factors by the SIRT1 deacetylase. Science 303:2011–2015

Chen D, Bruno J, Easlon E, Lin SJ, Cheng HL, Alt FW, Guarente L (2008) Tissue-specific regulation of SIRT1 by calorie restriction. Genes Dev 22:1753–1757

Clemmons DR (2007) Modifying IGF-1 activity: an approach to treat endocrine disorders, atherosclerosis and cancer. Nat Rev Drug Discov 6:821–833

Cohen HY, Miller C, Bitterman KJ, Wall NR, Hekking B, Kessler B, Howitz KT, Gorospe M, de Cabo R, Sinclair DA (2004) Calorie restriction promotes mammalian cell survival by inducing the SIRT1 deacetylase. Science 305:390–392

Datta SR, Brunet A, Greenberg ME (1999) Cellular survival: a play in three Akts. Genes Dev 13:2905–2927

Dkhar P, Sharma R (2014) Late-onset dietary restriction modulates protein carbonylation and catalase in cerebral hemispheres of aged mice. Cell Mol Neurobiol 34:307–313

Dupont J, Khan J, Qu BH, Metzler P, Helman L, LeRoith D (2001) Insulin and IGF-1 induce different pattern of gene expression in mouse fibroblast NIH-3T3 cells: identification by cDNA microarray analysis. Endocrinology 142:4969–4975

Fadini GP, Ceolotto G, Pagnin E, de Kreutzenberg S, Avogaro A (2011) At the crossroads of longevity and metabolism: the metabolic syndrome and lifespan determinant pathways. Aging Cell 10:10–17

Froesch ER, Schmid C, Schwander J, Zapf J (1985) Actions of insulin-like growth factors. Annu Rev Physiol 47:443–467

Frye RA (2000) Phylogenetic classification of prokaryotic and eukaryotic Sir2-like proteins. Biochem Biophys Res Commun 273:793–798

Gauguin L, Delaine C, Alvino CL, McNeil KA, Wallace JC, Forbes BE (2008) Alanine scanning of a putative receptor binding surface of insulin-like growth factor-I. J Biol Chem 283:20821–20829

Hadem IKH, Sharma R (2015) Age- and tissue-dependent modulation of IGF-1/PI3K/Akt protein expression by dietary restriction in mice. Metab Res, Horm. doi:10.1055/s-0035-1559770

Haigis MC, Guarente LP (2006) Mammalian sirtuins-emerging roles in physiology, aging, and calorie restriction. Genes Dev 20:2913–2921

Holzenberger M, Dupont J, Ducos B, Leneuve P, Geloen A, Even PC, Cervera P, Le Bouc Y (2003) IGF-1 receptor regulates lifespan and resistance to oxidative stress in mice. Nature 421:182–186

Huang H, Tindall DJ (2007) Dynamic FoxO transcription factors. J Cell Sci 120:2479–2487

Humbel RE (1984) Insulin-like growth factors, somatomedins, and multiplication stimulating activity: chemistry. In: Li CH (ed) Hormonal proteins and peptides, vol 12. Academic Press, Inc., New York, pp 57–79

Hussain MA, Schmitz O, Christiansen JS, Zapf J, Froesch RE (1995) Metabolic effects of insulin-like growth factor-I: a focus on insulin sensitivity. Metabolis 144:108–112

Jung HJ, Suh Y (2015) Regulation of IGF-1 signalling by microRNAs. Front Genet. 5:472

Jung KJ, Lee EK, Kim JY, Zou Y, Sung B, Heo HS, Kim MK, Lee J, Kim ND, Yu BP, Chung HY (2009) Effect of short term calorie restriction on pro-inflammatory NF-kB and AP-1 in aged rat kidney. Inflamm Res 58:143–150

Kaeberlein M, McVey M, Guarente L (1999) The SIR2/3/4 complex and SIR2 alone promote longevity in *Saccharomyces cerevisiae* by two different mechanisms. Genes Dev 13:2570–2580

Kanfi Y, Naiman S, Amir G, Peshti V, Zinman G, Nahum L, Bar-Joseph Z, Cohen HY (2012) The sirtuin SIRT6 regulates lifespan in male mice. Nature 483:218–221

Kenyon C (2011) The first long-lived mutants: discovery of the insulin/IGF-1 pathway for ageing. Philos Trans R Soc Lond B Biol Sci 366:9–16

Kenyon CJ (2010) The genetics of ageing. Nature 464:504–512

Kharbhih WJ, Sharma R (2014) Age-dependent increased expression and activity of inorganic pyrophosphatase in the liver of male mice and its further enhancement with short- and long-term dietary restriction. Biogerontology 15:81–86

Kim HJ, Jung KJ, Yu BP, Cho CG, Choi JS, Chung HY (2002) Modulation of redox-sensitive transcription factors by calorie restriction during aging. Mech Ageing Dev 123:1589–1595

Libina N, Berman JR, Kenyon C (2003) Tissue-specific activities of *C. elegans* DAF-16 in the regulation of lifespan. Cell 115:489–502

Lin L, Hron JD, Peng SL (2004) Regulation of NF-kappaB, Th activation, and auto inflammation by the forkhead transcription factor Foxo3a. Immunity 21:203–213

Longo VD, Fontana L (2010) Calorie restriction and cancer prevention: metabolic and molecular mechanisms. Trends Pharmacol Sci 31:89–98

Madrid LV, Wang CY, Guttridge DC, Schottelius AJG, Baldwin AS Jr, Mayo MW (2000) Akt suppresses apoptosis by stimulating the transactivation potential of the RelA/p65 subunit of NF-kappaB. Mol Cell Biol 20:1626–1638

Maier B, Gluba W, Bernier B, Turner T, Mohammad K, Guise T, Sutherland A, Thorner M, Scrable H (2004) Modulation of mammalian life span by the short isoform of p53. Genes Dev 18:306–319

Majaw T, Sharma R (2015) Arginase I expression is upregulated by dietary restriction in the liver of mice as a function of age. Mol Cell Biochem 407:1–7

Manning BD, Cantley LC (2007) AKT/PKB signaling: navigating downstream. Cell 129:1261–1274

Martin B, Mattsona MP, Maudsley S (2006) Caloric restriction and intermittent fasting: Two potential diets for successful brain aging. Ageing Res Rev 5:332–353

Masse I, Molin L, Billaud M, Solari F (2005) Lifespan and dauer regulation by tissue-specific activities of Caenorhabditis elegans DAF-18. Dev Biol 286:91–101

Mitsiades CS, Mitsiades N, Poulaki V, Schlossman R, Akiyama M, Chauhan D, Hideshima T, Treon SP, Munshi NC, Richardson PG, Anderson KC (2002) Activation of NF-kappaB and upregulation of intracellular anti-apoptotic proteins via the IGF-1/Akt signaling in human multiple myeloma cells: therapeutic implications. Oncogene 21:5673–5683

Montecino-Rodriguez E, Berent-Maoz B, Dorshkind K (2013) Causes, consequences, and reversal of immune system aging. J Clin Invest 123:958–965

Musarò A, Dobrowolny G, Rosenthal N (2007) Theneuroprotective effects of a locally acting IGF-1 isoform. Exp Gerontol 42:76–80

Nakagawa T, Guarente L (2011) Sirtuins at a glance. J Cell Sci 124:833–838

Nemoto S, Fergusson MM, Finkel T (2004) Nutrient availability regulates SIRT1 through a forkhead-dependent pathway. Science 306:2105–2108

O'Sullivan D, Szestak T, Pell J (2002) Regulation of hepatic insulin-like growth factor I leader exon usage in lambs: effect of immunization against growth hormone-releasing factor and subsequent growth hormone treatment. J Anim Sci 80:1074–1082

Oberbauer AM (2013) The regulation of IGF-1 gene transcription and splicing during development and aging. Front Endocrinol (Lausanne) 4:39

Obsil T, Obsilova V (2008) Structure/function relationships underlying regulation of FOXO transcription factors. Oncogene 27:2263–2275

Rincon M, Muzumdar R, Atzmon G, Barzilai N (2004) The paradox of the insulin/IGF-1 signaling pathway in longevity. Mech Ageing Dev 125:397–403

Salminen A, Kaarniranta K (2010) Insulin/IGF-1 paradox of aging: regulation via AKT/IKK/NF-kappaBsignaling. Cell Signal 22:573–577

Sansone L, Reali V, Pellegrini L, Villanova L, Aventaggiato M, Marfe G, Rosa R, Nebbioso M, Tafani M, Fini M, Russo MA, Pucci B (2013) SIRT1 silencing confers neuroprotection through IGF-1 pathway activation. J Cell Physiol 228:1754–1761

Schug TT, Li X (2010) Surprising sirtuin crosstalk in the heart. Aging 2:129–132

Sen R, Baltimore D (1986) Multiple nuclear factors interact with the immunoglobulin enhancer sequences. Cell 46:705–716

Sharma R, Dkhar P (2014) Biological basis of aging: theories and explanations. In: Sanchetee P (ed) Textbook of Geriatric Medicine. Indian Academy of Geriatrics (Paras Medical Publisher), Hyderabad, pp 24–31

Sharma R, Dutta D (2006) Age-dependent decreases in renal glucocorticoid receptor function is reversed by dietary restriction in mice. Ann NY Acad Sci 1067:129–141

Shavlakadze T, Winn N, Rosenthal N, Grounds MD (2005) Reconciling data from transgenic mice that overexpress IGF-I specifically in skeletal muscle. Growth Horm IGF Res 15:4–18

Solon-Biet SM, Mitchell SJ, de Cabo R, Raubenheimer D, Le Couteur DG, Simpson SJ (2015) Macronutrients and caloric intake in health and longevity. J. Endocrinol. 226:R17–R28

Sonntag WE, Steger RW, Forman LJ, Meites J (1980) Decreased pulsatile release of growth hormone in old male rats. Endocrinology 107:1875–1879

Speakman JR, Mitchell SE (2011) Caloric restriction. Mol Aspects Med 32:159–221

Suchiang K, Sharma R (2011) Dietary restriction regulates brain acetylcholinesterase in female mice as a function of age. Biogerontology 12:581–589

Sussenbach J, Steenbergh P, Holthuizen P (1992) Structure and expression of the human insulin-like growth factor genes. Growth Regulat 2:1–9

Tanno M, Sakamoto J, Miura T, Shimamoto K, Horio Y (2007) Nucleocytoplasmic shuttling of the NAD^+-dependent histone deacetylase SIRT1. J Biol Chem 282:6823–6832

Tilstra JS, Clauson CL, Niedernhofer LJ, Robbins PD (2011) NF-kB in Aging and Disease. Aging Dis. 2:449–465

Tran D, Bergholz J, Zhang H, He H, Wang Y, Zhang Y, Li Q, Kirkland JL, Xiao ZX (2014) Insulin-like growth factor-1 regulates the SIRT1-p53 pathway in cellular senescence. Aging Cell 13:669–678

Van Der Heide LP, Hoekman MF, Smidt MP (2004) The ins and outs of FoxO shuttling: mechanisms of FoxO translocation and transcriptional regulation. Biochem J 380:297–309

Van Heemst D (2010) Insulin, IGF-1 and longevity. Aging Dis 1:147–157

Vogt PK, Jiang H, Aoki M (2005) Triple layer control: phosphorylation, acetylation and ubiquitination of FOXO proteins. Cell Cycle 4:908–913

Walford RL, Liu RK, Gerbased M, Mathies M, Smith GS (1973) Long term dietary restriction and immune function in mice—response to sheep red blood-cells and to mitogenic agents. Mech Ageing Dev 2:447–454

Yang H, Adamo ML, Koval AP, McGuinness MC, Ben-Hur H, Yang Y (1995) Alter- native leader sequences in insulin- like growth factor I mRNAs modulate translational efficiency and encode multiple signal peptides. Mol Endocrinol 9: 1380–1395

Yang HW, Youm YH, Dixit VD (2009) Inhibition of thymic adipogenesis by caloric restriction is coupled with reduction in age-related thymic involution. J Immunol 183:3040–3052

Yeung F, Hoberg JE, Ramsey CS, Keller MD, Jones DR, Frye RA, Mayo MW (2004) Modulation of NF-kappaB-dependent transcription and cell survival by the SIRT1 deacetylase. EMBO J 23:2369–2380

Zoncu R, Efeyan A, Sabatini DM (2011) mTOR: from growth signal integration to cancer, diabetes and ageing. Nat Rev Mol Cell Biol 12:21–35

Arginase I Regulation by Dexamethasone in the Liver of Aging Mouse

Teikur Majaw and Ramesh Sharma

Abstract Arginase I is a metalloenzyme that catalyses arginine into urea and ornithine. It plays an essential role in hepatic metabolism of most ureotelic organisms as a principal component of urea cycle responsible for elimination of nitrogenous waste (ammonia) in the form of urea. Arginase I is predominantly expressed in the cytoplasm of liver but is less expressed in other tissues. Changes in the expression of arginase I is regulated by various factors including dietary and various hormones. We were interested to know the regulation of arginase I by dexamethasone in the liver of mice at two different ages: young (1-month) and old (18-month). Our results indicate that dexamethasone significantly upregulates the activity of arginase I in both young and old mice, albeit to a varying degree with young mice showing a higher magnitude of upregulation as compared to the old mice. The enzyme protein level ascertained by Western blotting also matched that of the activity. Arginase I mRNA level paralleled that of protein level on such treatments. These results show that dexamethasone mediated induction of liver arginase I expression declines with age of mice.

Keywords Arginase I · Dexamethasone · Liver · Mice

1 Introduction

Aging is characterized by a gradual buildup of diverse deleterious changes in every part of the cells, tissues and the organism as a whole that enhance the predisposition to diseases and ultimately death of an organism with advancing age (Harman 2000). These transitions in biological processes occurring in tissues and organs may lead to the increasing risk of age-associated diseases and to a disturbed body homeostasis. The variability in the rate of aging seen among individuals is generally due to genetic makeup and various other intrinsic and extrinsic factors. Another important

T. Majaw · R. Sharma (✉)
Department of Biochemistry, North-Eastern Hill University, Shillong 793022, India
e-mail: sharamesh@gmail.com

© Springer Science+Business Media Singapore 2017
P.C. Rath et al. (eds.), *Topics in Biomedical Gerontology*,
DOI 10.1007/978-981-10-2155-8_6

characteristic of aging is that even though several functions deteriorate after adulthood, the time of onset of deterioration of each function is different, the rate at which one function declines is different from that of another and among organs (Kanungo et al. 1997). While the definite mechanism of aging at the molecular level is unknown, a progressive increase with age in the accumulation of damaged macromolecules, chiefly proteins, may be critical in the role of senescence. In particular, the increased amount of damaged proteins is important because it could lead to malfunction of nearly all biological processes (Ryazanov and Nefsky 2002).

Liver is essential in the regulation of various cellular processes like glucose homeostasis, metabolism and detoxification of xenobiotics. Its pivotal function in maintaining metabolic homeostasis makes it an interesting target for aging study. Protein metabolism results in production of ammonia that is toxic if accumulated and must be removed via the urea cycle in the liver. Among urea cycle enzymes, the only enzyme known to exist in two distinct isoforms is arginase. These isozymes are classified as arginase I (liver-type) and arginase II (kidney-type) (Stewart and Caron 1977). Arginase I is located in cytoplasm and predominantly expressed in the liver as a component of the urea cycle while arginase II is located in matrix of mitochondria and is expressed highly in kidney and intestine but is less expressed or absent in liver (Mori 2006). They are similar with regard to the enzymatic properties and requirement of manganese ions but dissimilar with respect to sub-cellular location, tissue distribution, regulation of expression and immunological reactivity (Wu and Morris 1998). Expression of arginase I is regulated by metabolism of dietary protein that alter amino acid nitrogen flux into the urea cycle and by various hormones like glucagon, insulin, and glucocorticoids (Morris 1992).

Since arginase I plays an important regulatory role in physiology and development of an organism, we became interested to understand the regulation of arginase I in liver of mice as a function of age. In this study, we report the regulation of arginase I in mouse liver by a synthetic glucocorticoid, dexamethasone. We employed enzyme activity assays, immunoblotting and real-time PCR techniques to determine the arginase I activity, protein level and mRNA level, respectively.

2 Materials and Methods

2.1 Animals

Animal used were male albino mice (BALB/c). Normal laboratory conditions for animals were kept at 25 ± 2 °C with 12 h light/dark cycle. Standard pellet diet and water were fed ad libitum. Young (1-month) and old (18-month) mice were used for determination of the effect of dexamethasone on arginase I activity and expression. Guidelines of the institution for the use of animals were followed.

2.2 Dexamethasone Treatment

Pilot experiments were conducted to determine the optimal dose- and time-response of arginase I to dexamethasone administration. Dexamethasone (10 μg/100 g body weight) was dissolved in ethanolic saline and administered intraperitoneally. Age-matched control received only ethanolic saline.

2.3 Enzyme Preparation

Mice were sacrificed at a fixed time of the day by cervical dislocation method. Livers were excised immediately, wash in normal saline and blotted dry. Tissue homogenate preparation was followed according to Majaw and Sharma (2015). It was centrifuged and cytosolic fraction was used as enzyme source.

2.4 Determination of Arginase I Activity

Activity of arginase I was determined spectrophotometrically using diacetyl-monoxime to measure urea production (Satoh and Ito 1968; Majaw and Sharma 2015). Fifty microlitres of enzyme preparation was used for the reaction and the specific activity of enzyme was defined as micromole of urea formed per min per milligram protein.

2.5 Protein Estimation

The dye binding method of Bradford (1976) was used to determine the protein concentration of the enzyme preparation. Bovine serum albumin (BSA) was used as standard.

2.6 Western Blotting

Western blotting was performed to determine the protein level of arginase I following procedures as described (Majaw and Sharma 2015). Briefly, 60 μg of proteins was separated on SDS-PAGE. They were blotted onto nitrocellulose membrane. After blocking non-specific binding with skimmed milk in Tris buffer saline, the membrane was incubated with 1:500 diluted primary antibodies (rabbit anti-Arginase I). Unbound antibodies were removed by washing with Tween-20

Tris buffer saline. It was further incubated for 3 h in secondary antibody (goat anti-rabbit IgG-ALP conjugate). BCIP/NBT substrate was added and the color intensity of the band was quantified using KDS software.

2.7 Total RNA Isolation and Quantitative Real-Time PCR

RNA isolation and quantitative real-time PCR was performed as described (Majaw and Sharma 2015). In brief, total RNA was extracted using 1 ml of ice cold TRIsure and 2 μg was reverse transcribed to cDNA. Quantitative real-time PCR was performed with primers (arginase I: forward: 5′-AACACGGCAGTCGCTTTAACC-3′, reverse: 5′-GGTTTTCATGTGGCGC ATTC-3′; GAPDH: forward: 5′-CAGGTTGTCTCCTGCGACTT-3′, reverse: 5′-CCCTGTTGCTGTAGCCGTA-3′) at a final concentration of 300 nM and 2.5 μl of cDNA template. Data were normalized to the endogenous control (GAPDH) and the fold change of arginase I mRNA was calculated by $2^{-\Delta\Delta Ct}$ method (Livak and Schmittgen 2001) and expressed relative to the control group.

2.8 Statistical Analysis

One-way ANOVA was used for statistical analysis using GraphPad Prism 6.03. When significant, post hoc Tukey's Multiple Comparison Test was performed. Values are reported as mean ± SEM with 4–5 mice in a group and p value ≤ 0.05 was set for statistical significance.

3 Results and Discussion

From our previous studies on the endogenous arginase I activity and protein level in liver of mice of different age groups viz., young, adult and old, we observed that the activity of arginase I in the liver of mice was highest in young which significantly declines as the animal ages. The protein level of arginase I expression estimated by Western blotting also correlated with the activity showing a decreased level with the advance of age (Majaw and Sharma 2015). Our observations from previous findings suggest that aging is accompanied by decreased efficiency in the urea cycle and the maintenance of the level of arginase activity and expression is necessary for the proper physiological function of the animal during development. Modulation by various hormones may provide an insight into better understanding the age-related regulation of arginase expression.

Glucocorticoids are crucial hormones secreted from adrenal glands during the stressful situations and are named for their ability to promote the conversion of proteins and lipids into glucose during stress. Studies reported that adrenalectomy prevents the development features characteristic of aging suggesting that glucocorticoids play important roles in the process of aging (Landfield et al. 1981). The glucocorticoids' action on the organism is multitudinal and is mediated by binding to their respective cytosolic/nuclear receptors. In our experiment, the effect of dexamethasone on arginase I activity and expression was studied in liver of 1- and 18-month old mice. We observed that dexamethasone increases the activity and protein level of arginase I in both young and old mice as compared to the age-matched control (Fig. 1a, b), although the magnitude of induction is higher in young mice as compare to the old mice. Two immune reactive bands of arginase I was observed in Western blotting. Arginase I mRNA contains two methionine codons and translation can be initiated from one or both of the codons resulting into two immune reactive bands (Ohtake et al. 1998; Akiba et al. 2002). At the mRNA level, a similar trend of arginase I mRNA expression was also detected in both the ages (Fig. 2). Reports by Okun et al. (2015) have also indicated the upregulation of arginase I by dexamethasone which is consistent with our present findings. Glucocorticoids are known to induce lean tissue wasting and protein turnover (Watson et al. 2012) and the increased arginase I may be of functional importance to support the disposal of ammonia produced from protein metabolism during such intervention.

The difference in the magnitude of response to dexamethasone suggests the possibility of the different level of glucocorticoid receptors (GR) expression at different stages of development of mice. It is also possible that the receptor's

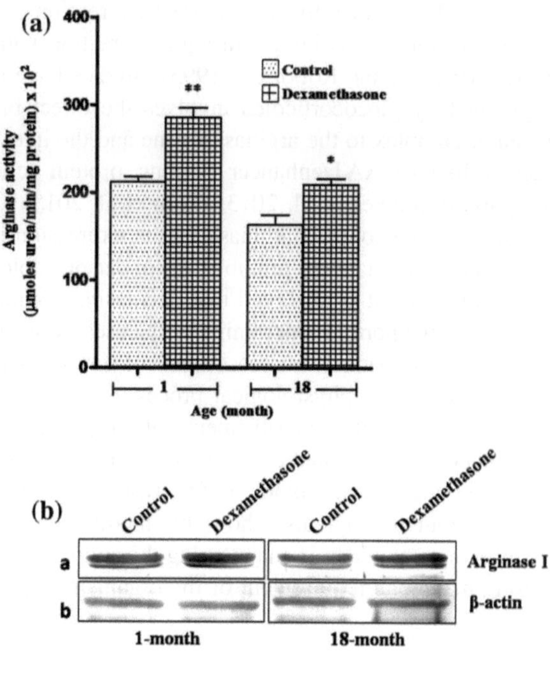

Fig. 1 **a** Arginase I activity expressed as urea formed per min per mg protein **b** western blot of (**a**) arginase I protein and (**b**) β-actin from liver of young (1-month) and old (18-month) mice treated with dexamethasone. Values are reported as mean ± SEM (4/5 mice). *Asterisks* (*) and (**) indicate statistical significant value at $p < 0.05$ and $p < 0.01$, respectively compared to age-matched control mice. β-actin was used as a loading control

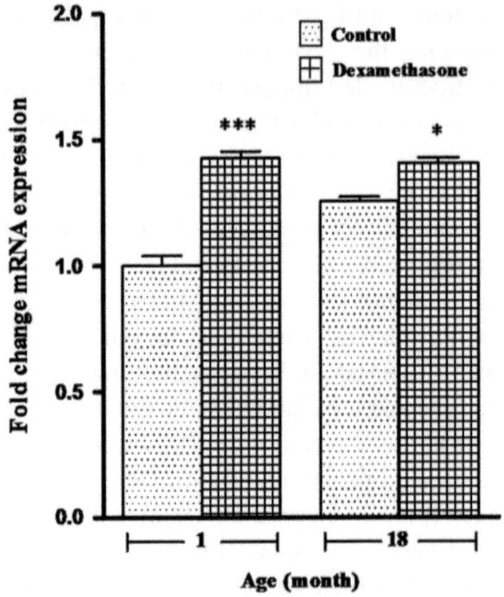

Fig. 2 Arginase I mRNA level ascertained by real-time PCR from the liver of 1-month and 18-month old treated with dexamethasone. Data are reported as fold change over age-matched control mice ($2^{-\Delta\Delta Ct}$ method). *Asterisks* (*) and (***) indicate statistical significant value at $p < 0.05$ and $p < 0.001$, respectively when compared to age-matched control mice. Values are reported as mean ± SEM (n = 4/5)

affinity towards glucocorticoids altered as the animal aged. From the perspective of glucocorticoids receptor level, it has been reported that receptors level in liver was found to significantly decrease gradually in old mice while the affinity remained unaltered (Dutta and Sharma 2004). Since dexamethasone acts by binding to the glucocorticoids receptors, the higher induction of arginase I expression in young mice corresponded with the high receptor level at the young age as compared to the old mice. Up-regulation of arginase I expression is mediated either by the direct action of glucocorticoids or through interaction with cytokines and other growth factors (Pannen and Robotham 1995). Studies have illustrated that upregulation of arginase I by glucocorticoids involves the direct binding of the glucocorticoids-receptor complex to the arginase I gene and the interaction with other transcription factors like CCAAT/enhancer binding protein (c/EBP) may enhance the transcription (Grontved et al. 2013; Okun et al. 2015). Our results, thus, indicate that dexamethasone controls arginase I gene expression in liver of mice and may help maintain homeostasis of amino acids under physiological conditions.

In summary, it is observed that decreasing arginase I activity and protein level may lower the normal functioning of the urea cycle and may cause increase level of ammonia. Exposure to increase level of ammonia may perturb the normal functioning of various physiological processes in various organs. This may, although partly, explain the involvement of arginase I in the process of aging. Simultaneously, elevated arginine due to decreased level of arginase I in aged mice could augment nitric oxide (NO) synthesis and lead to age-related oxidative damage. Our findings also show that modulation of arginase I expression by glucocorticoids may help in maintaining the endogenous level of arginase I expression during metabolic requirement of the organism (Fig. 3).

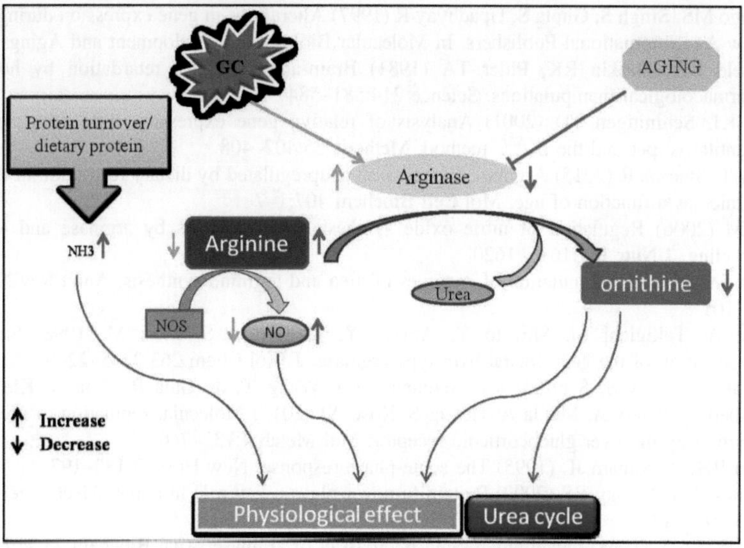

Fig. 3 Schematic diagram showing the effect of aging and dexamethasone on the liver arginase I in mice. Aging decreases the arginase I expression and dexamethasone acts by modulating the upregulation of arginase I. The possible downstream effects due to decreased arginase I caused by aging are the perturbed urea cycle, excess NO production and the other related physiological processes. Glucocorticoids, like dexamethasone, however, act as by attenuating the negative effect of aging

Acknowledgments Authors express gratitude to the Department of Biochemistry, North-Eastern Hill University, Shillong for providing research facilities and infrastructures under UGC-UPE, DRS and DST-FIST, New Delhi. TM is also grateful to CSIR, New Delhi for the support financially under JRF and SRF fellowship (F.No: 09/347(0194)/2010-EMR-I).

References

Akiba T, Kuroiwa N, Shimizu-Yabe A, Iwase K, Hiwasa T, Yokoe H, Kubosawa H, Kageyama R, Dahrington GJ, Mori M, Tanzawa H, Takiguchi M (2002) Expression and regulation of the gene for arginase I in mouse salivary glands: requirement of CCAAT/enhancer-binding protein α for the expression in the parotid gland. J Biochem 132:621–627

Bradford MM (1976) A rapid and sensitive method for the quantification of microgram quantities of protein using the principle of protein-dye binding. Anal Biochem 72:248–254

Dutta D, Sharma R (2004) Age-dependent dietary regulation of glucocorticoid receptor in the liver of mice. Biogerontology 5:177–184

Grontved L, John S, Baek S, Liu Y, Buckley JR, Vinson C, Aguilera G, Hager GL (2013) C/EBP maintains chromatin accessibility in liver and facilitates glucocorticoid receptor recruitment to steroid response elements. EMBO J 32:1568–1583

Harman D (2000) Antioxidant supplements: effect on disease and aging in the United States population. J Am Aging Assoc 23:25–31

Kanungo MS, Singh S, Gupta S, Upadhyay R (1997) Alterations in gene expression during aging. New Age International Publishers, In Molecular Biology of Development and Aging
Landfield PW, Baskin RK, Pitler TA (1981) Brain-age correlates retardation by hormonal pharmacological manipulations. Science 214:581–584
Livak KJ, Schmittgen TD (2001) Analysis of relative gene expression data using real-time quantitative pcr and the $2^{-\Delta\Delta ct}$ method. Methods 25:402–408
Majaw T, Sharma R (2015) Arginase I expression is upregulated by dietary restriction in the liver of mice as a function of age. Mol Cell Biochem 407:1–7
Mori M (2006) Regulation of nitric oxide synthesis and apoptosis by arginase and arginine recycling. J Nutr 137:1616–1620
Morris SM Jr (1992) Regulation of enzymes of urea and arginine synthesis. Annu Rev Nutr 12: 81–101
Ohtake A, Takigichi M, Shigeto Y, Amaya Y, Kawamoto S, Mori M (1998) Structural organization of the gene for rat-liver type arginase. J Biol Chem 263:2245–2249
Okun JG, Conway S, Schmidt KV, Schumacher J, Wang X, de Guia R, Zota A, Klement J, Seibert O, Peters A, Maida A, Herzig S, Rose AJ (2015) Molecular regulation of urea cycle function by the liver glucocorticoid receptor. Mol Metab 4:732–740
Pannen BH, Robotham JL (1995) The acute-phase response. New Horiz 3:183–197
Ryazanov AG, Nefsky BS (2002) Protein turnover plays a key role in aging. Mech Ageing Dev 123:207–213
Satoh PS, Ito Y (1968) Quantitative rapid microassay of arginase. Anal Biochem 23:219–229
Stewart JA, Caron H (1977) Arginases of mouse brain and liver. J Neurochem 29:657–663
Watson ML, Baehr LM, Reichardt HM, Tuckermann JP, Bodine SC, Furlow JD (2012) A cell-autonomous role for the glucocorticoid receptor in skeletal muscle atrophy induced by systemic glucocorticoid exposure. Am J Physiol Endocr Metab 302:1210–1220
Wu G, Morris SM Jr (1998) Arginine metabolism and beyond. Biochem J 336:1–7

Inorganic Pyrophosphatase of Cardiac and Skeletal Muscle is Enhanced by Dietary Restriction in Mice During Aging

Wanaiborlang Jala Kharbhih and Ramesh Sharma

Abstract Hydrolysis of ATP generates inorganic pyrophosphate (PPi) in the cells, extracellular matrix, cartilage and body fluids. Inorganic pyrophosphatase (iPPase) converts PPi to orthophosphate; the extra energy from this reaction can substitute for ATP-derived energy under certain conditions. We studied the activity expression of iPPase in cardiac and skeletal muscle of young and old mice subjected to short- and long-term dietary restriction. Western blot analysis using anti-iPPase and differential polymerase chain reaction using iPPase specific primers were used to ascertain iPPase level. Significant increase in both expression and activity of iPPase were observed in older mice as compared to younger ones. Twenty four hours of fasting enhanced the expression and activity of iPPase in the cardiac and skeletal muscle of both young and old mice which were abrogated upon 24 h of re-feeding them. However, both groups of mice on long-term dietary restriction (DR) showed an additive enhancement in the level of iPPase when compared with their respective age-matched controls. This might bring about metabolic reprogramming in replenishing energy scarcity of long-term dietary restricted mice.

Keywords Inorganic pyrophosphatase · Aging · Dietary restriction · Cardiac muscle · Skeletal muscle

1 Introduction

Many biosynthetic reactions produce inorganic pyrophosphate (PPi) as a metabolic by-product. The advantage of this lies on a greater $\Delta G'^0$ stored in α, β phosphoanhydride bond of nucleoside triphosphates compared to that in the β and γ and that the PPi is rapidly hydrolyzed to orthophosphate, driving both the kinetics and energetics of anabolic reactions towards biosynthesis (Heinonen 2001). For this reason, PPi homeostasis is exceedingly important for the cell. It is a potent inhibitor of crystal

W.J. Kharbhih · R. Sharma (✉)
Department of Biochemistry, North-Eastern Hill University, Shillong 793022, India
e-mail: sharamesh@gmail.com

© Springer Science+Business Media Singapore 2017

P.C. Rath et al. (eds.), *Topics in Biomedical Gerontology*,
DOI 10.1007/978-981-10-2155-8_7

nucleation and growth (Fleisch et al. 1966), found in the extracellular matrix of tissues and in body fluids (Fleisch and Bisaz 1962; Russell et al. 1971). Most detailed studies of PPi metabolism have been done in cultured hepatocytes and chondrocytes (Davidson and Halestrap 1988; Rosen et al. 1997; Rosenthal and Henry 1999). PPi is generated in mitochondria and its concentration in both intra- and extracellular is regulated by mitochondrial energy metabolism (Ryan et al. 1999; Johnson et al. 2000). Accordingly, PPi produces mitochondrial membrane potential in an inorganic pyrophosphatase (iPPase)-dependent manner (Pereira-da-Silva et al. 1993). It serves as a donor of phosphate in protein phosphorylation in yeast and mammalian cells derived from ATP (da Silva et al. 1991). Removal of PPi from the cytosol is carried out by two main non-homologous enzyme types: soluble inorganic pyrophosphatases (iPPases) and proton-translocating membrane-bound pyrophosphatases (H+-PPases) (Pérez-Castiñeira et al. 2001; Serrano et al. 2007). The first of these is found as the sole enzyme that keeps cytosolic PPi at very low levels in animals.

In mammals, several pathophysiological conditions related to calcium phosphate homeostasis are linked to deregulation of iPPase levels (Weissen-Plenz et al. 2008; Polewski et al. 2010). Up-regulation of iPPase is found in several proteomic studies dealing with cancer tissues from different origins, mostly in lung, gastric and colon (Sang-Ho et al. 2012). Extracellular PPi acts partly as a key inhibitor of hydroxyapatite crystal growth. However, intracellular PPi acts as a biochemical intermediate interacting with certain ATPases, and thus regulating the transport of iron and mitochondrial function, leading to the regulation of protein synthesis (Terkeltaub 2001).

Studies conducted in rodents showed mammalian lifespan could be altered through dietary restriction, but without malnutrition, extending the average and the maximum lifespan (McCay et al. 1935). Until 1990s that DR became extensively viewed as a scientific model that could present insights into the retardation of aging process and thereby recognize underlying mechanisms of aging (Kennedy et al. 2007). The converse relationship between calorie intake and prolonged lifespan of mice suggests a role for regulators of energy metabolism in the mechanisms of DR. It's induced reprogramming of metabolic pathways that may be a key event in the causation of healthy aging and longevity (Dhahbi et al. 2001; Anderson and Weindruch 2007). A role for nutrient-responsive signalling molecules like SIRT1, mTOR and PGC-1a, in model animals (yeast, worms, flies, and mice) point to interconnections in aging and DR (Mair and Dillin 2008). Knowing the role of iPPase in the energy metabolism, we tried to correlate the effect of age and DR on the cardiac and skeletal muscle iPPase of mice and observed that iPPase expression levels increase significantly during aging and the long-term intermittent fasting and feeding has an additive effect in maintaining enhanced level of iPPase in these tissues of mice.

2 Experimental Procedures

2.1 *Animal Studies*

Male albino (Balb/c strain) mice were purchased from Pasteur institute, Shillong for experimental study. They were divided into two groups: young age of one month and old age of eighteen months. These were further divided into group one (Y-AL, n = 6) young ad libitum fed and (Y-DR, n − 6) young DR fed mice. Group two was further divided into (O-AL, n = 6) old ad libitum fed and (O-DR, n = 5) old DR fed mice. DR mice were subjected to food restriction (feeding on alternate days) as described in Goyary and Sharma (2008). They were sacrificed by cervical dislocation at a fixed time (11 AM) following institutional guidelines on use of animals. All chemicals were of analytical grade and the antibodies used were detailed in our earlier publication (Kharbhih and Sharma 2014).

2.2 *Enzyme Activity, Western Blotting and PCR Assays*

iPPase activity was determine by the protocol obtained from (Shatton et al. 1983; Syiem and Sharma 1996) with some modifications. The specificity activity was expressed as μmole PPi hydrolyzed mg protein^{-1} min^{-1}. Protein estimation was done by Bradford (1976) and standard plot obtained by using bovine serum albumin. Western blot analyses and differential display PCR assays were as according to the procedure described earlier (Kharbhih and Sharma 2014). The data were analyzed using unpaired t-test and non-parametric Mann-Whitney test of both the GraphPad Prism 6.03 and SPSS 16.0 with $p < 0.05$.

3 Results and Discussion

The specific activity of iPPase in cardiac and skeletal muscle tissues increase (50 %) in older mice compared with young ones (Figs. 1a and 2a). Western blot analysis (Figs. 1b and 2b) of iPPase and its mRNA expression (Figs. 1c and 2c) of the cardiac and skeletal muscle show a similar pattern as of the enzyme activity.

The body weight of both groups of mice was found to significantly decrease when subjected to long-term DR as compared with ad libitum fed mice as reported earlier (Kharbhih and Sharma 2014). DR intervention results in enhanced activity of cardiac and skeletal muscle iPPase in young (19 and 38 %) and old (18 and 14 %) mice when compared with their age-matched counterpart, respectively (Figs. 1a and 2a). Western blot analysis of this protein confirms the above observation (Figs. 1b and 2b) and similarly, iPPase mRNA expression analysis also follows the same

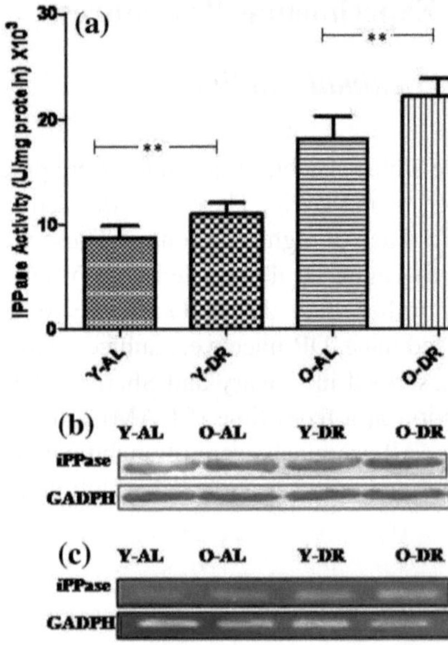

Fig. 1 Effect of DR on **a** activity of iPPase in the cardiac muscle of young and old mice. Statistically significant p values (**, <0.01) between compared groups are shown. **b** Immunoblot analysis. **c** iPPase mRNA expression in young (Y-AL, Y-DR) and old (O-AL, O-DR) mice. GAPDH mRNA was used as a control

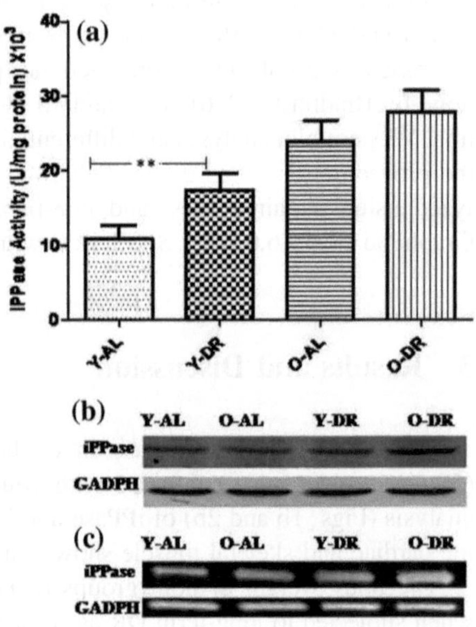

Fig. 2 Effect of DR **a** on the activity of iPPase in the skeletal muscle of young and old mice. Values are expressed as mean data with standard deviation in each group. Statistical significance was as described above. **b** Immunoblot analysis. **c** Expression of iPPase mRNA level. GAPDH mRNA was used as a control

pattern in both young (22 and 30 %) and old (20 and 18 %) DR mice in comparison to the ad libitum fed mice, respectively (Figs. 1c and 2c).

The iPPase is a key enzyme that has a potential to regulate energy homeostasis in mammalian metabolism. In this paper, we found that young mice have significantly lower level of iPPase in the cardiac and skeletal muscle when compared to the older ones. These results corroborate with others and our earlier reports (Panda et al. 2007; Kharbhih and Sharma 2014) of increased level of iPPase the liver of aged mice. Though the magnitude of enhancement in iPPase is lesser in cardiac and skeletal muscle compared to the liver. It has been reported that enzymes like nucleoside triphosphate pyrophosphohydrolase, involved in PPi generation and coupling of glucose oxidation and fatty acid metabolism increase with advance in age (Johnson et al. 2001; Tollet-Egnell et al. 2001). But when animals are subjected to DR, it alters the metabolic programming and tries to maintain iPPase at an elevated level in both the muscles studied to suit the metabolic demand of these tissues. Due to the lower energy status during DR, the reprogramming to increase iPPase expression is crucial for extra energy needed by the cells (Higami et al. 2006; Everitt et al. 2010). DR, thus, enhances the level of iPPase to meet the energy requirements in cardiac and skeletal muscle, though the magnitude of enhancement is lesser in older mice.

Acknowledgments Authors would like to thank the Department of Biochemistry, North-Eastern Hill University, Shillong for providing the necessary research facilities under UGC-UPE, DRS and DST FIST, New Delhi. WJK thanks CSIR, New Delhi for the Fellowship support as JRF and SRF (F.No: 9/347(0185)/2009-EMR-1).

References

Anderson RM, Weindruch R (2007) Metabolic reprogramming in dietary restriction. Interdiscip Top Gerontol 35:18–38

Bradford MM (1976) A rapid sensitive method for the quantitation of microgram quantities of protein utilizing the principle of protein dye binding. Anal Biochem 72:248–254

da Silva LP, Lindahl M, Lundin M, Baltscheffsky H (1991) Protein phosphorylation by inorganic pyrophosphate in yeast mitochondria. Biochem Biophys Res Commun 178:1359–1364

Davidson AM, Halestrap AP (1988) Inorganic pyrophosphate is located primarily in the mitochondria of the hepatocyte and increases in parallel with the decrease in light-scattering induced by gluconeogenic hormones, butyrate and ionophore A23187. Biochem J 254:379–384

Dhahbi JM, Mote PL, Wingo J, Rowley BC, Cao SX, Walford RL, Spindler SR (2001) Caloric restriction alters the feeding response of key metabolic enzyme genes. Mech Ageing Dev 122:1033–1048

Everitt AV, Rattan SIS, Couteur DG, Cabo R (2010) Calorie Restriction. Aging and Longevity, Springer, New York

Fleisch H, Bisaz S (1962) Isolation from urine of pyrophosphate, a calcification inhibitor. Am J Physiol 203:671–675

Fleisch H, Russell RG, Straumann F (1966) Effect of pyrophosphate on hydroxyapatite and its implications in calcium homeostasis. Nature 212:901–903

Goyary D, Sharma R (2008) Late onset of dietary restriction reverses age-related decline of malate-aspartate shuttle enzymes in the liver and kidney of mice. Biogerontology 9:11–18

Heinonen JK (2001) Biological role of inorganic pyrophosphate. Kluwer Academic Publishers, Dordrecht

Higami Y, Tsuchiya T, Chiba T, Yamaza H, Muraoka I, Hirose M, Komatsu T, Shimokawa I (2006) Hepatic gene expression profile of lipid metabolism in rats: impact of caloric restriction and growth hormone/insulin-like growth factor-1 suppression. J Gerontol A Biol Sci Med Sci 61:1099–1110

Johnson K, Jung A, Murphy A, Andreyev A, Dykens J (2000) Mitochondrial oxidative phosphorylation is a downstream regulator of nitric oxide effects on chondrocyte matrix synthesis and mineralization. Arthritis Rheum 43:1560–1570

Johnson K, Pritzker K, Goding J, Terkeltaub R (2001) The nucleoside triphosphate pyrophosphohydrolase isozyme PC-1 directly promotes cartilage calcification through chondrocyte apoptosis and increased calcium precipitation by mineralizing vesicles. J Rheumatol 28:2681–2691

Kennedy BK, Steffen KK, Kaeberlein M (2007) Ruminations on dietary restriction and aging. Cell Mol Life Sci 64:1323–1328

Kharbhih WJ, Sharma R (2014) Age-dependent increased expression and activity of inorganic pyrophosphatase in the liver of male mice and its further enhancement with short- and long-term dietary restriction. Biogerontology 15:81–86

Mair W, Dillin A (2008) Aging and survival the genetics of life span extension by dietary restriction. Annu Rev Biochem 277:727–754

McCay MF, Crowell MF, Maynard LA (1935) The effect of retarded growth upon the length of life span and upon the ultimate body size. Nutrition 5:155–172

Panda H, Pandey RS, Debata PR, Supakar PC (2007) Age-dependent differential expression and activity of cytosolic inorganic pyrophosphstase gene. Biogerontology 8:517–525

Pereira-da-Silva L, Sherman M, Lundin M, Baltscheffsky H (1993) Inorganic pyrophosphate gives a membrane potential in yeast mitochondria, as measured with the permeant cation tetraphenylphosphonium. Arch Biochem Biophys 304:310–313

Pérez-Castiñeira JR, Gómez-García R, López-Marqués RL, Losada M, Serrano A (2001) Enzymatic systems of inorganic pyrophosphate bioenergetics in photosynthetic and heterotrophic protists: remnants or metabolic cornerstones? Int Microbiol 4(3):135–142

Polewski MD, Johnson KA, Foster M, Millan JL, Terkeltaub R (2010) Inorganic pyrophosphatase induces type I collagen in osteoblasts. Bone 46:81–90

Rosen F, McCabe G, Quach J, Solan J, Terkeltaub R (1997) Differential effects of aging on human chondrocyte responses to transforming growth factor beta: increased pyrophosphate production and decreased cell proliferation. Arthritis Rheum 40:1275–1281

Rosenthal AK, Henry LA (1999) Thyroid hormones induce features of the hypertrophic phenotype and stimulate correlates of CPPD crystal formation in articular chondrocytes. J Rheumatol 26:395–401

Russell RG, Bisaz S, Donath A, Morgan DB, Fleisch H (1971) Inorganic pyrophosphate in plasma in normal persons and in patients with hypophosphatasia, osteogenesis imperfecta, and other disorders of bone. J Clin Invest 50:961–969

Ryan LM, Kurup IV, Cheung HS (1999) Transduction mechanisms of porcine chondrocyte inorganic pyrophosphate elaboration. Arthritis Rheum 42:555–560

Sang-Ho J, Gyung HK, Young HC (2012) Pyrophosphatase overexpression is associated with cell migration, invasion, and poor prognosis in gastric cancer. Tumor Biol. 33:1889–1898

Serrano A, Perez-Castineira JR, Baltscheffsky M, Baltscheffsky H (2007) H + -PPases: yesterday, today and tomorrow. IUBMB Life 59:76–83

Shatton JB, Ward C, Willams A, Weinhouse S (1983) A microcolorimetric assay of inorganic pyrophosphatase. Anal Biochem 130: 114–119

Syiem D, Sharma R (1996) Purification and kinetic characterization of chicken liver inorganic pyrophosphatase. Indian J Biochem Biophys 33:363–370

Terkeltaub RA (2001) Inorganic pyrophosphate generation and disposition in Pathophysiology. Am J Physiol Cell Physiol 281:C1–11

Tollet-Egnell P, Flores-Morales A, Stahlberg N, Malek RL, Lee N, Norstedt G (2001) Gene expression profile of the aging process in rat liver: normalizing effects of growth hormonereplacement. Mol Endocrinol 15:308–318

Weissen-Plenz G, Nitschke Y, Rutsch F (2008) Mechanisms of arterial calcification: spotlight on the inhibitors. Adv Clin Chem 263–293 (Elsevier Ltd., Amsterdam)

Oxidative Stress and the Brain: An Insight into Cognitive Aging

Sambe Asha Devi and Abhijit Satpati

Abstract The last two decades has witnessed accumulating evidences in favour of oxidative stress as a causative link between normal brain aging and several neuropathological conditions. Reactive oxygen species (ROS) is considered a major factor contributing to decline in brain function with aging. Cognitive impairment and oxidative stress are common occurrences in old age and are often traceable to events such as increased lipid peroxidation and protein oxidations in specific cognitive regions of the brain, the hippocampus and cerebral cortex. With the advent of various cognitive tests, oxidative pathology is further confirmed by behavioural changes in animal and humans as well. While much of the biomedical research is concentrating on methods to cure aging related neurodegeneration and cognitive decline, biogerontologists are seeking something larger with a view to finding ways to prevent it at its root. Here we discuss and highlight the possibilities of curtailing the cognitive deficits during normal and pathological aging through evidences in favour of vitamin and non-vitamin supplements —in an effort to run the biological clock backwards and extend healthy brain function.

Keywords Aging brain · Antioxidants · Cognition · Lipofuscin · Neurodegeneration · Oxidative stress

1 Introduction

It is known that although the brain makes up 2 % of the total body weight and uses almost 20 % of the cardiac output to meet its ATP demands, is most vulnerable to the increased production of superoxide ($O_2^{\bullet -}$) because of its weak antioxidant defences compared to the heart that is one other post-mitotic organ. Paradoxically,

S. Asha Devi (✉) · A. Satpati
Laboratory of Gerontology, Department of Zoology, Bangalore University, Bangalore 560056, India
e-mail: sambeashadevi@bub.ernet.in

A. Satpati
e-mail: asuba@vsnl.com

© Springer Science+Business Media Singapore 2017

P.C. Rath et al. (eds.), *Topics in Biomedical Gerontology*,
DOI 10.1007/978-981-10-2155-8_8

the various chemicals referred to as neurotransmitters that the brain uses for transmitting messages also participate in reactions that generate many important free radicals (FRs). Although in an healthy brain these FRs are neutralized by multiple antioxidant defences, superoxide dismutase (SOD), glutathione (GSH), catalase (CAT) and glutathione peroxidase (GSH-Px), the generation of reactive peroxyl, alkoxyl and hydroxyl radicals from hydrogen peroxide and many of the lipid peroxides are very high with aging because of the increased accumulation of reactive iron. One of the most significant findings in aging brain's function is a gradual decline in spatial learning, memory and attention.

Because of the recent efforts in reducing oxidative stress (OS) through antioxidants, this review focuses on the normal aging brain and the impact of antioxidants on the cognitive loss with evidences from animal and human studies. Concomitantly, a brief review of the involvement of antioxidants in the pathology of progression of neurodegenerative diseases will be discussed with an emphasis on alleviating cognitive decline.

1.1 Oxidative Stress in Normal Brain Aging

By definition, FRs are atoms or molecules that have unpaired electrons and hence highly reactive in search of another free electron. Most of the FRs that damage biological systems are oxygen radicals and reactive oxygen species (ROS) resulting from aerobic metabolism. The sites of generation of the radicals are the mitochondria, peroxisomal fatty acid, cytochrome P-450, and phagocytic cells. Most of the ROS involve the oxidation of NADH for energy production which in turn is utilized to phosphorylate ADP. The mitochondrial electron transport chain (ETC) is about four-electron reduction of O_2 to H_2O with NADH and succinate transferring their electrons to complex I (NADH dehydrogenase) and complex II (succinate dehydrogenase) of the ETC. Coenzyme Q accepts electrons from complexes I and II, and in turn donates electrons to cytochrome b in complex III (ubiquinone-cytochrome c reductase) wherein the electrons are donated to cytochrome c1, and so cytochrome c to complex IV (cytochrome c oxidase) that finally reduces O_2 to H_2O. The one-electron reduction of O_2 results in superoxide radical ($O_2^{\bullet-}$). The mitochondria can convert 1 to 2 % of O_2 molecules to super oxide ($O_2^{\bullet-}$) under supra-physiological concentration of oxygen (Beckman and Ames 1998; Yang et al. 2008b) and 0.2 % under normal physiological concentrations of oxygen (Staniek and Nohl 2000). However, the superoxide anion is converted to less toxic hydrogen peroxide (H_2O_2) by superoxide dismutase (SOD). Although H_2O_2 is not a FR, it is further converted to highly reactive FR, the hydroxyl radical ($OH^\bullet$) through the Fenton reaction. The $OH^\bullet$ is a toxic radical damaging the lipids, proteins and nucleic acids. Also a high iron content causes increased generation of $OH^\bullet$ from H_2O_2 via Fenton reaction and depressed scavenging ability for ROS (Deloncle et al. 2006). In brief, the mitochondrial free radical theory of aging (MFRTA) is supported by several evidences on: (i) increased ROS generation with aging,

(ii) decreased activity of many antioxidant enzymes with age, (iii) increased occurrences of mitochondrial mutations during aging, and (iv) reduced energy output (Bratic and Larsson 2013).

The brain has high concentrations of polyunsaturated fatty acids (Floyd and Hensley 2002; Shulman et al. 2004) and OS is a major factor in a normal aging brain and is linked to the observed reduction in neurological activities and negatively correlated to lifespan (Navarro et al. 2002). Most importantly, oxidative damage leads to cognitive impairments in the aged humans. In fact, most of the OS effects are traceable to mitochondrial dysfunction and production of ROS that contribute to brain aging (Miquel et al. 1980). Additionally, the endogenous antioxidants, SOD, CAT and GSH-Px and glutathione (GSH) decline in several critical sites for cognition such as the hippocampus and cerebral cortex (Siqueira et al. 2005; Hasan et al. 2009). Studies from this laboratory have demonstrated increased H_2O_2 levels in the HC and CC of aging rats (Jolitha et al. 2006).

1.1.1 Lipid Peroxidation and Brain Aging

It is known that lipofuscin is a product of lipid peroxidation and that there occurs a progressive accumulation of these granules with age. Interestingly, its biochemical composition is reported to vary in different animal species and different brain regions at varying ages (Benavides et al. 2002). Since mitochondria are sites of ROS generation, the organelle contains peroxidized non-degradable granules. Spectrofluorimetric studies from our laboratory on *Wistar* rat brain have shown increased levels of lipofuscin–like fluorescent-like substance and malondialdehyde levels in the cerebral cortex and hippocampus with age (Jolitha et al. 2006). Electron microscopic studies show appearance of more complex and fused lipofuscin granules in the nuclear periphery of hippocampal neurons from old rats (Asha Devi et al. 2011) and this is in contrast to the young wherein lipofuscin is more often scattered than clustered (Fig. 1). Experiments from other labs have indicated variegated lipofuscin granules in these regions in the aged rat brain (Brunk and Terman 2002).

Overall, lipofuscin is considered as a matrix that in fact reflects on the insults and damage that a neuron has undergone during the life of an individual, and more grievously as a risk factor for the development of neurodegenerative conditions such as Alzheimer's disease.

1.1.2 Protein Oxidation and Brain Aging

Proteins that increase with aging in the mice hippocampus are mainly enzymes that mediate energy production and OS (Yang et al. 2008a). In addition, one of the cellular processes that is altered specifically in the aging hippocampus is that of protein processing (VanGuilder and Freeman 2011). A wide proteomic analysis of brains from 4 days to 15 months of age in mice has identified 40 proteins that exhibit changes in during the mouse lifespan showing that six proteins increase and

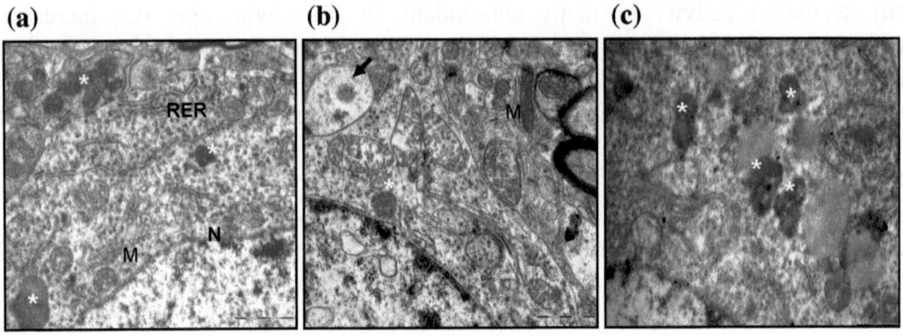

Fig. 1 Accumulation of lipofuscin (*asterisk*) in the perikaryon of hippocampal neuron in **a** adult (4–5mo) ×18,500; **b** late-adult (13–14 mo) and **c** middle-aged Wistar rats (21–22mo) ×30,000. *Arrow* indicates enlarged mitochondrium with disrupted cristae. Note the cluster of lipofuscin granules(asterix) intermingled and almost fused with mitochondria that lack cristae in the middle-aged hippocampus. *N* nucleus; *RER* rough endoplasmic reticulum; *M* mitochondrium

27 decrease. Collectively, an analysis of such proteins encompasses the biological processes in which those proteins have pivotal role and correspond mainly to more than one third to protein metabolic processes, one third to transport, one fourth to nucleic acid metabolic process, almost one fifth to intracellular signal cascade and finally more than one sixth to stress proteins. Interestingly, about one fifth of all these identified proteins is related to energy metabolism and is localized in the mitochondria (Yang et al. 2008a). To detect protein oxidation, carbonyls are the most commonly employed marker, and the use of 2D gel electrophoresis has provided very useful results for the study of specific carbonylated protein spots during OS and senescence (Baraibar et al. 2011; Anand et al. 2014).

One other cell type that is more abundant in the central nervous system, the glial cells, is more resistant than neurons to OS and protects the neurons from oxidative insults (Hollensworth et al. 2002). This forms an important avenue for further proteomic studies on glial cells during aging, or disease states. Miura et al. has proposed that astrocytes respond to OS and its capacity to do so is, however, not depressed with age. The authors found that α-tubulin on exposure to H_2O_2 undergoes tyrosyl phosphorylation and aging actually enhances this phosphorylation and prevents the synthesis of microtubules, although aging does not inhibit cell protection against severe OS (Miura et al. 2005). Therefore, it is widely accepted that proteomic studies on the response of glial cells to OS with aging and neurodegenerative diseases contribute to an understanding of neuron metabolism.

1.2 Brain Aging and Cognition

Since the brain needs high energy, a gradual decline in many of the respiratory enzymes as a function of aging not only impact the cognitive function but to appalling conditions of neurodegenerative diseases (Lin and Beal 2006; Boveris and Navarro 2008). Studies have demonstrated an interrelation between mitochondrial and behavioural functions in animals (Navarro and Boveris 2007; Boveris and Navarro 2008; Bratic and Larsson 2013; Petrosillo et al. 2013) and humans (Ames and Liu 2004; Aliev et al. 2009; Shankar 2010). To determine the neural mechanisms of learning and memory various types of animal models are of great assistance since they recapitulate certain symptoms in neurological, neurodegenerative and psychiatric disorders.

Until recently, age-related memory deficits were connected to loss of neurons in specific areas such as the prefrontal cortex and hippocampus. However, with the advent of sophisticated cell-counting techniques, unbiased stereology, it is now known that loss of function in these areas are not because of loss of neurons but low sensitivity in the receptors for neurotransmitters and their associated second messengers besides the altered calcium-sensitive signalling agents that are associated with conversion of short-term memories to long-term memories.

Tables 1 and 2 lists few examples of existing methods and their recent advancements in the methods for assessing cognitive behavior.

2 Brain Aging and Oxidant Pathology

Lysosomal dysfunctions in senescent cells often results in important markers of senescence- accumulation of beta-galactosidase and lipofuscin (Carnero 2013). Geng et al. (2010) have reported increased β-galactosidase activity in rat hippocampus with aging. Intriguingly, aging and neurodegenerative diseases share common pathways related to FR damage and lowered energy synthesis (Perluigi et al. 2010). It is known from animal experiments, the involvement of oxygen free radicals in cognitive decline and antioxidant vitamins to increase the cholinergic and glutaminergic neurotransmitters thereby improving the survival of neurons and their synaptic responses. Alzheimer's (AD) and Parkinson's disease (PD) are associated with OS although oxidative injury is speculated to develop secondary to excessive OS as a result of mitochondrial abnormalities, insufficient energy supply, decreased antioxidant defence (Grundman and Delaney 2002) and increased ROS generation (Demetrius and Driver 2013). Evidences for oxidative processes in AD have come from experimental as well as several neuropathological studies (Markesberry and Carney 1999). Lipofuscin is now perceived as a strong link between cognitive loss and oxidative stress in AD since lipofuscin contains Aβ and its precursor, and its transfer from intra-to extracellular space is more disastrous because of its modifying action on the morphological and biochemical functions (Gómez-Ramos and Morán 2007). In fact, Giaccone and his co-scientists view the released lipofuscin into the extracellular space could act as a potential source of Aβ oligomers over a period of time to form senile plaques (Giaccone et al. 2011).

In AD and PD patients, native mitochondria are replaced by enlarged or giant mitochondria, altered calcium homeostasis when related to their age-matched controls (Trimmer et al. 2000). It is possible that genetic factors that are strong risk

Table 1 Traditional/conventional behavioral tests

Test	Type of learning	Brain region involved	Type of memory	Application in Clinical cases	References
Morris water maze Rats and mice 	Spatial and non-spatial learning	HC	Spatial working memory	AD	Morris (1984) Bromley-Brits et al. (2011)
				PD with dementia	Alvin and Terry (2009) Astur et al. (2002)
Barens Maze (Alternate of MWM) Rats and mice 	Spatial and non-spatial learning	HC and ErC	Spatial reference memory	AD	Barnes (1979) O'Leary and Brown (2009)
Radial arm maze Rats and mice  Humans	Spatial learning	HC, PFC	Spatial working and reference memory	AD	Olton and Samuelson (1976) Ganguly and Guha (2008)
				HD	Ji et al. (2015) Astur et al. (2004)
T-maze Rats and mice 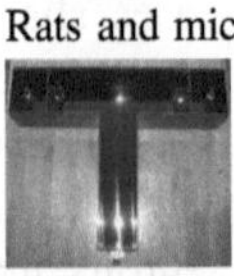	Spatial learning	HC, PFC	Spatial working memory	AD	Deacon and Rawlins (2005)
				HD	Faizi et al. (2012), Lione et al., (1999) Carroll et al., (2011)
Y-maze Rats and mice 	Spatial learning	HC, PFC	Spatial and working memory	AD	Glickman and Jensen (1961) Faizi et al. (2012)
Delayed matching to position Rats, Baboon	Associative learning	PC and vmPFC	Spatial and working memory	AD	Yamada et al. (2005)
				HD	Yhnell et al. (2015) Rodriguez et al. (2011)
Delayed non-matching to position Rats and mice	Associative learning	PC and vmPFC	Spatial and working memory	HD	Wiig and Burwell (1998) Yhnell et al. (2015)

(continued)

Table 1 (continued)

Test	Type of learning	Brain region involved	Type of memory	Application in Clinical cases	References
Passive avoidance Rats and mice  Human	Associative learning	HC, PFC and AMY	Fear conditioning	Psychopathic behaviour	Ogren and Stidel (2015) Blair and Mitchell (2009)
				AD	Hosseini et al. (2013) Finger et al. (2007)
Novel object recognition Rats, mice and monkey 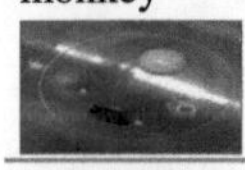	Associative learning	HC, PC	Recognition memory, STM, LTM and intermediate memory		Bevins and Besheer (2006).
				Schiz	Rajagopal et al. (2014)
Set shifting Task (ID/ED) Rat and mice  Humans	Discrimination Learning	PFC, Frontal lobe	Working memory and attentional set	Schiz	Scheggia et al. (2014) Egerton et al. (2005) Roger et al. (2000)
Open field test rats and mice 	Motor activity	Limbic system	CC, CB, HC, Septum, Thalamic reticular nucleus,		Walsh and Cummins (1976)
				AD associated Locomotory behaviour	Faizi et al. (2012)
Elevated plus maze Rats, mice, guinea-pigs, gerbils, voles and wild mice 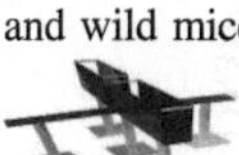	Anxiety related behaviour	Limbic system, HC, AMY, dorsal raphe Nucleus			Handley and Mithani (1984) Pellow et al. (1985)
				Anxiety like behaviour	Walf and Frye (2007)

AD Alzheimer's Disease; *AMY* Amygdala; *CC* cerebral cortex; *DMP* Delayed matching to position; *DNMP* Delayed non-matching to position; *ED* Extra dimensional; *EPM* Elevated Plus maze; *ErC* Entorhinal cortex; *HC* Hippocampus; *HD* Huntington disease; *ID* intra dimensional; *LTM* Long term memory; *mPFC*, Medial Prefrontal cortex; *MWM* Morris water maze; *NOR* Nobel object recognition; *OFM* Open field maze; *PAL* Passive avoidance learning; *PC* Perirhinal cortex; *PFC* Prefrontal cortex; *RAM* Radial arm maze; Schiz, Schizophrenia; *STM* Short term memory; *VmPFC* Ventromedial PFC

Table 2 Modification over traditional/conventional behavioral tests

Test	Modification over existing training paradigm	Type of learning	Brain region involved	Type of memory	References
Oasis Maze 	MWM	Spatial learning	HC	Remote spatial memory	Clark et al. (2005)
Annular Maze 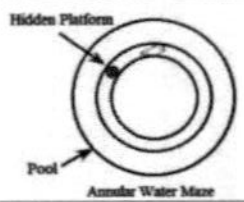	MWM	Spatial learning	HC	Remote spatial memory	Clark et al. (2005)
Modified T maze (for spatial and reference memory) 	T maze	Spatial learning	PFC	PFC dependent working and reference memory	Shoji et al. (2012)
Radial arm water maze Alamed et al. (2006) 	MWM and RAM	Spatial learning	HC, PFC	Working along with reference memory	Fujisaki et al. (2014)
Trial-unique, delayed nonmatching-to-location 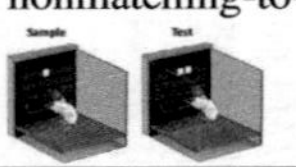	DNMP	Associative learning	HC	HC dependent learning in novel search task as well as a novel test of paired associate learning	Talpos et al. (2010)

(continued)

B & W IN PRINT

B & W IN PRINT

Table 2 (continued)

Test	Modification over existing training paradigm	Type of learning	Brain region involved	Type of memory	References
The touch screen operant platform 	Operant conditioning	Working memory	PFC, HC, AMY, distinct thalamic nuclei, subthalamic nucleus, fornix, subiculum, medial septal/vertical limb of diagonal band and nucleus basalis magnocellularis	Working memory and pattern separation	Horner et al. (2013)
Visual Cognitive Tasks using the floor projection maze 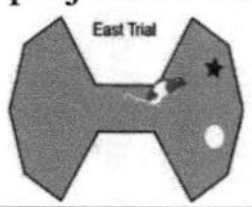	NOR and T Maze (alternation)	Spatial learning	HC, PC, PFC	Spatial working and Recognition memory	Jacobson et al. (2014)

AMY Amygdala; *DMP* Delayed matching to position; *DNMP* Delayed non-matching to position; *HC* Hippocampus; *MWM* Morris water maze; *NOR* Nobel object recognition; *PC* perirhinal cortex; *PFC* prefrontal cortex; *RAM* Radial arm maze factors for AD function through FR mechanisms e.g. ApoE genotype, that a major genetic marker for AD, is correlated to increased OS markers in animal and human studies (Ramassamy et al. 2000). Additionally, reactive nitrogen species could play a role in the synthesis of neurofibrillary tangles (Giasson et al. 2000) resulting in neuronal death. It is also known that aging and Alzheimer's are intricately related because the CA1 site being susceptible to neuronal loss in both the situations and that evidences exist on neuronal dysfunction and calcium accumulation leading to glucocorticoid toxicity during aging and Alzheimer's (Landfield et al. 1992). Additionally, Weinreb et al. (2007) found a significant parallelism in the protein profile affected between aging and neurodegenerative diseases in the hippocampus of rats. They have demonstrated that in the aged hippocampus, OS and mitochondrial dysfunction are important and on treatment with an anti-AD drug, Ladostigil, or with an anti-Parkinson drug, Rasagiline, can reverse the effect of aging on various mitochondrial and key regulator genes involved in neurodegeneration, cell survival, synaptogenesis, oxidation, and metabolism.

The long struggle against OS-related normal brain aging has spurred a pursuit for means that might not only protect the brain from the progressive diminution of neuronal communication but also from the awful destruction of the brain cells leading to AD and PD, conditions that affect the cognitive skills and movements with the AD being the gravest disease of old age. Memory deficits of normal aging occur primarily in secondary memory systems, the ones that can store newly acquired information. Among these the hippocampus probably mediates our learning about place, and the prefrontal cortex for acquiring the capacity to perform the specific tasks.

3 Vitamin and Non-vitamin Supplements for Normal Aging and Pathological Aging

Numerous studies have associated supplements, vitamin and non-vitamins; with cognition because of their potential to alleviate OS and inflammation, more so with age.

3.1 Vitamins

Data obtained have indicated that the detoxification mechanisms decline with age and that vitamin E is effective in slowing the storage of lipofuscin in vitro (Kan et al. 1991). Further, studies indicate the existence of regional differences in the lipid peroxidation and protein oxidation with a decline in the antioxidant enzymes, SOD, CAT and GSH-Px with age while an up-regulation of these enzymes are evident in the vitamin E supplemented ones (Vohra et al. 2001; Asha Devi and Ravi kiran 2004). The beneficial effect of vitamin E alone in improved spatial learning has been demonstrated by Dolu et al. (2015) in adult rats that were subjected to Morris water maze (MWM). Dietary antioxidants are regarded as being important in modulating OS in aging and age-associated diseases. Supplementation to the elderly with vitamin E has shown to enhance immune response, delay onset of AD, and increase resistance to oxidative injury associated with exercise (Meydani 2002). Previous studies from our laboratory (Asha Devi et al. 2006) as well as of Takatsu et al. (2009) have shown a synergistic role of vitamin E and physical exercise in enhancing cognition especially in older rats. The benefits of vitamin E and C in lowering OS with aging has been evidenced from studies on lipofuscin-like substances in the stress-responsive hypothalamus (Manjula et al. 2013) and cognitive sites, cerebral cortex (Asha Devi et al. 2012) and hippocampus (Asha Devi and Manjula 2014).

A significant inverse relation between vitamin E and cognitive decline has also been established in 2889 older subjects, aged 65–102 years during a period of 7 years (Morris et al. 2002). Grodstein and his collaborators (2003) have shown modest cognitive benefits from the use of vitamins E and C in their long-term study spanning a period of 5 years on some 14,968 women who were 70–79 years of age. Similarly, studies from the Honolulu-Asia Aging Study (HAAS) involving the Japanese-American males have revealed a relation between past intakes of antioxidant supplements at least four and ten years prior to cognitive tests and found that in men on both vitamin E and C supplements had a significantly lowered risk for cognitive dysfunction. (Masaki et al. 2000). Further evidences come from population studies such as Rotterdam study wherein an inverse relation between the risk of dementia and intake of vitamin E and C (Engelhart et al. 2002).

3.2 *Pyrrolo-Quinoline Quinine*

Recapitulating the evidences in favour of mitochondrial aging and the brain as a highly vulnerable organ to OS generated in the mitochondria, it is notable to mention pyrrole-quinoline quinine (PQQ) for preventing brain degeneration with aging. PQQ reduces the excitotoxicity effects that are one of the causes of age-related cognitive impairment seen in AD and PD by protecting the brain from toxic proteins, beta-amyloid and alpha-synuclein (Kim et al. 2010). Studies in rats subjected to extreme oxidant stress to mimic aging have shown better memory when PQQ was supplemented (Ohwada et al. 2008). Further, old rats supplemented with PQQ, not only exhibited better navigation in a maze but also showed superior memory.

3.3 *Polyphenols*

Of the many potential and dietary phyto-nutrients, flavonoids have attracted attention because of evidences from humans (Van Dongen et al. 2000) and animals (Youdim et al. 2000) of their potential role in preventing neurodegeneration that occurs with aging. Grape seed proanthocyanidin (GSPE) has beneficial effects on alleviating neuropathology in experimental animals (Ferruzzi et al. 2009; Deshane et al. 2004). Grape seed extract contains lipids, proteins, carbohydrates and phenols and proanthocyanidins (PACs) which are the most abundant phenolic compounds contain polymers of dimmers or trimers of (+) catechin and (−) epicatechin. Studies from our laboratory on adult *Wistar* rats that were supplemented with GSPE showed higher acetyl choline (ACh) levels and choline acetyltransferase (ChAT) activity with lower acetylcholine esterase activity in the CC and HC sites (Jolitha et al. 2009). Older rats had a remarkable improvement in cognition in a T-maze (Fig. 2), and when their hippocampus were examined for indicators of OS, the rats showed significantly lower levels of MDA (Fig. 3a) and H_2O_2 (Fig. 3b) compared

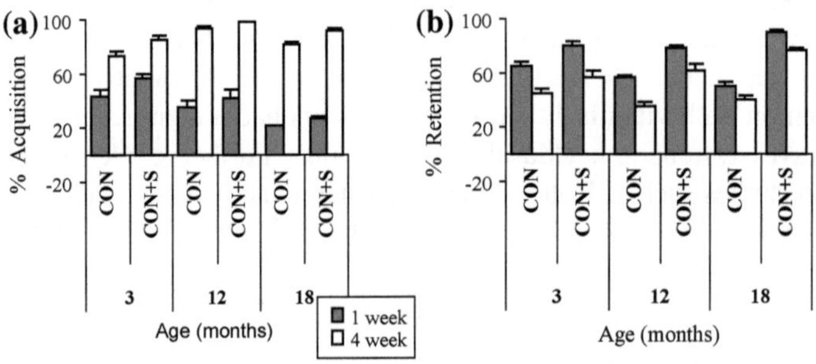

Fig. 2 Fewer errors made by GSPE supplemented (CON+S) adult, late-adult and middle-aged Wistar rats when tested for % acquisition in a T-maze **a** at 1st and 4th of learning. **b** represents % retrieval of the acquired task at the end of 1st and 4th week

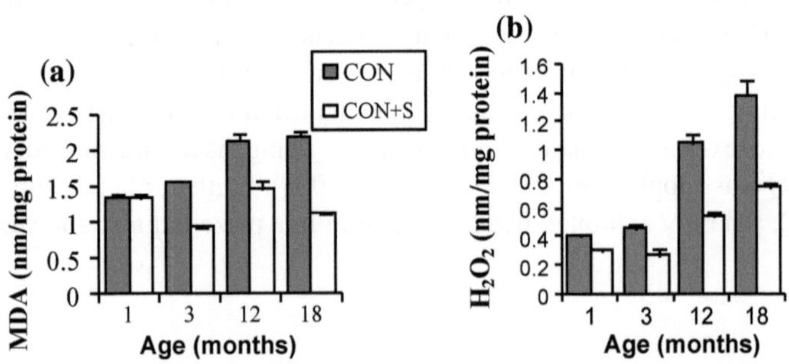

Fig. 3 MDA and H_2O_2 levels in the hippocampus of young, adult, late-adult and middle-aged rats

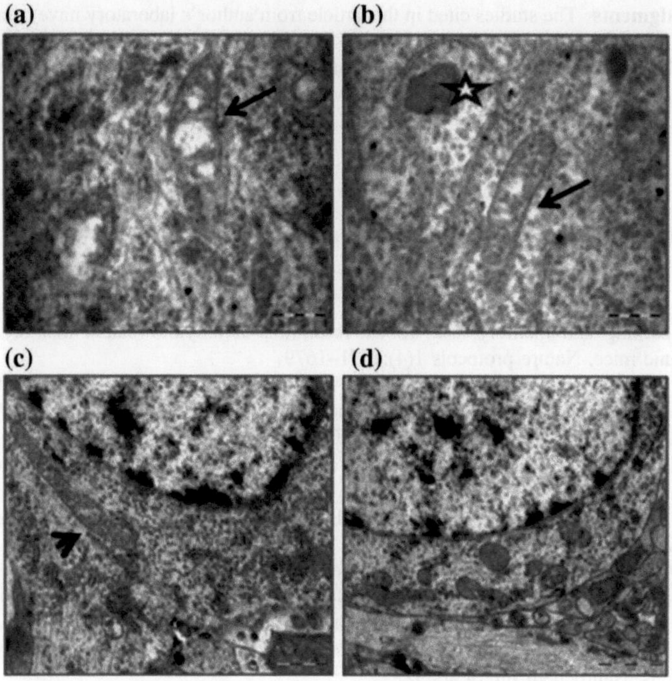

Fig. 4 Electron micrographs of hippocampus in 22-mo-old rats. **a** vacuolated mitochondria (*arrow*) ×30,000; **b** disrupted cristae, a lipofuscin granule (*asterix*) ×18,500; **c** a giant mitochondria (*arrow head*); **d** numerous normal mitochondria distributed around the perinuclear area of the neuron in rats supplemented with grape seed proanthocyanidin for 12 weeks at 150 mg/kg body weight ×18,500

to their age-matched controls. On further examination of the HC across age, rats that were on GSPE supplement had a reduction in LF granules besides healthy mitochondria (Fig. 4).

4 Conclusion

We have reviewed here, many and growing evidences that support the hypothesis that OS does play a role in the cognitive decline and the associated pathogenesis with aging. Studies in animals and humans have increased the significance of further development and testing of antioxidants as a primary strategy for the prevention of cognitive decline in normal aging and prevention of onset of oxidant-related pathology. In conclusion, it is the tip of the iceberg that has been put forth on the significance of specific antioxidant supplements that can decelerate cognitive decline. Much needs to be unravelled from large prospective studies for a better understanding of cognitive aging.

Acknowledgments The studies cited in this article from author's laboratory have been supported from research grants from the University Grants Commission (F.No. F.3-196/2001, SR-II and F. No.3-356/2007, SR 2008) and Indian Council of Medical Research (54/9/CFP/GER/20th NCD-II dt.10/7/2012). We are thankful to DST for providing Junior Research Fellowship to Mr. Abhijit Satpati under the PURSE programme (SR/S9/Z-23/2010/38© dt. 27.06.2011).

References

Alamed J, Wilcock DM, Diamond DM, Gordon MN, Morgan D (2006) Two-day radial-arm water maze learning and memory task; robust resolution of amyloid-related memory deficits in transgenic mice. Nature protocols 1(4):1671–1679

Aliev G, Palacios HH, Walrafe B, Lipsitt AE, Obrenovich ME, Morales L (2009) Brain mitochondria as a primary target in development of treatment strategies for Alzheimer's disease. Int J Biochem Cell Biol 41:1989–2004

Alvin V, Terry Jr. (2009) Spatial navigation (Water Maze) tasks. In: JJ Buccafusco (ed) Methods of behavior analysis in neuroscience, Chap. 13. CRC Press, Boca Raton FL

Ames BN, Liu J (2004) Delaying the mitochondrial decay of aging with acetylcarnitine. Ann N Y Acad Sci 1033:108–116

Anand S, Asha Devi S, Ravi kiran T (2014) Differential expression of the cerebral cortex proteome in physically trained adult rats. Brain Res Bull 104:88–91

Asha Devi S, Manjula KR (2014) Intermittent cold-induced hippocampal oxidative stress is associated with changes in the plasma lipid composition and is modifiable by vitamins C and E in old rats. Neurochem Int 74:46–52

Asha Devi S, Manjula KR, Subramanyam MVV (2012) Protective role of vitamins E and C against oxidative stress caused by intermittent cold exposure in aging rat's frontoparietal cortex. Neurosci Letts 529:155–160

Asha Devi S, Ravi Kiran T (2004) Responses of antioxidant enzymes of the cerebral cortex and hippocampus to exercise and vitamin E. Neurobiol Aging 25:501–508

Asha Devi S, Jolitha N, Ishii N (2006) Grape seed proanthocyanidin extract (GSPE) and antioxidant defense in the brain of adult rats. Med Sci Monitor 12:124–129

Asha Devi S, Sagar Chandrasekhar BK, Manjula KR, Ishii N (2011) Grape seed proanthocyanidin lowers brain oxidative stress in adult and middle-aged rats. Exp Gerontol 46:958–964

Astur RS, Taylor LB, Mamelak AN, Philpott L, Sutherland RJ (2002) Humans with hippocampus damage display severe spatial memory impairments in a virtual Morris water task. Behavioural Brain Res 132:77–84

Astur RS, Tropp J, Sava S, Constable RT, Markus EJ (2004) Sex differences and correlations in a virtual Morris water task, a virtual radial arm maze, and mental rotation. Behavioural Brain Res 151:103–115

Baraibar MA, Hyzewicz J, Rogowska-Wrzesinska A, Laudouce R, Roepstorff P, Mouly V, Friguet B (2011) Oxidative stress-induced proteome alterations target different cellular pathways in human myoblasts. Free Radic Biol Med 51:1522–1532

Barnes CA (1979) Memory deficits associated with senescence: a neurophysiological and behavioral study in the rat. J Comp Physiol Psychol 93:74–104

Beckman KB, Ames BN (1998) The free radical theory of aging matures. Physiol Rev 78:547–581

Benavides SH, Monserrat AJ, Farina S, Porta EA (2002) Sequential histochemical studies of neuronal lipofuscin in human cerebral cortex from the first to ninth decade of life. Arch Gerontol Geriatr 34:219–231

Bevins RA, Besheer J (2006) Object recognition in rats and mice: a one-trial non-matching-to-sample learning task to study 'recognition memory. Nat Protoc 1:1306–1311

Blair RJR, Mitchell DGV (2009) Psychopathy, attention and emotion. Psychological Med. 39:543–555

Boveris A, Navarro A (2008) Brain mitochondrial dysfunction in aging. IUBMB Life 60:308–314
Bratic A, Larsson N (2013) The role of mitochondria in aging. J Clin Invest 123:951–957
Bromley-Brits K, Deng Y, Song W (2011) Morris water maze test for learning and memory deficits in Alzheimer's disease model mice. J Vis Exp 53:e2920
Brunk UT, Terman A (2002) The mitochondrial-lysosomal axis theory of aging: accumulation of damaged mitochondria as a result of imperfect autophagocytosis. Eur J Biochem 269:1996–2002
Carnero A (2013) Markers of cellular senescence. Methods Mol Biol 965:63–81
Carroll JB, Southwell AL, Graham RK, Lerch JP, Ehrnhoefer DE, Cao LP, Zhang WN, Deng Y, Bissada N, Henkelman RM, Hayden MR (2011) Mice lacking caspase-2 are protected from behavioral changes, but not pathology, in the YAC128 model of Huntington disease. Mol Neurodegener 6:59–62
Clark RE, Broadbent NJ, Squire LR (2005) Hippocampus and remote spatial memory in rats. Hippocampus 15:260–272
Deacon RM, Rawlins JN (2005) Hippocampal lesions, species-typical behaviours and anxiety in mice. Behav Brain Res 156:241–249
Deloncle R, Guillard O, Bind JL, Delaval J, Fleury N, Mauco G, Lesage G (2006) Free radical generation of protease resistant prion after substitution of manganese for copper in bovine brain homogenate. Neurotoxicol 27:437–444
Deshane J, Chaves L, Sarikonda KV, Isbell S, Wilson L Kirk M, Grubbs C, Barnes S, Meleth, S, Kim H (2004) Proteomic analysis of rat brain protein modulations by grape seed extract. J Agric Food Chem 52:7872–7883
Demetrius LA, Driver J (2013) Alzheimer's as a metabolic disease. Biogeron 14:641–649
Dolu N, Khan A, Dokutan S (2015) Effect of vitamin E administration on learning of the young male rats. J Exp Neurosci 9:81–85
Egerton A, Reid L, McKerchar CE, Morris BJ, Pratt JA (2005) Impairment in perceptual attentional set-shifting following PCP administration: a rodent model of set-shifting deficits in schizophrenia. Psychopharmacol 179:77–84
Engelhart MJ, Geerlings MI, Ruitenberg A, van Swieten JC, Hofman A, Witteman JC, Breteler MM (2002) Dietary intake of antioxidants and risk of Alzheimer disease. JAMA 287(24):3223–3229
Faizi M, Bader PL, Saw N, Nguyen TV, Beraki S, Wyss-Coray T, Longo FM, Shamloo M (2012) Thy1-hAPPLond/Swe+mouse model of Alzheimer's disease displays broad behavioral deficits in sensorimotor, cognitive and social function. Brain Behav 2:142–154
Ferruzzi MG et al. (2009) Bioavailability of gallic acid and catechins from grape seed polyphenol extract is improved by repeated dosing in rats: implications for treatment in Alzheimer's disease. J Alzheimer's Dis 18(1):113–124.
Finger EC, Marsh AA, Buzas B, Kamel N, Rhodes R, Vythilingham M, Pine DS, Goldman D, Blair JR (2007) The Impact of tryptophan depletion and 5-HTTLPR genotype on passive avoidance and response reversal instrumental learning tasks. Neuropsychopharmacol 32:206–215
Floyd RA, Hensley K (2002) Oxidative stress in brain aging. Implications for therapeutics of neurodegenerative diseases. Neurobiol Aging 23:795–807
Fujisaki K, Tsuruya K, Yamato M, Toyonaga J, Noguchi H, Nakano T, Taniguchi M, Tokumoto M, Hirakata H, Kitazonol T (2014) Cerebral oxidative stress induces spatial working memory dysfunction in uremic mice: neuroprotective effect of tempol. Nephrol Dial Transplant. 29:529–538
Ganguly R, Guha D (2008) Alteration of brain monoamines & EEG wave pattern in rat model of Alzheimer's disease & protection by Moringa oleifera. Ind J Med Res 128:744–751
Geng YQ, Guan IT, Xu XH, Fu YC (2010) Senescence-associated beta-galactosidase activity expression inn aging hippocampal neurons. Biochem Biophys Res Commun 396:866–869
Giaccone G, Orsi L, Cupidi C, Tagliavini F (2011) Lipofuscin hypothesis of Alzheimer's disease. Dement Geriatr Cogn Disord 1:292–296

Giasson BI, Duda JE, Murray IV, Chen Q, Souza JM, Hurtig HI, Ischiropoulos H, Trojanowski JQ, Lee VM (2000) Oxidative damage linked to neurodegeneration by selective alpha-synuclein nitration in synucleinopathy lesions. Science 290:985–989

Glickman SE, Jensen GD (1961) The effects of hunger and thirst on Y-maze exploration. J Comp Physiol Psychol 54:83–85

Gómez-Ramos P, Morán A (2007) Ultrastructural localization of intracellular Aß-peptide in Alzheimer disease brains. J Alzheimer's Dis 11:53–59

Grodstein F, Chen J, Willett WA (2003) High dose antioxidant supplements and cognitive function in community-dwelling elderly women. Am J Clin Nutr 77:975–985

Grundman M, Delaney P (2002) Antioxidant strategies for Alzheimer's disease. Proc Nutr Soc 61:191–202

Handley SL, Mithani S (1984) Effects of α-adrenoreceptor agonists and antagonists in a maze-exploration model of 'fear'-motivated behaviour. Naunyn-Schmeideberg's Arch Pharmacol 327:1–5

Hasan M, Tripathi S, Mahdi AA, Mitra K, Negi MPL (2009) Lipofuscin, lipid peroxidation and antioxidant status in discrete regions of the aged rat brain. Proc Indian Natl Acad Sci 75:173–181

Hollensworth SB, Shen CC, Sim JE, Spitz DR, Wilson GL, Ledoux SP (2002) Glial cell type-specific responses to menadione-induced oxidative stress. Free Radic Biol Med 28:1161–1174

Horner AE, Heath CJ, Hvoslef-Eide M, Kent BA, Kim CH, Nilsson SRO, Alsiö J, Charlotte A Oomen CA, Lisa M Saksida, Bussey TJ (2013) The touchscreen operant platform for testing learning and memory in rats and mice. Nat Protoc 8:1961–1984

Hossein N, Alaei H, Reisi P, Radahmad M (2013) The effect of treadmill running on passive avoidance learning in animal model of Alzheimer disease. Int J Prevent Med. 4:187–192

Jacobson T, Kent BQ, Yang F, Burwell RD (2014) Automated Visual Cognitive Tasks for Recording Neural Activity Using a Floor Projection Maze. J Vis Exp Feb 20(84):e51316. doi:10.3791/51316

Ji ES, Kim YM, Shin MS, Kim CJ, Lee KS, Kim K, Ha J, Chung YR (2015) Treadmill exercise enhances spatial learning ability through suppressing hippocampal apoptosis in Huntington's disease rats. J Exerc Rehab 11:133–139

Jolitha AB, Subramanyam MVV, Asha Devi S (2006) Modification by vitamin E and exercise of oxidative stress in regions of aging rat brain: Studies on superoxide dismutase isoenzymes and protein oxidation status. Exp Geront 41:753–763

Jolitha AB, Subramanyam MVV, Asha Devi S (2009) Age-related responses of the rat cerebral cortex; influence of vitamin E and exercise on the cholinergic system. Biogeron 10:53–63

Kan S, Asha Devi S, Kawashima S (1991) Effect of vitamin E on the accumulation of fluorescent material in cultured cerebral cortical cells of mice. Exp Gerontol 26:365–374

Kim J, Harada R, Kobayashi M, Kobahasi N, Soda K (2010) The inhibitory effect of pyrroloquinoline quinine on the amyloid formation and cytotoxicity of truncated alpha-synnuclein. Mol Neurodegener 5:20–28

Landfield PW, Thibault O, Mazzanti ML, Porter NM, Kerr DS (1992) Mechanisms of neuronal death in brain and alzheimer's disease: Role of endocrine-mediated calcium dyshomeostasis. J Neurobiol 23:1247–1260

Lin MT, Beal MF (2006) Mitochondrial dysfunction and oxidative stress in neurodegenerative diseases. Nature 443:787–795

Lione LA, Carter RJ, Hunt MJ, Bates GP, Morton AJ, Dunnett SB (1999) Selective discrimination learning impairments in mice expressing the human Huntington's disease Mutationm. J Neurosci 19:10428–10437

Manjula KR, Subramanyam MVV, Asha Devi S (2013) Protection against oxidative stress caused by intermittent cold exposure by combined supplementation with vitamin E and C in the aging rat hypothalamus. Neurochem Res 38:876–885

Markesberry WE, Carney JN (1999) Oxidative stress in Alzheimer's disease. Brain Pathol 9:133–146

Masaki KH, Losonezy KG, Izmirlian G, Foley DJ, Ross GW, Petrovitch H, Havlik R, White LR (2000) Association of vitamin E and C supplement use with cognitive function and dementia in elderly men. Neurology 54:1265–1272

Meydani M (2002) Nutrition interventions in aging and age associated disease. Proc Nutr Soc 61:165–171

Miura Y, Kano M, Abe K, Urano S, Suzuki S, Toda T (2005) Age-dependent variations of cell response to oxidative stress: proteomic approach to protein expression and phosphorylation. Electrophoresis 26:2786–2796

Miquel J, Economos AC, Fleming J, Johnson JE (1980) Mitochondrial role in cell aging. Exp Gerontol 15:575–591

Morris R (1984) Developments of a water-maze procedure for studying spatial learning in the rat. J Neurosci Methods 11:47–60

Morris MC, Evans DA, Bienias JL, Tangney CC, Wilson RS (2002) Vitamin E and cognitive decline in older persons. Arch Neurol 59:1125–1132

Navarro A, Boveris A (2007) The mitochondrial energy transduction and the aging process. Am J Cell Physiol. 292:C670–C686

Navarro A, Sanchez del Pino MJ, Omez CG, Peralta JL, Boveris (2002) Behavioral dysfunction, brain oxidative stress, and impaired mitochondrial electron transfer in aging mice. Amer J Physiol 282:R985–R992

Ogren SO, Stidel O 2015. Passive avoidance. In: Stolerman IP (ed) Encyclopedia of psychopharmacology. Springer-Verlag Berlin, Heidelberg, pp 1220–1228

Ohwada K, Takeda H, Yamazaki M (2008) Pyrroloquinoline (PQQ) prevents cognitive deficit caused by oxidative stress in rats. J Clin Biochem Nutr 42:29–34

O'Leary TP, Brown RE (2009) Visuo-spatial learning and memory deficits on the Barnes maze in the 16-month-old APPswe/PS1dE9 mouse model of Alzheimer's disease. Behav Brain Res 201:120–127

Olton DS, Samuelson RJ (1976) Remembrance of places passed: spatial memory in rats. J Exp Psych: Anim Behav Process 2(2):97–116. doi:10.1037/0097-7403.2.2.97

Pellow S, Chopin P, File SE, Briley M (1985) Validation of open: closed arm entries in an elevated plus-maze as a measure of anxiety in the rat. J Neurosci Methods 14:149–167

Petrosillo G, Benedictis VD, Ruggiero FM, Paradies G (2013) Decline in cytochrome c oxidase activity in rat-brain mitochondria with aging. Role of peroxidized cardiolipin and beneficial effect of melatonin. J Bioenerg Biomembr 45:431–440

Perluigi M, di Domenico F, Giorgia A, Schinina ME, Coccia R, Cini C, Bellia F, Cambria MT, Cornelius C, Butterfield DA, Calabrese V (2010) Redox proteomics in aging rat rain: involvement of mitochondrial reduced glutathione status and mitochondrial protein oxidation in the aging process. J Neurosci Res 88:3498–3507

Rajagopal L, Massey BW, Huang M, Oyamada Y, Meltzer HY (2014) The novel object recognition test in rodents in relation to cognitive impairment in Schizophrenia. Curr Pharmaceutical Design 20:1–11

Ramassamy C, Averill D, Beffert U, Therous L, Lussier-Cacan S, Cohn JS, Christen Y, Schoofs A, Davignon J, Poirier J (2000) Oxidative insults are associated with apolipoprotein E genotype in Alzheimer's disease brain. Neurobiol Dis 7:23–37

Rodriguez JS, Zürcher NR, Bartlett TQ, Nathanielsz PW, Nijland MJ (2011) Delayed matching to sample task performance in juvenile baboons. J Neurosci Methods 196:258–263

Roger RD, Andrews TC, Graasby PM, Brooks DJ, Robbins TW (2000) Contrasting cortical and subcortical activations produced by attentional-set shifting and reversal learning in humans. J Cogn Neurosci 12:142–162

Scheggia D, Bebensee A, Weinberger DR, Papaleo F (2014) The ultimate intra-/extra-dimensional attentional set-shifting task for mice. Biol Psychiatry 75:660–670

Shulman RG, Rothman DL, Behar KL, Hyder F (2004) Energetic basis of brain activity implications for neuro imaging. Trends Neuro Sci 27:489–495

Shankar SK (2010) Biology of aging brain. Indian J Pathol Microbiol 53:595–604

Shoji H, Hagihara H, Takao K, Hattori S, Miyakawa T (2012) T-maze forced alternation and left-right discrimination tasks for assessing working and reference memory in mice. J Vis Exp 60:e3300

Siqueira IR, Fochesatto C, de Andrade A, Santos M, Hagen M, Bello-Klein A, Netto CA (2005) Total antioxidant capacity is impaired in different structures from aged rat brain. Int J Dev Neurosci 23:663–671

Staniek K, Nohl H (2000) Are mitochondria a permanent source of reactive oxygen species? Biochem Biophys Acta 146:268–275

Takatsu H, Owada K, Abe K, Nakano M, Urano S (2009) Effect of vitamin E on learning and memory deficit in aged rats. J Nutr Sci Vitaminol (Tokyo) 55:389–393

Talpos JC, McTighe SM, Dias R, Saksida LM, Bussey TJ (2010) Trial-unique, delayed nonmatching-to-location (TUNL): A novel, highly hippocampus-dependent automated touch-screen test of location memory and pattern separation. Neurobiol learning and Mem 94:341–352

Trimmer PA, Swerdlow RH, Parks JK, Keeney P, Bennett JP, Miller SW, Davis RE, Parker WD (2000) Abnormal mitochondrial morphology in sporadic Parkinson's and Alzheimer's disease cybrid cell lines. Exp Neurol 162:37–50

Van Dongen MC, Van Rossum E, Kessels AG, Sielhorst HJ, Knipschild PG (2000) The efficacy of Ginkgo for elderly people with dementia and age-associated memory impairment: new results of a randomized clinical trial. J Am Geritarics Soc 48:1193–1194

VanGuilder HD, Freeman WM (2011) The hippocampal neuroproteome with aging and cognitive decline: past progress and future directions. Front in Aging Neurosci 3, article 8

Vohra BP, Sharma SP, Kansai VK (2001) Age-dependent variations in mitochondrial and cytosolic antioxidant enzymes and lipid peroxidation in different regions of central nervous system of guinea pigs. Indian J Biochem Biophys 38:321–326

Walsh RN, Cummins RA (1976) The open-field test: a critical review. Psychological Bull 83:482–504

Walf AA, Frye CA (2007) The use of the elevated plus maze as an assay of anxiety-related behavior in rodents. Nat Protocol. 2:322–328

Weinreb O, Drigues N, Sagi Y, Reznick AZ, Amit T, Youdim MBH (2007) The application of proteomics and genomics to the study of age-related neurodegeneration and neuroprotection. Antioxid Redox Signal 9:169–179

Wiig KA, Burwell RD (1998) Memory impairment on a delayed non-matching-to-position task after lesions of the perirhinal cortex in the rat. Behav Neurosci 112:827–838

Yamada K, Takayanagi M, Kamei H, Naga T, Dohniwa M, Kobayash K, Yoshida S, Ohhara T, Takauma K, Nabeshima T (2005) Effects of memantine and donepezil on amyloid β-induced memory impairment in a delayed-matching to position task in rats. Behav Brain Res 162:191–199

Yang A, Liu T, Li S, Zhang X, Ding Q, Que H, Yan X, Wei k, Liu S 2008a Comparative proteomic analysis of brains of naturally aging mice. Neurosci 154:1107–1120

Yang JL, Weissman L, Bohr VL, Mattson MP (2008b) Mitochondrial DNA damage and repair in neurodegenerative disorders. DNA Repair 7:1110–1120

Yhnell E, Dunnett SB, Brooks SP (2015) The utilisation of operant delayed matching and non-matching to position for probing cognitive flexibility and working memory in mouse models of Huntington's disease. J Neurosci Methods. doi:10.1016/j.jneumeth.2015.08.022

Youdim KA, Shukitt-hale B, Martin A, Wang H, Denisova NA, Bickford PC (2000) Short-term dietary supplementation of blueberry polyphenolics: beneficial effects on aging brain performance and peripheral tissue function. Nutr Neurosci 3:383–397

Age-Related Changes in Antioxidant Enzymes of Rat Kidney and Oxidative Stress Parameters with Special Reference to Methylation of the Catalase Promoter

Srikanta Jena, Suresh K. Bunker, Jagneshwar Dandapat and G.B.N. Chainy

Abstract Oxidative stress plays an important role in etiology of age-associated decline in renal physiological functions. The purpose of the present study was to determine the profiles of antioxidant enzymes (AOEs) and oxidative stress parameters in rat kidney from birth to old age with a particular reference to catalase promoter methylation. Results revealed that the levels of oxidative stress indices (lipid peroxides and protein carbonyls) were significantly higher, while levels of small antioxidants (reduced glutathione and ascorbic acid) were significantly lower in kidneys of old rats than that of adult ones. Also the results suggest lack of a correlation among the expression levels of antioxidant enzymes and their respective activity levels in rat kidney with relation to aging process. For example, when SOD1 activity was significantly low in old age than adult ones, its mRNA and protein levels were higher in old age. On the contrary, SOD2 activity and its mRNA level were higher in old age in comparison to adult rats without any alteration in its translated products. GPx activity and its mRNA did not change in old age in comparison to adult ones. Although GR activity remained unaltered during aging, its mRNA level was reduced in old age than in adult ones. On the other hand, CAT activity and levels of its mRNA and protein were reduced in old rats than that of adult ones. The methylation level in CpG dinucleotides of CAT promoter (from −268 to +52) was higher in old rats than in adult ones and was inversely proportional to its expression. The results suggest that (i) rates of prooxidant generation exceed the overall antioxidant defenses during aging in kidney and (ii) emphasize important role of methylation in the expression of CAT during aging.

Keywords Oxidative stress · Antioxidant enzyme expression · Aging
Rat kidney · Catalase-promoter methylation

S. Jena · S.K. Bunker · J. Dandapat · G.B.N. Chainy (✉)
Department of Biotechnology, Utkal University, Bhubaneswar 751004, Odisha, India
e-mail: chainyg@gmail.com

S. Jena
Department of Zoology, Ravenshaw University, Cuttack 753003, India

© Springer Science+Business Media Singapore 2017

P.C. Rath et al. (eds.), *Topics in Biomedical Gerontology*,
DOI 10.1007/978-981-10-2155-8_9

1 Introduction

Aging is associated with impairment of cellular functions as a consequence of reactive oxygen species (ROS)-damaged macromolecules in cells (Harman 1956). ROS (superoxide radicals ($O_2^{\bullet -}$), hydroxyl radicals (•OH) and hydrogen peroxide (H_2O_2)) are produced as by-products of cellular metabolism. Being highly reactive in nature, they can oxidize biomolecules present in their vicinity that leads to impairment of cellular functions. However, cellular system is equipped with both enzymatic as well as non-enzymatic antioxidant defences to counter act ROS. The major cellular antioxidant system is comprised of superoxide dismutases (SOD), catalase (CAT), glutathione peroxidase (GPx) and glutathione reductase (GR). While SOD dismutates $O_2^{\bullet -}$, a product of incomplete reduction of oxygen during metabolism, to H_2O_2, CAT and GPx neutralize H_2O_2. At low concentration hydrogen peroxide acts as a cellular signaling molecule to regulate expression of several transcriptional factors, whereas at higher concentration it acts as an oxidant in cellular system, leading to oxidative stress (Gough and Cotter, 2013). Catalase breaks down H_2O_2 to water and molecular oxygen, whereas GPx reduces H_2O_2 to water at the cost of oxidation of reduced glutathione (GSH). The reduced glutathione is recovered back from oxidized glutathione (GSSG) by the enzyme GR. There are also some cellular small antioxidant molecules like GSH and ascorbic acid (ASA), which directly scavenge the ROS. In general, a delicate balance is maintained between the rate of production of ROS and its neutralization in cells. When the balance tilts towards the production of ROS, the organism is pushed to a stress condition which is commonly referred to as "Oxidative Stress" (OS) (Halliwell and Gutterige 2001). Although role of oxidative stress in aging is supported by numerous data, yet a direct cause-and -effect relationship between oxidative stress and aging needs more evidences (Kregel and Zang 2007).

Kidney is one of the vital organs of the body that is responsible for the maintenance of homeostasis. Several age related pathological states/diseases of kidney are considered as cause or effects of oxidative stress (Baud and Ardajlou 1993; Ichikawa et al. 1993). Although age-related alterations in histology and physiology of kidney is well documented for several species of vertebrates including that of rodents (Lindeman 1990; Martin and Sheaff 2007), information with reference to antioxidant defenses is limited from birth to old age. A careful analysis of previous reports on age-related changes in antioxidant enzymes and oxidative stress parameters in kidney suggests that studies are limited because they are either confined to alterations in activities of SOD, CAT, GPX and LPX (Farooqui et al. 1987; Tian et al. 1998; Gunduz et al. 2004) or confined to changes in expression of some selected renal antioxidant enzymes with age (Rao et al. 1990; Martin et al. 2002). Further, they were restricted either between adult and old age or among young, adult and old ages. None of them had simultaneously measured oxidative stress parameters, small antioxidants and activities of antioxidant enzymes along with their respective expressions at mRNA and/or protein levels from birth to old age. Consequently, the picture emerged out of above studies is not complete and therefore, it is difficult to deduce meaningful conclusion on the relationship between antioxidant defence mechanism of the kidney with age.

With above background, we have reported the status of major antioxidant enzymes, two small antioxidant molecules, and two oxidative stress parameters in kidney from day 1 of birth to 730 days of age in order to construct a dynamic composite picture of renal antioxidant defense mechanism during aging. Since it is reported that methylation of CpG dinucleotides present in promoter regions of genes plays a crucial role in their expression during development and ageing (Richardson 2003; Van Otterdijk et al. 2013), we have also compared methylation status of CpG dinucleotides in a specific region of renal catalase promoter in young, adult and old rats and correlated the values with catalase activity and expression. Also an attempt was made to assess the physiological status of kidney during aging by monitoring serum creatinine and urea nitrogen levels. It is presumed that information on antioxidant profile with age will have important implications in understanding the changing renal functions during aging.

2 Materials and Methods

2.1 Animals and Experimental Design

All experiments were carried out in Wistar strain male rats (*Rattus norvegicus*). Animal care, maintenance, breeding, sacrifice of animals, collection and storage of tissue (−80 °C) and serum samples (−20 °C) have been described in details elsewhere (Sahoo et al. 2008a) and explained in brief as follows. Wistar strain rats (*Rattus norvegicus*) were supplied by National Institute of Nutrition (Hyderabad, India). Animals were maintained at 25 ± 2 °C and controlled 12:12 h light and dark cycles with free access to food (Sahoo et al. 2008a) and drinking water. Animal care, maintenance and experiments were done under the supervision of the Institutional Animal Ethics Committee (IAEC) regulated by the Committee for the Purpose of Control and Supervision of Experiments on Animals (CPCSEA), Govt. of India. Animals of different ages (1, 7, 15, 30, 90 and 730 days), were sacrificed by decapitation and blood samples were collected to obtain sera. The sera were stored at −20 °C until the measurement of functional indices (serum urea nitrogen and creatinine) of kidney. Kidneys were dissected out quickly, cleaned in 0.9 % (w/v) cold normal saline, pat dried and weighed. Kidneys were kept at −80 °C until use.

Experiments:

Experiment I: In the first experiment, activities of antioxidant enzymes (SOD, CAT, GPx and GR), oxidative stress parameters (LPX and PC) and small antioxidants molecules (GSH and ascorbic acid) and serum levels of urea nitrogen and creatinine were estimated in kidney of 1, 7, 15, 30, 90 and 730 days old rats. Expression levels of antioxidant enzymes were determined in 15, 90 and 730 days old rats.

Experiment II: In the second experiment, methylation status of kidney catalase promoter and its protein expression were investigated in 30, 90 and 700 days old rats.

2.2 Sample Preparation

A 20 % (w/v) homogenate of kidney was prepared in ice-cold phosphate buffer (50 mM, pH 7.4) with a Potter-Elvehjem type motor-driven homogenizer. The crude homogenate was centrifuged at 600×g for 10 min to pellet down nuclei and other cell debris. The resultant supernatant (S1) was used to estimate oxidative stress and non enzymatic parameters. The supernatant (S1) was treated with Triton X-100 (final conc. 0.1 %, v/v) and centrifuged at 10,000×g for 15 min. The supernatant (S2) so obtained was used to assay enzyme activities. The whole tissue processing was done at 4 °C.

2.3 Oxidative Stress Parameters

Lipid peroxidation level was estimated by monitoring the formation of thiobarbituric acid-reactive substances (TBARS) according to the method of Ohkawa et al. (1979) in the presence of 0.02 % (w/v) butylated hydroxyl toluene to prevent peroxidation during the subsequent boiling. Concentration of TBARS was calculated as n mol TBARS formed per mg protein from its extinction co-efficient of $1.56 \times 10^5\ M^{-1}\ cm^{-1}$. Protein carbonyl content in samples was estimated by the method of Levine et al. (1990) and calculated as nmoles carbonyl/mg protein.

2.4 Non-enzymatic Small Antioxidant Parameters

Samples were precipitated in ice-cold 5 % TCA containing 0.01 N HCl and the clear supernatant was collected after centrifuging the samples at 100×g for 15 min. GSH level was measured using the method of Sedlak and Lindsay (1968) and expressed as µmoles/g tissue. Ascorbic acid content in samples was measured by the method of Mitsui and Ohta (1961) and expressed as µg/g tissue.

2.5 *Antioxidant Enzyme Activities*

The supernatants (S2) of the samples were passed through a 1 ml of Sephadex G-25 column and elutes were used to assay the activities of SOD, GPx and GR. The activity of SOD was determined following the modified nitrite method of Das et al. (2000) and expressed as units/mg protein, where one unit of enzyme activity is defined as the amount of enzyme capable of inhibiting 50 % of nitrite formation under assay conditions. GPx activity was assayed by measuring the oxidation rate of NADPH in the presence of cumine-hydroperoxide, GSH and GR (Paglia and Valentine 1967). The enzymatic activity of GR was assayed by measuring the oxidation rate of NADPH in the presence of GSSG as previously described method of Massey and Williams (1965). The activity of GPx and GR were expressed as nmoles of NADPH oxidized/min/mg protein. For measuring the activity of CAT, the supernatants (S2) of the samples were treated with ethanol (0.17 M) and the break down of 12 mM H_2O_2 was followed at 240 nm (Cohen et al. 1970; Aebi 1974). The activity was calculated taking 43.6 M^{-1} cm^{-1} as the molar extinction co-efficient of H_2O_2 and finally expressed as μkatal/mg protein.

2.6 *Serum Analyses*

Serum urea nitrogen (SUN) and creatinine concentration were determined by modified Berthelot method and alkaline picrate method, respectively, using standard protocols provided by the kit (Crest Biosystem, India).

2.7 *Estimation of Protein*

Protein concentration was estimated by the method of Bradford (1976) and bovine serum albumin was taken as standard.

2.8 *DNA Isolation and Quantification*

Total genomic DNA was isolated from tissues by RNase and Proteinase K treatment followed by standard phenol/chloroform extraction method (Sambrook and Russell 2001), quantified by UV-Visible spectrophotometer (Cary 100, Varian) and purity was checked by calculating the ratio between absorbance at 260–280 nm.

2.9 RNA Isolation and Synthesis of cDNA

Total RNA was isolated from kidney tissues using Trizol reagent (Molecular Research Center Inc., USA). RNA samples were subjected to DNase1 treatment to remove genomic DNA contamination in the presence of RNase inhibitor. Primer sequences (Table 1) of antioxidant genes were custom made by Integrated DNA technology, USA. G3PDH served as the internal standard. Reverse transcription of 5 μg total RNA was performed, using 200 ng random hexamer, 20 U RNase inhibitor, 1.0 mM dNTPs, and 40U of M-MuLV reverse transcriptase (Fermentas), to make ss-cDNA at 37 °C for 1 h.

2.10 Semi-quantitative RT-PCR

The reverse transcribed products (cDNA) were then subjected to PCR at 94 °C for 3 min, 3-step cycling, each cycle consist of denaturation at 94 °C for 30 s, annealing (Table 1) for 30 s, extension at 72 °C for 1 min (30 cycle for GAPDH, SOD1, SOD2, CAT and GPx1, and 33 cycle for GR) followed by final extension at 72 °C for 5 min. The PCR mixture of 25 μl included 1 μl RT product as templet, 2.5 μl 10X buffer, 0.2 mM dNTPs, 1.5 mM $MgCl_2$, 1U *Taq* DNA polymerase (Fermentas) and 25 pmol of each primer. The PCR products were electrophoresed on ethidium bromide stained 1.2 % agarose gels. The expression levels were measured densitometrically.

2.11 Methylation Specific Primer PCR (MSP-PCR)

One μg of genomic DNA was treated with sodium bisulfite using the Epitect Bisulfite kit according to manufacturer's protocol. To analyze the methylation level of the catalase promoter, primers for MSP were designed using the meth primer software-The Li Lab (www.urogene.org/methprimer). In brief, 50 ng bisulfite-treated genomic DNA was amplified using either a methylated-specific or unmethylated specific primers. PCR was performed in a Thermal cycler (My Cycler, BIO RAD) using the following conditions: 95 °C for 5 min; 35 cycles of 95 °C for 30 s 54 °C for 30 s, 72 °C for 60 s and 7 min final extension at 72 °C. The product sizes were 150 bp for both methylated and unmethylated products. PCR products were analysed on 1.5 % agarose gels. Gels were visualized by ethidium bromide staining and analyzed under gel documentation system (BIO RAD, Universal Hood II). Methylated and unmethylated status were determined by the band intensity of the MSP-PCR products.

Table 1 Accession numbers, primer sequences, number of cycles, annealing temperature, binding positions product sizes for amplification of G3PDH, SOD1, SOD2, CAT, GPx, GR genes and CAT-MSP-PCR

Gene	Accession no.	Primer sequence (5′–3′) S = sense, AS = antisense	Cycles	Annealing temp. (°C)	Binding position	Product size (bp)
G3PDH	DQ 403053	S-TCCCTCAAGATTGTCAGCAA AS-AGATCCACAACGGATACATT	30×	48	352–371 659–640	308
SOD1	NM_017150	S-GCAGAAGGCAAGCGGTGAAC AS-TAGCAGGACAGCAGATGAGT	30×	52	159–178 605–586	447
SOD2	BC070913	S-CTGAGGAGAGCAGCGGTCGT AS-CTTGGCCAGCGCCTCGTGGT	30×	58	51–70 308–289	258
CAT	BC081853	S-GCGAATGGAGAGGCAGTGTAC AS-GAGTGAGTTGTCTTCATTAGCACTG	30×	52	702–722 1353–1328	652
GPx1	S41066	S-CTCTCCGCGGTGGCACAGT AS-CCACCACCGGGTCGGACATAC	30×	56.8	367–385 656–636	290
GR	NM_053906	S-TCACTGCTCCGCACATCC AS-CTCAACACCGCCAGCGTTCTCC	33×	54.1	419–436 714–693	296
CAT-MSP	AH004967.1	S-GAGTTTTAGTGGTTAATTAGGAGGC AS-GTAAAACAAAAAAACCGAACGAA	35×	55	−162 to −184 −34 to −11	150
CAT-USP	AH004967.1	S-GTTTTAGTGGTTAATTAGGAGGTGG AS-ACATAAAACAAAAAAACCAAACAA	35×	54	−158 to −182 −34 to −9	150

2.12 Western Blotting

A 10 % homogenate of kidney was prepared in 50 mM Tris-HCl buffer, pH 7.4, containing 150 mM NaCl, 104 mM PMSF, 100 μM E-64, 80 μM aprotinin, 100 μM leupeptin, 1 % triton X-100 and 0.1 % SDS to avoid protein degradation. Homogenates were centrifuged at 1000×g for 20 min at 4 °C. Fifty micrograms of total cellular proteins were resolved in 12 % SDS-PAGE and transferred to PVDF membrane (0.4 μm, PALL Life Sciences) at 23 mA current for 1 h. The membrane was blocked in 5 % blocking solution for 1 h at room temperature. The blot was then incubated with rabbit polyclonal anti-G3PDH (1:1000), anti-CAT (1:5,000), anti-SOD1 (1:5,000) or anti-SOD2 (1:2500) for 1 h at room temperature. Anti-G3PDH, anti-CAT and SOD1 primary antibodies were obtained from Imgenex India Pvt. Ltd., India, while SOD2 was obtained from Santa Cruz Biotechnology, Inc., USA. The membrane was washed 3 times (5 min each) with washing solution and subsequently incubated with HRP-conjugated anti-rabbit goat IgG (1:7500, Santa Cruz Biotechnology, Inc., USA) for 1 h at room temperature. After washing, specific immunoreactive proteins were detected with ECL kit (Santa Cruz Biotechnology, Inc., USA) in X-ray film and their expression levels were measured by densitometry.

2.13 Bioinfomatics

Rat catalase gene (Gene Bank AH 004967.1), 320 bp was analyzed for CpG island using the CpG Island Searcher programme http://cpgislands.usc.edu/cpg.aspx. The CpG dinucleotides was defined as DNA sequences of 200 bp with GC content more than 50 %. Promoter region of catalase gene (−268 to +52) has GC content 65.7 % and observed CpG and expected CpG ratio is 0.65. To recognize putative binding sites for transcription factor we used the TFSEARCH programme http://www.cbrc.jp/research/db/TFSEARCH.html).

2.14 Statistical Analyses

Quantitation of Western blot and RT-PCR bands were done using computer assisted densitometry Image-Quant TL, Image Analysis Software v2003. Relative densities were determined as the ratio of AOEs band/G3PDH band. The data were presented as mean ± S.D. Data were analysed by one-way analysis of variance (ANOVA) followed by Duncan's new multiple range tests to find out the level of significance among the mean values. The minimal statistical significance was considered at $p \leq 0.05$ levels.

3 Results

3.1 *Results of Experiment I*

3.1.1 Body Weight (BW), Kidney Weight (KW), Serum Urea Nitrogen and Creatinine Levels

The body and kidney weights of rats increased significantly with age (Table 2). The relative weights of kidney increased (41 %) in 7 days old rats in comparison to 1 day, then a gradual decrease in relative weights of kidney were recorded with age. The level of SUN increased from 7 days of age to 90 days (about twofolds) linearly and then decreased significantly in old age (23 % in comparison to 90 days old rats). The serum creatinine level increased linearly from 7 days of age to 730 days (Table 2).

3.1.2 Oxidative Stress Parameters

LPx levels of kidney did not change till 30 days of age, then increased by 193 % in 90 days old rats. Its level further increased by 41 % in 730 days old rats in comparison to 90 days old rats. Protein carbonyl content of kidney did not change till 15 days of age, then increased by 72 % in 30 days old rats and remain unchanged till adulthood (90 days). Further a small but significant increase in its level (10 % in comparison to 90 days old rats) was recorded in old rats (Table 3).

3.1.3 Non-enzymatic Small Antioxidant Molecules

GSH content of the kidney was significantly higher (48 %) in 90 day old rats than 7 day old rats, then decreased in old age. The magnitude of decrease was 10 % in old age in comparison to 90 day old rats. ASA content of kidney did not change from 7 days of age to 90 days but decreased significantly by 12 % in old age (Table 3).

3.1.4 Expressions and Activities of Antioxidant Enzymes

Total SOD

Total SOD activity of kidney increased gradually from 7 days of age to 90 days (about 16-folds) and did not change further in old age (Table 4).

Table 2 Kidney weight (g), body weight (g) and renal functional indices (SUN (mg/dL) and creatinine (mg/dL)) of rat kidney of different ages

Age (days)	1	7	15	30	90	730
KW (g)	0.05 ± 0.002^{a}	0.16 ± 0.001^{b}	0.26 ± 0.01^{c}	0.57 ± 0.06^{d}	$0.75 \pm 0.02e$	1.68 ± 0.09^{f}
BW (g)	6.14 ± 0.17^{a}	12.75 ± 0.21^{b}	24.64 ± 1.01^{c}	63 ± 4^{d}	111 ± 2^{e}	324 ± 13.92^{f}
Relative KW	0.92 ± 0.04^{a}	1.3 ± 0.02^{b}	1.08 ± 0.04^{c}	0.9 ± 0.05^{a}	0.68 ± 0.02^{d}	0.51 ± 0.01^{e}
SUN	–	6.65 ± 0.18^{a}	9.56 ± 0.5^{b}	10.87 ± 0.79^{c}	14.73 ± 0.35^{d}	11.23 ± 0.71^{c}
Creatinine	–	0.23 ± 0.02^{a}	0.48 ± 0.01^{b}	0.56 ± 0.03^{c}	0.7 ± 0.05^{d}	0.88 ± 0.04^{e}

Data are expressed as means ± S.D. of five animals. Within each row, data having different superscripts differ significantly ($p \leq 0.05$) from each other

Table 3 Age-related changes in oxidative stress parameters (LPx (nmoles TBARS/mg protein) and PC (nmoles/mg proteins)) and levels of small antioxidant molecules (GSH (μmoles/g tissue) and ASA (μg/g tissue)) of rat kidney of different ages

Age (days)	7	15	30	90	730
LPx	0.43 ± 0.01^{a}	0.44 ± 0.03^{a}	0.45 ± 0.02^{a}	1.32 ± 0.03^{b}	1.91 ± 0.05^{c}
PC	3.4 ± 0.75^{a}	3.5 ± 0.86^{a}	6.04 ± 0.71^{b}	6.09 ± 0.22^{b}	6.6 ± 0.25^{c}
GSH	1.01 ± 0.03^{a}	1.21 ± 0.01^{b}	1.25 ± 0.09^{b}	1.5 ± 0.05^{c}	1.35 ± 0.04^{d}
ASA	459.2 ± 14^{a}	443.9 ± 18^{a}	446.6 ± 21^{a}	443.9 ± 9^{a}	387.3 ± 27^{b}

Data are expressed as means ± S.D. of five animals. Within each row, data having different superscripts differ significantly ($p \leq 0.05$) from each other

SOD1

SOD1 activity of kidney increased gradually from day 1 to day 90 (about 19-folds) and then decreased significantly in old age (26 %) in comparison to 90 day old rats, (Table 4). The transcript level of SOD1 elevated gradually with age, (Fig. 1a). The translational level of SOD1 was significantly higher in 90 day old rats in comparison to 15 day old rats and did not change in old age (Fig. 1b).

SOD2

The activity of SOD2 increased gradually from day1 to old age (about 19-folds), (Table 4). The transcript level of SOD2 increased with age (Fig. 2a), whereas its translational level was significantly higher in 90 day old rats in comparison to 15 day old and 730 day old rats (Fig. 2b).

CAT

CAT activity of kidney gradually increased from day 1 to 90 days of age (about 20-folds) and then its activity decreased in old age (27 %) in comparison to 90 day old rats, (Table 4). The transcript and translational levels of CAT was significantly higher in 90 day old rats in comparison to 15 day old and 730 day old rats (Fig. 3a, b).

GPx

GPx activity of kidney increased twofolds on 30 days of age in comparison to 1 day and then decreased (twofolds) in 90 days old rats and remain more or less same in old age (Table 4). The translate level of GPX did not change with age (Fig. 4a).

Table 4 Age-related changes in antioxidant enzymes (SOD (total, SOD1 and SOD2 in units/mg protein), CAT (µkatal/mg protein), GPx (µmoles NADPH oxidized/min/mg protein), SOD/CAT+GPx (arbitrary unit) and GR (µmoles NADPH oxidized/min/mg protein)) in rat kidney of different ages

Age (days)	1	7	15	30	90	730
SOD(total)	0.6 ± 0.04^{a}	0.64 ± 0.03^{a}	1.33 ± 0.12^{b}	2.42 ± 0.14^{c}	10.08 ± 0.54^{d}	9.66 ± 0.51^{d}
SOD1	0.37 ± 0.02^{a}	0.43 ± 0.06^{a}	0.91 ± 0.13^{b}	1.4 ± 0.14^{c}	7.09 ± 0.56^{d}	5.18 ± 0.46^{e}
SOD2	0.23 ± 0.02^{a}	0.198 ± 0.03^{a}	0.44 ± 0.07^{b}	1.02 ± 0.1^{c}	3.14 ± 0.16^{d}	4.48 ± 0.13^{e}
CAT	0.36 ± 0.03^{a}	1.97 ± 0.16^{b}	2.16 ± 0.17^{b}	$3.43 \pm 0.^{c}$	7.16 ± 0.29^{d}	5.23 ± 0.31^{e}
GPx	0.22 ± 0.01^{a}	0.35 ± 0.01^{b}	0.4 ± 0.03^{c}	0.47 ± 0.01^{d}	0.31 ± 0.01^{e}	0.32 ± 0.01^{be}
SOD/CAT+GPx	1.04 ± 0.14^{a}	0.27 ± 0.02^{b}	0.51 ± 0.04^{c}	0.62 ± 0.05^{c}	1.35 ± 0.1^{d}	1.73 ± 0.06^{e}
GR	52.8 ± 2.05^{a}	77.1 ± 7.3^{b}	71.58 ± 5.64^{b}	61.65 ± 4.1^{c}	76.16 ± 2.13^{b}	75.96 ± 7.28^{b}

Data are expressed as means ± S.D. of five animals. Within each row, data having different superscripts differ significantly ($p \leq 0.05$) from each other

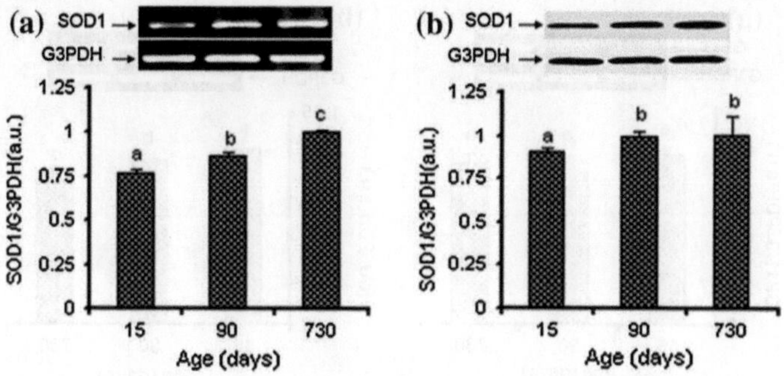

Fig. 1 Age-related changes in the expression of SOD1 in rat kidney of different age groups. Quantitative representations of SOD1/G3PDH mRNA **a** and proteins **b** obtained by densitometric analyses, in arbitrary units (a.u.). Data are expressed as means ± S.D. of five animals. Data having different superscripts differ significantly ($p \leq 0.05$) from each other

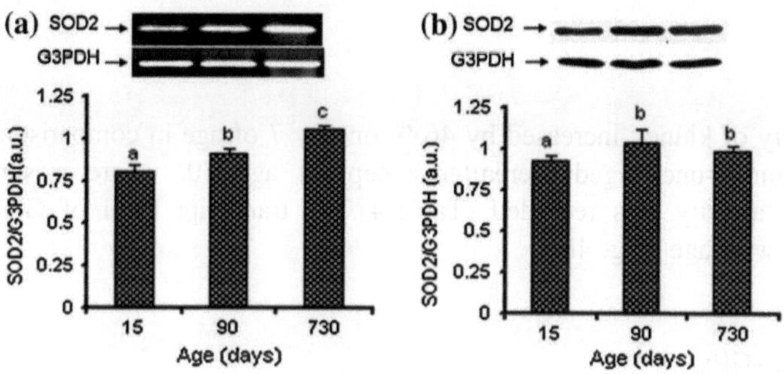

Fig. 2 Age-related changes in the expression of SOD2 in rat kidney of different age groups. Quantitative representations of SOD2/G3PDH mRNA **a** and proteins **b** obtained by densitometric analyses, in arbitrary units (a.u.). Data are expressed as means ± S.D. of five animals. Data having different superscripts differ significantly ($p \leq 0.05$) from each other

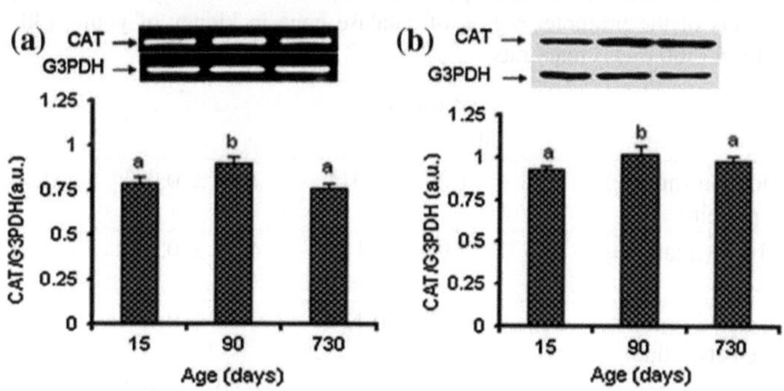

Fig. 3 Age-related changes in the expression of CAT in rat kidney of different age groups. Quantitative representations of CAT/G3PDH mRNA **a** and proteins **b** obtained by densitometric analyses, in arbitrary units (a.u.). Data are expressed as means ± S.D. of five animals. Data having different superscripts differ significantly ($p \leq 0.05$) from each other

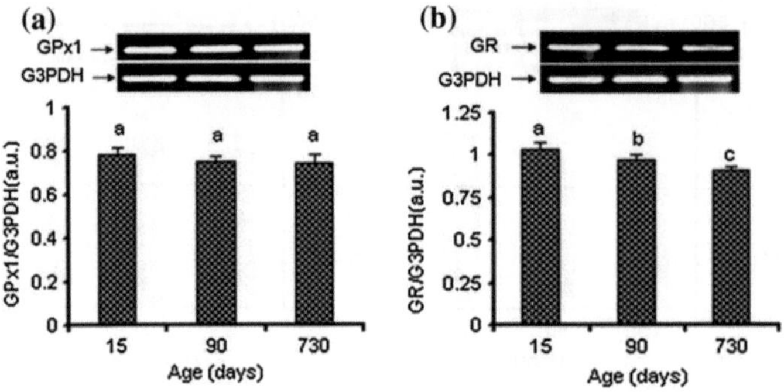

Fig. 4 Age-related changes in mRNA levels of GPx1 and GR in rat kidney of different age groups. Quantitative representations of GPx1/G3PDH **a** and GR/G3PDH **b** obtained by densitometric analyses, in arbitrary units (a.u.) Data are expressed as means ± S.D. of five animals. Data having different superscripts differ significantly ($p \leq 0.05$) from each other

GR

GR activity of kidney increased by 46 % on day 7 of age in comparison to day 1 and remained unchanged thereafter except at age 30, where a significantly decreased activity was recorded (Table 4).The transcript level of GR declined gradually with age (Fig. 4b).

SOD/CAT+GPx

The ratio of SOD/CAT+GPx was high on day 1 of age and then decreased by 74 % in day 7. Thereafter the ratio gradually increased with age (Table 4).

Table 5 Age related changes in lipid peroxidation, catalase activity, translated product and methylation status of the promoter region of catalase gene in kidney of young (30 days), adult (90 days) and old (700 days) male rats

Parameters	Age (days)		
	30	90	700
Lipid peroxidation (nmol TBARS/mg protein)	1.20 ± 0.044^{a}	1.65 ± 0.07^{b}	2.30 ± 0.06^{c}
Catalase activity (μkatal/mg protein)	2.92 ± 0.18^{a}	4.78 ± 0.18^{b}	2.5 ± 0.08^{c}
Catalase translated product (Integrated density value)	1.49 ± 0.12^{a}	2.03 ± 0.14^{b}	1.04 ± 0.09^{c}
Catalase MSP product (Integrated density value)	121.74 ± 8.10^{a}	86.40 ± 8.80^{b}	157.62 ± 10.66^{c}

Data are expressed as mean ± S.D. of five animals. Western blotting and MSP-PCR products are presented as integrated values. Data having different superscripts in same row differ significantly ($p \leq 0.05$) from each other

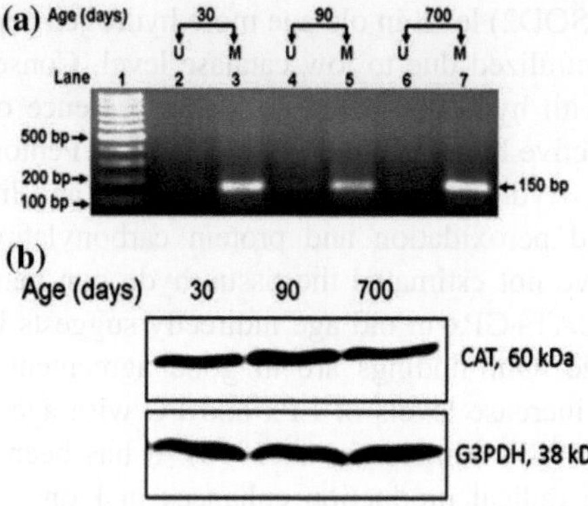

Fig. 5 Representative MSP-PCR products of promoter region and western blot of rat kidney catalase of young (30 days), adult (90 days) and old (700 days) rats. **a** MSP-PCR products (lane 2 to 3), adult (lane 4 to 5) and old (lane 6 to 7) and loaded against 100 bp DNA ladder (lane 1). U- represent unmethylated specific PCR product amplified with unmethylated primers, M- represent methylated specific PCR product amplified with methylated primers. **b** Representative western blots

3.2 Results of Experiment II

3.2.1 MSP-PCR

Band intensity of MSP-PCR products of catalase promoter was significantly low in 90 day old rats than 30 day old and 700 day old rats (Table 5 and Fig. 5a). Catalase activity and expression of its translated products were significantly low in 30 day old and 700 day old rats in comparison to 90 day old animals (Table 5 and Fig. 5b). The level of LPx increased gradually with respect to age (Table 5).

4 Discussion

Findings of the present study suggest augmentation of oxidative stress in the kidney of old rats as evident by the enhancement in the levels of LPx and PC. However, their increase cannot be explained to changing activities of antioxidant enzymes with age. It is evident from the present study that levels of antioxidant enzymes increased gradually from birth to adult hood but their respective activities in old age in compare to adult rats differ from each other. For example, in old age SOD1 and CAT activities decreased, SOD2 activity enhanced while that of GPx and GR remained unaltered. Changes in the enzyme levels along with decrease in levels of small antioxidant molecules such as GSH and ASA indicate an overall decrease in antioxidant capacity of the tissue to neutralize ROS which resulted in augmentation of levels of LPx and PC in old age. It is possible that due to increase in mitochondrial SOD (SOD2) level in old age more hydrogen peroxides are produced that could not be neutralized due to low catalase level. Consequently, superoxide radicals may react with hydrogen peroxide in the presence of Fe^{2+} causing production of highly reactive hydroxyl radicals by classical Fenton reaction (Halliwell and Gutterige 2001). Hydroxyl radicals attack proteins and lipids present in their vicinity causing lipid peroxidation and protein carbonylation. Although in the present paper we have not estimated the tissue hydrogen peroxide level, the elevated ratio of SOD/CAT+GPx in old age indirectly suggests increased production of hydrogen peroxide. Our findings are in good agreement with earlier reports which demonstrated increase levels of LPx and PC with age in several tissues of rats (Farooqui et al. 1987; Almeida et al. 1998). It has been observed that mitochondrial superoxide radical production enhances in kidney of mice along with protein carbonyl content without any change in mitochondrial SOD activity with aging (Miyazawa et al. 2009). Choksi et al. (2007) reported that mouse kidney mitochondrial electron transport chain complexes were oxidatively damaged with age. GPx and CAT are two key enzymes responsible for the metabolism of hydrogen peroxides. It is apparent from the present study that CAT exhibited a dominant role than GPx in kidney with aging. The fact is substantiated by observing unaltered activities of GPx and GR during development and aging. It is difficult to explain the reasons for observed decrease in GR activity on 30 days of age. This phase of neonatal period may be a critical window where the renal cells may be in a vulnerable state due to low GSH availability in the cellular environment. GSH plays a key role in cellular defence by acting as a non-enzymatic antioxidant. Besides, it also plays a major role as a substrate for GPx activity (Halliwell and Gutterige 2001) (Fig. 6).

Our results on age-related alterations in renal antioxidant defences are in good agreement with earlier reports on other tissues (brain (Sahoo and Chainy 1997), heart (Van der loo et al. 2005), liver (Martin et al. 2002), skeletal muscle (Lambertucci et al. 2007), testis (Sahoo et al. 2008b) and kidney (Martin et al. 2002; Thiab et al. 2015)). Thiab et al. (2015) confined their studies to biochemical measurement of antioxidant enzymes in different regions of rat kidney during aging. However, authors have restricted their investigation up to 60 week old rats. It is possible that in old age imbalance among the activities of antioxidant enzymes in tissues may be responsible for the increase in tissue oxidative stress as a consequence of proper clearance of ROS. Recently, through proteomic analyses, it is observed that elevated levels of oxidative and proteolytic proteins are associated with decline in glycolytic proteins and antioxidant enzymes in kidney of aging mice (Amelina and Cristobal 2009).

Our investigation on expression of antioxidant enzymes at mRNA and protein levels suggest that the levels of transcripts and proteins of antioxidant enzymes did not necessarily reflect the changes observed in their respective activities in the tissue. It has been reported that mRNA expression of antioxidant enzymes such as Cu/Zn-SOD, GPx and GR in kidney are elevated in old age with the exception of Mn-SOD and catalase (Martin et al. 2002). The discrepancies between our results and that of Martin et al. (2002) may be due to use of different rat strains and

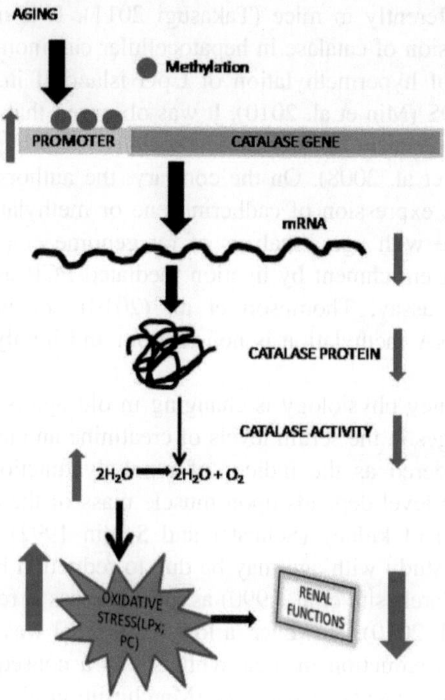

Fig. 6 A schematic diagram of regulation of catalase expression and its implication during aging process

adoption of different methods to assess the level of transcripts. We have used semi quantitative approach to quantify transcripts while Martin et al. (2002) have used real time PCR to quantitate mRNA levels of antioxidant enzymes. On the other hand, our results are in good agreement with that of Rao et al. (1990) where authors have reported a decrease in mRNA levels of SOD, GPX and catalase in kidney of male Fischer 344 rats in relation to age. It is here to mention that expression of mRNA and protein levels of antioxidant enzymes in Leydig cells of testis do not reflect their respective activities with age (Luo et al. 2006).

The region of catalase promoter (−268 to +52) selected for the present study is characterized by having seventeen CpG dinucleotides and several CCAAT sites. The region is also characterized by the presence of binding sites for several transcriptional factors such as AP1, SP1, C/EBPα, C/EBPβ and GATA1 and GATA2 (Nakashima et al. 1989; Taniguchi et al. 2005) and as evident by bioinformatic study mentioned in materials and methods. Observed inverse relationship between expression of catalase and methylation of its promoter in tissues in relation to age implies an important role of methylation in expression of the gene. Several evidences suggest that DNA methylation affects gene expression in age dependent manner (Issa 2000). It is reported that DNA methylation status progressively moves towards maturation status throughout the lifespan and modulates transcriptional status of tissues differently in mice (Takasugi 2011). It is not out of context to mention that expression of catalase in hepatocellular carcinoma cells is reported to reduce as a result of hypermethylation of CpG island II in its promoter due to augmentation of ROS (Min et al. 2010). It was observed that loss of expression of rat kidney cadherin gene in old age was associated with hypermethylation of its promoter (Akintola et al. 2008). On the contrary, the authors failed to notice any change either in the expression of cadherin gene or methylation status of its promoter in liver tissue with age. Analysis of rat genome employing genome-wide *Hpa*II tiny fragment enrichment by ligation mediated PCR assay and the luminometric methylation assay, Thompson et al. (2010) concluded that epigenetic dysregulation of DNA methylation is non-random and highly tissue specific with respect to aging.

The fact that kidney physiology is changing in old age is further supported by the significant changes in the serum levels of creatinine and urea. Serum creatinine and urea are considered as the indices of renal dysfunction (Lindeman 1990). Generally creatinine level depends upon muscle mass of the body and glomerular filtration rate (GFR) of kidney (Schuster and Seldin 1992). Increased creatinine level in the present study with age may be due to reduction in GFR of the kidney (Lindeman 1990; Marchesini et al. 1990) as muscle mass is reported to decrease in old age (Paturi et al. 2010). However, a low SUN level was recorded in old rats. This may be due to reduction in urea synthesis as a consequence of age-related decline in functional capacity of the liver (Marchesini et al. 1990).

Taken together the results of the present study suggest that deterioration in functional aspect of kidney in old age is a consequence of redox imbalance contributed by both the elevation of oxidative stress indices and declined antioxidant potential. And epigenetic changes in promoter region of antioxidant enzymes in general and catalase in particular may play a role in their expression, thereby reducing the antioxidant potential of the cells and tissues.

Acknowledgments This work was supported by Department of Biotechnology, Government of India. The authors are thankful to the Head, Department of Biotechnology, Utkal University, India for providing the laboratory facilities. The author is also gratified to the Department of Biotechnology, Government of India for providing the fellowship (Ref. No. DBT-JRF/05-06/123). SKB is grateful to CSIR, India for providing fellowship (Ref. No. CSIR/JRF/09/173(0128)/2010-EMR-I.

References

Aebi H (1974) Catalase. In: Bergmayer HU (ed) Methods of enzymatic analysis, vol II. Academic press, New York, pp 673–683

Akintola AD, Crislip ZL, Catania JM, Chen G, Zimmer WE, Burghardt RC, Parrish AR (2008) Promoter methylation is associated with the age-dependent loss of N-cadherin in the rat kidney. Am J Physiol Renal Physiol 294:F170–F176

Almeida H, Magalhaes MC, Magalhaes MM (1998) Age-related changes in lipid peroxidation products in adrenal gland. Age 21:119–121

Amelina H, Cristobal S (2009) Proteomic study on gender differences in aging kidney of mice. Proteome Sci 7:16

Baud L, Ardajlou R (1993) Involvement of reactive oxygen species in kidney damage. Br Med Bull 49:621–629

Bradford MM (1976) A rapid and sensitive method for the quantitation of microgram quantities of protein utilizing the principle of protein-dye binding. Anal Biochem 72:248–254

Choksi KB, Nuss JE, Boylston WH, Rabek JP, Papaconstntinou J (2007) Age related increase in oxidatively damaged proteins of mouse kidney mitochondrial electron transport chain complexes. Free Radic Biol Med 43:1423–1438

Cohen G, Dembiec D, Marcus J (1970) Measurment of catalase activity in tissue extracts. Anal Biochem 34:30–38

Das K, Samanta L, Chainy GBN (2000) A modified spectrophotometric assay of superoxide dismutase using nitrite formation of superoxide radicals. Indian J Biochem Biophys 37: 201–204

Farooqui MY, Day WW, Zamorano DM (1987) Glutathione and lipid peroxidation in the aging rat. Comp Biochem Physiol B 88:177–180

Gough DR, Cotter TG (2013) Hydrogen peroxide: a jekyll and hyde signaling molecule. Cell Death and Dis 2:e213

Gunduz F, Senturk UK, Kuru O, Aktekin B, Aktekin MR (2004) The effect of one year's swimming exercise on oxidative stress and antioxidant capacity in aged rats. Physiol Res 53:171–176

Halliwell B, Gutterige JMC (2001) Free radicals in biology and medicine, 3rd edn. Oxford University press Inc., New York

Harman D (1956) Aging: a theory based on free radical and radiation chemistry. J Gerontol 11:298–300

Ichikawa I, Kiyama S, Yoshioka T (1993) Renal antioxidant enzymes: their regulation and function. Kidney Int 45:1–9

Issa JP (2000) CpG-island methylation in aging and cancer. Curr Top Microbiol Immunol 249:101–118

Kregel KC, Zang HJ (2007) An integrated view of oxidative stress in aging: basic mechanisms, functional effects and pathological considerations. Am J Physiol Regul Integr Comp Physiol 292:18–36

Lambertucci RH, Levada-Pires AC, Rossoni LV, Curi R, Pithon-Curi TC (2007) Effects of aerobic exercise training on antioxidant enzyme activities and mRNA levels in soleus muscle from young and aged rats. Mech Ageing Dev 128:267–275

Levine RL, Garland CD, Oliver CN, Amici A, Climent I, Lenz AG, Ahn BW, Shaltiel S, Stadtman ER (1990) Determination of carbonyl content in oxidatively modified proteins. Methods Enzymol 186:464–478

Lindeman RD (1990) Overview: renal physiology and pathophysiology of aging. Am J Kidney Dis 4:275–282

Luo L, Chen H, Trush MA, Show MD, Anway MD, Zirkin BR (2006) Aging and the Brown Norway rat Leydig cell antioxidant defense system. J Androl 27:240–247

Marchesini G, Bianchi GP, Fabbri A, Lolli R, Bugianesi E, Zoli M, Pisi E (1990) Synthesis of urea after a protein-rich meal in normal man in relation to ageing. Age and Ageing 19:4–10

Martin JE, Sheaff MT (2007) Renal aging. J Pathol 211:198–205

Martin R, Fitzl G, Mozet C, Martin H, Welt K, Wieland E (2002) Effect of age and hypoxia/reoxygenation on mRNA expression of antioxidative enzymes in rat liver and kidneys. Exp Gerontol 37:1481–1487

Massey V, Williams CH (1965) On the reaction mechanism of yeast glutathione reductase. J Biol Chem 40:4470–4480

Min JY, Lim SO, Jung G (2010) Downregulation of catalase by reactive oxygen species via hypermethylation of CpG island II on the catalase promoter. FEBS Lett 584:2427–2432

Mitsui A, Ohta T (1961) Photooxidative consumption and photoproductive formation of ascorbic acid in green leaves. Plant Cell Physiol 2:31–44

Miyazawa M, Ishii T, Yasuda K, Noda S, Onouchi H, Hartman PS, Ishii N (2009) The role of mitochondrial superoxide anio (O^{2-}) on physiological aging in C57BL/6 J mice. J Radiat Res 50:73–82

Nakashima H, Yamamoto M, Goto K, Osumi T, Hashimoto T, Endo H (1989) Isolation and characterization of the rat catalase-encoding gene. Gene 79:279–288

Ohkawa H, Ohishi N, Yagi K (1979) Assay for lipid peroxides in animal tissues by thiobarbituric acid reaction. Anal Biochem 95:352–358

Paglia DE, Valentine WN (1967) Studies on the quantitative and qualitative characterization of erythrocyte glutathione peroxidase. J Lab Clin Med 70:158–169

Paturi S, Gutta AK, Katta A, Kakaria SK, Arvapalli RK et al (2010) Effects of aging and gender on muscle mass and regulation of Akt-mTOR-p70s6k related signalling in the F344BN rat model. Mech Ageing Dev 131:202–209

Rao G, Xia E, Richardson A (1990) Effect of age on the expression of antioxidant enzymes in male Fischer F344 rats. Mech Ageing Dev 53:49–60

Richardson B (2003) Impact of aging on DNA methylation. Ageing Res Rev 2:245–261

Sahoo A, Chainy GBN (1997) Alterations in the activities of cerebral antioxidant enzymes of rat are related to aging. Int J Dev Neurosci 15:939–948

Sahoo DK, Roy A, Chainy GBN (2008a) Protective effects of vitamin E and curcumin on L-thyroxine-induced rat testicular oxidative stress. Chem Biol Interact 176:121–128

Sahoo DK, Roy A, Chainy GBN (2008b) Rat testicular mitochondrial antioxidant defense system and its modulation by aging. Acta Biol Hung 59:413–424

Sambrook J, Russell DW (2001) Molecular cloning: a laboratory manual. Cold Spring Harbor Laboratory Press, Cold Spring Harbor, NY

Schuster VL, Seldin DW (1992) Renal clearance. In: Seldin DW, Giebisch G (eds) The kidney: physiology and pathology, 2nd edn. Raven Press, New York

Sedlak J, Lindsay RH (1968) Estimatioin of total, protein-bound and nonprotein sulfhydryl groups in tissues with Ellman's reagent. Anal Biochem 25:192–205

Takasugi M (2011) Progressive age dependent DNA methylation changes starts before adulthood in mouse tissue. Mech Ageing Dev 132:65–71

Taniguchi M, Hashimoto M, Hori N, Sato K (2005) CCAAT/ enhancer binding protein-β (C/EBP-β), a pivotal regulator of the TATA-less promoter in the rat catalase gene. FEBS Lett 579:5785–5790

Thiab NR, King N, Jones GL (2015) Effect of ageing and oxidative stress on antioxidant enzyme activity in different regions of rat kidney. Cell Biochem, Mol. doi:10.1007/s11010-015-2503-2

Tian L, Cai Q, Wei H (1998) Alterations of antioxidant enzymes and oxidative damage to macromolecules in different organs of rats during aging. Free Radic Biol Med 24:1477–1484

Thompson RF, Atzmon G, Gheorghe C, Liang HQ, Lowes C, Greally JM, Barzilai N (2010) Tissue-specific dysregulation of DNA methylation in aging. Aging Cell 9:506–518

Van der Loo B, Bachschmid M, Labugger R, Kilo J, Hahn R, Palacios-Callender M, Lüscher TF (2005) Expression and activity patterns of nitric oxide synthases and antioxidant enzymes reveal a substantial heterogeneity between cardiac and vascular aging in the rat. Biogerontol 6:325–334

Van Otterdijk SD, Mathers JC, Strathdee G (2013) Do age-related changes in DNA methylation play a role in the development of age-related diseases? Biochem Soc Trans 41:803–807

Differential Expression of Long Noncoding RNA in the Rat Brain During Aging

Sukhleen Kour and Pramod C. Rath

Abstract The discovery of genome-wide transcription through high-throughput sequencing in the mammals has revealed that only ~2 % of the genome is expressed into protein-coding mRNAs and the rest ~98 % makes different types of intergenic, intronic and repeat sequence-rich, small and long regulatory noncoding RNAs with multitude of biological functions. The complexity of mammalian brain has been largely attributed to diverse region-specific transcriptomes, a major portion of which has been recently found to consist of innumerable forms of long (>200 nt.) noncoding RNAs (lncRNAs), implicated in various functions such as brain development, cell-lineage specification, learning and memory. However, their relative association with processes involved in aging and age-related disorders has not been sufficiently explored. Here, we have characterized a repeat sequence containing long intergenic noncoding RNA (lincRNA), *LINC-RBE* (rat brain expressed) from the rat genome, which is differentially expressed in the brain during maturation and aging. Through expression analysis, *LINC-RBE* was shown to express in specific cell types and neuroanatomical compartments, e.g., cortex, hippocampus and cerebellum of the rat brain in an age-dependent manner. Thus, our study showed the possible interrelationship between lincRNAs and various brain functions during aging, which may provide an alternative basis to study various age-related neurological diseases and disorders.

Keywords Aging · Brain · Noncoding DNA · Long noncoding RNA · Bioinformatics · RNA expression

S. Kour · P.C. Rath (✉)
Molecular Biology Laboratory, School of Life Sciences,
Jawaharlal Nehru University, New Delhi 110067, India
e-mail: pcrath@mail.jnu.ac.in

© Springer Science+Business Media Singapore 2017

P.C. Rath et al. (eds.), *Topics in Biomedical Gerontology*,
DOI 10.1007/978-981-10-2155-8_10

1 Introduction

Multicellular diploid organisms develop either from single diploid cells by asexual methods such as vegetative growth of totipotent cells or from haploid cells by sexual method by a process of fertilization of a female gonadal cell, the mature oocyte by a male gonadal cell, the sperm. The latter mechanism has been selected for mammals during evolution. In this process, the oocyte or the mother cell contributes the essential genomic information along with the mitochondria, while the sperm or the father cell contributes the variations in the genomic information to the next generation and thereby to the gene pool of the population. Along with this, two more major genetic contributions are also made during the process of fertilization. The first is by the transposable elements (TEs) and the second is by the noncoding RNAs present in the haploid cells. These two components of the genome are now known to contribute to both genetic and epigenetic aspects of the zygotic cell, which determines certain aspects of the subsequent steps of the embryonic development. Following embryonic development, the multicellular organism undergoes functional specialization and maturation to adulthood functions. Reproduction is one of the major functions of adulthood. Soon after attainment of adulthood or sexual maturity to reproduce itself, the organism slowly starts experiencing the aging process. Aging sets in slowly but surely. Aging is manifested in all types of cells, organs and organisms, which finally expresses into progressive and cumulative decline in all physiological functions of the cells and tissues. Prominent manifestations among them are the following: decrease in production of reproductive cells ultimately leading to cessation of reproductive ability; weakening of the bones and muscles; wrinkling of the skin; greying of the hairs; decreased physical ability for alertness, movement and capacity to work; decreased mental ability often leading to loss of memory and cognition; lowered immune system; lowered metabolism in almost all the organs; decreased level of confidence and overall loss of physiological balance and coordination among various organ systems of the body. Thus old age makes a person susceptible to many diseases. Often old patients feel less cared for medical treatments. The geriatric medicine is usually either less developed or less cared for in many countries. With all these difficulties, we still notice many old persons keep themselves alive with joyful personality, effective interaction with others and substantial human contribution to their family, immediate neighbourhood and the society at large. Until they remain useful to others, they get attention for their own life. However, many other elderly, due to poor conditions of their health, often become less fortunate in the later part of life. Generally, old age is prolonged for a considerable period of years, thereby adding problems due to lack of quality of life, healthcare issues and family as well as societal responsibilities towards the elderly. Thus a biological natural process of aging of a living organism ultimately gives rise to a multitude of challenges for scientists, clinicians, policy makers and organizations as well as the government. Many individuals and organizations are actively thinking to convert these challenges in the field of aging into opportunities for research on aging, drug-development for elderly and old age-associated diseases, making use of traditional medicine system for healthy aging, innovations in pharmaceutical industry, advancements in clinical practices in geriatric medicine and patient care in hospitals as well as winning the good will and blessings of the elderly populations of the world for the policy makers, organizations, government and society. As biologists, we need to understand the basic mechanism(s) of the aging process in mammals at cellular, biochemical and molecular levels. The present research article is an example of a piece of basic or fundamental research on one of the aspects of the aging process in the rat brain as an experimental model system.

In the mid-20th century, with the discovery of DNA as the genetic material (Watson and Crick 1953), one of the greatest challenges of biology had emerged as how to explain the decoding of information in the form of nucleotide sequence in DNA, a double helical nucleic acid macromolecule, to the stretch of amino acid sequences in proteins, presumed to perform a diversity of cellular functions in all organisms. This was answered following comprehensive experiments and collective insights about the role of a special class of ribonucleic acid molecule, called as 'messenger RNA' (mRNA). It is synthesized from DNA (transcription) and carries information to ribosomes for protein synthesis (translation). This proposed the concept of "central dogma of molecular biology" (DNA makes RNA makes protein) (Amiel 1965; Geiduschek and Haselkorn 1969). Subsequently, it was realized that a rather small fraction of mammalian genome was assigned for all proteins, the rest of the DNA especially the repetitive ($\sim$40 %), intergenic ($\sim$40 %) and intronic ($\sim$15 %) DNA was not represented in proteins. They were inadvertently called as "Junk DNA" with no function. With the sequencing of mammalian genomes during the last two decades (Ananda et al. 2014; Gibbs et al. 2004; Venter et al. 2001; Waterston et al. 2002) together with the advancements in the high-throughput genomic screening techniques such as microarrays for global gene expression analysis and next generation sequencing (NGS) (Capobianco 2014; Kawaji et al. 2014) and data acquired from ENCODE (Encyclopedia of DNA elements) (Birney et al. 2007; Siggens and Ekwall 2014) and FANTOM (Functional annotation of mammalian cDNAs) consortia (Carninci et al. 2005; Liang et al. 2015; Lizio et al. 2015), a whole new revelation was made. The $\sim$98 % of the non-protein coding DNA or "dark matter" of the genome is now known to pervasively transcribe into multifunctional forms of non-protein coding RNA molecules, called as noncoding RNAs (ncRNAs), and only $\sim$1.5–1.8 % consists of protein-coding sequences (Denas et al. 2015; Guttman et al. 2009; Iyer et al. 2015; Quek et al. 2015). Thus, instead of having only protein-coding genes with multiple transcription start sites, alternative promoter and enhancer elements, and variable 5′ and 3′-upstream regions, maximum portion of mammalian genome coding for myriad of complex ncRNAs of intergenic and intronic origins define the complexity and flexibility of the genome with respect to its evolution (Liang et al. 2015; Marques and Ponting 2014). Until now, many of these ncRNAs have been functionally associated with almost all biological processes from development, lineage specification, differentiation (Ballarino et al. 2015; Cajigas et al. 2015; Gong et al. 2015; Shahryari et al. 2015), cellular reprogramming (Flynn and Chang 2014; Kim et al. 2015; Loewer et al. 2010) to cell proliferation, growth and cell death. Repetitive sequences, besides being prevalently associated with heterochromatin and telomeric organization (Arning et al. 2015; Chow et al. 2010; Lee 2011; Redon et al. 2013; Saksouk et al. 2014), X-chromosome inactivation (dosage compensation), translational regulation, neo-centromere formation and generation of small regulatory RNAs (Faulkner et al. 2009; Hoffman et al. 2014; Kapusta et al. 2013), have also emerged as an important source of noncoding RNAs in the mammalian genome (Singh and Rath 2012), however, still their functional relevance in the genome is largely unknown (Faulkner et al. 2009). The revised "central dogma of molecular biology" now includes the ncRNAs as regulatory molecules (Fig. 1) (Wahlestedt 2013).

Noncoding RNAs, based on their expression and function, are divided into two major categories: (a) housekeeping or structural ncRNAs, that include well-characterized infrastructural RNAs such as ribosomal RNAs, transfer RNAs, and small nuclear/spliceosomal RNAs; and (b) regulatory ncRNAs, which comprise of various types of small noncoding RNAs (sncRNAs) such as microRNAs or miRNAs ($\sim$19–22 nt.), PIWI-interacting RNAs or piRNAs ($\sim$30–32 nt.), small nucleolar RNAs or snoRNAs, promoter-associated small RNAs or PASRs (Hu et al. 2012) and transcription initiation RNAs or tiRNAs (Zaramela et al. 2014), among others, and a wide range of thousands of long noncoding RNAs (lncRNAs). SncRNAs comprise of substantially characterized and highly conserved fraction of transcripts, primarily implicated in functions such as post-transcriptional modulation of mRNA stability and degradation (Olive et al. 2015; Philippe et al. 2015), dosage compensation of X chromosome (Song et al. 2009) and chromatin organization (Chen et al. 2014). Besides this, recently many of them have been studied for their involvement in aging and age-related diseases in *C. elegans* (*lin-14, mir-71, mi-238 and mi-246*) (Boehm and Slack 2005; de Lencastre et al. 2010) and animal (mouse, human) models (*miR-21, miR-130a* and *miR-494*) (Serna et al. 2012) (Table 1). LncRNAs, on the other hand, are $\geq$200 to several thousand nt. in size (e.g., *XACT*, $\sim$250 kb; *Xist*, $\sim$17 kb) (Mattick and Rinn 2015; Ponting et al. 2009; St Laurent et al. 2015). Also, some recently discovered class of heterogeneous, polyadenylated and 5′-methyl-capped, regulatory RNAs that are spatio-temporally expressed in sense, antisense, overlapping and bi-directional manners from the intronic, intergenic, repeat sequence-rich regions of mammalian genome (Guttman et al. 2009; Porro et al. 2014; Kurokawa 2011; Milligan and Lipovich 2014; Mortimer et al. 2014; Pnueli et al. 2015; St Laurent et al. 2015; Vucicevic et al. 2015). Although most of the lncRNAs are pervasively transcribed in the genome, they show poor sequence conservation and are highly unstable or present in few copies per cell. However, the dynamics, developmental and tissue/cell type-specific expression patterns of lncRNAs are regulated by transcriptional factors (e.g., Oct-4, Nanog, p53 etc.) and epigenetic modifications of histones and DNA methylation, this together with conservation of their promoter sequences strongly propose for their biological function and evolutionary significance (Ponting et al. 2009; Wang and Chang 2011). Their intrinsic tendency to form various secondary structures for functional aspects, is a prerequisite for

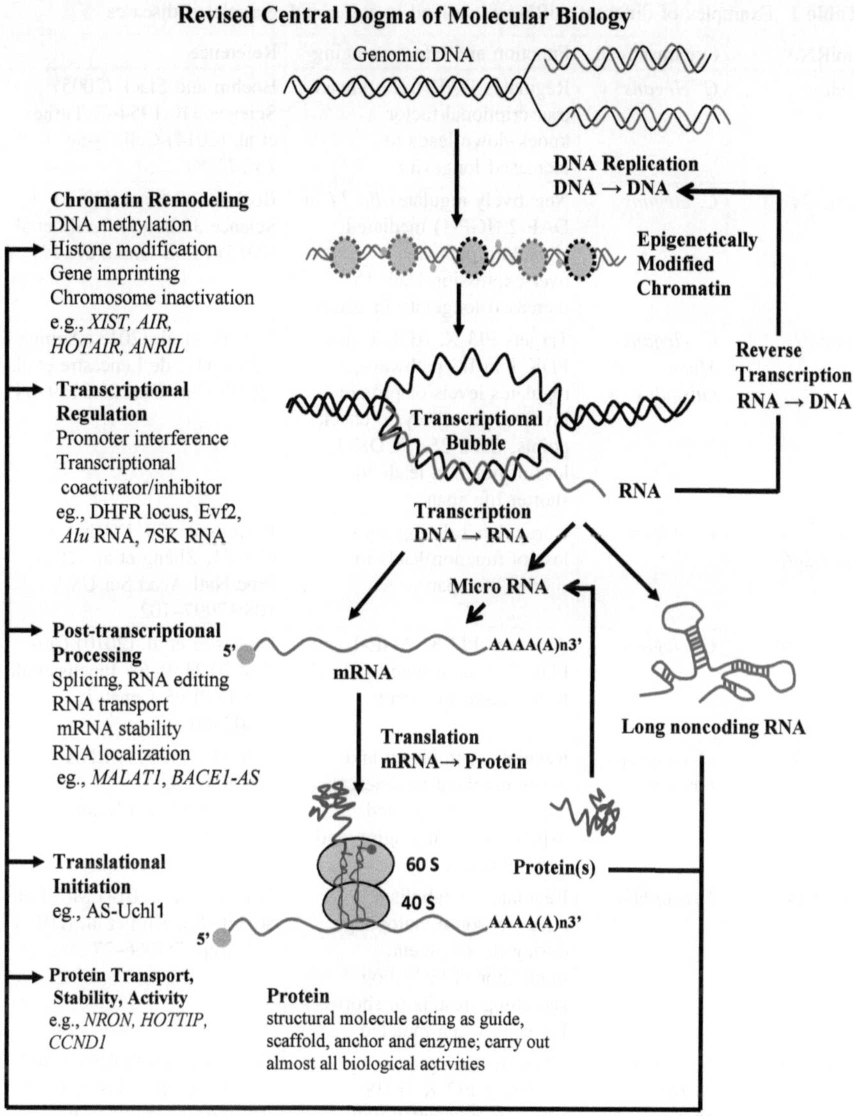

Fig. 1 Schematic representation of the revised "central dogma of molecular biology" depicting the flow of genetic information in eukaryotes and various roles of lncRNAs along with proteins in regulating the steps. The generalized view of the flow of genetic information includes: DNA is the source of all genetic information and from DNA the information is decoded in the form of messenger RNA (mRNA) and then to protein through the ribosomal machinery. The earlier assumption was that proteins perform all functions in a cell for sustenance of life. This concept has been revised with the discovery of various regulatory noncoding RNAs including microRNAs (miRNAs) acting on the mRNAs and long noncoding RNAs (lncRNAs). LncRNAs together with various proteins are involved in maintenance and epigenetic modification of chromatin, transcriptional and post-transcriptional gene regulation, translational initiation, protein metabolism and transport and many others. This gives a whole new concept of exploring the genome and cellular functions in a new dimension (Mattick and Rinn 2015)

Table 1 Examples of different miRNAs involved in aging and age-related diseases

miRNA	Organism/cell	Function and effect on aging	Reference
linc-4	*C. elegans*	Regulates DAF16, Forkhead transcriptional factor, knock-down leads to increased longevity	Boehm and Slack (2005) Science 310:1954–7, Turner et al. (2014) Cell Cycle 13:772–81
linc-14	*C. elegans*	Negatively regulates *lin-14* in DAF 2 (IGF-1) mediated signalling pathway, over-expression leads to increased longevity in adults	Boehm and Slack (2005) Science 310:1954–7, Shi et al. (2013) PLoS One 8:e75475
mir-71	*C. elegans* *Mus musculus*	Targets PI3 K, AGE-1 and PDK 1 in IIS pathways, regulates levels of protein involved in cell cycle check points, CDC-25.1, CDK-1, loss of function leads to shorter life span	Lucanic et al. (2013) Aging 5:394–411, de Lencastre et al. (2010) Curr Biol 20:2159–68
mir-238 *mir-246*	*C. elegans*	Upregulated during aging, loss of function leads to shorter life span	Karp et al. (2011) RNA 17, 639–51, Zhang et al. (2011) Proc Natl Acad Sci USA 108:17997–102
mir-239	*C. elegans*	Activates PI3 K, AGE-1 and PDK 1, loss of function leads to increased longevity	Lencastre et al. (2010) Curr Biol 20:2159–68, Pincus et al. (2011) PLoS Genet 7: e1002306
mir-34	*C. elegans* *Drosophila*	Regulates autophagy and genes involved in senescence and aging, upregulated expression during aging and early dormancy	Kato et al. (2011) RNA 17:1804–20, Liu et al. (2012) Nature 482:519–23
miR-14	*Drosophila*	Regulates metabolism, stress-response, autophagy during development, modulator of hedgehog signalling, results in shorter life span when mutated	Nelson et al. (2014) Mol Cell 56:376–88, Kim et al. (2014) Cell Rep. 7:2066–77
miR-8/miR-200	*Drosophila* *Homo sapiens*	Regulates cell growth by inhibiting PI3 K in IIS pathway, associated with neurogenesis and synapse structure maintenance, implicated in aging by promoting cell growth	Trümbach and Prakash (2015) Cell Tissue Res 359:161–77, Hyun et al. (2009) Cell 139:1096–108
miR-669c	*Mus musculus*	Regulates oxidative defence by targeting glutathione-*S*-transferases activity, elevated levels during mid-age	Maes et al. (2008) Mech Ageing Dev 129:534–41

(continued)

Table 1 (continued)

miRNA	Organism/cell	Function and effect on aging	Reference
miR-709	*Mus musculus*	Targets cytochrome C complex, GSK3β of Wnt/β-catenin signalling, elevated levels during mid-age	Maes et al. (2008) Mech Ageing Dev 129:534–41, Chen et al. (2014) Cell Signal 26:2583–9
miR-93	*Mus musculus*	Targets glutathione-*S*-transferase cytochrome complex and SIRT1, activates PI3 K/Akt signaling pathway, elevated levels in extremely old age	Li et al. (2011a) Mech Ageing Dev 132:75–85, Jiang et al. (2015) Oncotarget 6:8286–99
miR-214	*Mus musculus*	Targets glutathione-*S*-transferases and cytochrome complex in liver, elevated levels in extremely old age, has cardioprotective effect in myocardial ischemia	Maes et al. (2008) Mech Ageing Dev 129, 534–41, Wan et al. (2015) Cell Mol Biol 61:1–6
miR-34a	*Drosophila Mus musculus Homo sapiens*	Targets many pathways: glutathione-*S* transferase, MGST1 and SIRT1, targets transcriptional factors SP-1 and NRF2 in liver, post-transcriptionally regulates p53, increased expression with age and in patients with Alzheimer's disease, induces senescence in endothelial progenitor cells	Liu et al. (2012) Nature 482:519–23, Yu et al. (2015) Cell Death Differ 22:1170–80, Boon et al. (2013) Nature 495:107–10, Rokavec et al. (2014) J Mol Cell Biol 6:214:30, Li et al. (2011b) Aging 3:985–02
miR-217	*Mus musculus Homo sapiens*	Inhibits SIRT1 and deacetylation of FOXO1, increased levels in human atherosclerotic plaques, targets PTEN, upregulated levels induce premature endothelial cell senescence	Menghini et al. (2009) Circulation 120:1524–32, Deng et al. (2014) Cancer Letters 355:184–91
miR-29, miR-22, miR-30	*Mus musculus, Homo sapiens*	Activates Rb tumour suppressor pathway and represses Myb-related protein B encoding gene, upregulated in elderly mouse and during cellular senescence	Dimmeler and Nicotera (2013) EMBO Mol Med 5:180–90, Takahashi et al. (2012) PLoS One 7:e48974
miR-17-92 cluster	*Homo sapiens, Mus musculus*	Downregulates PTEN and suppress IIS pathway, upregulated in many cancers, down regulated in many aging models and several cell types such as human umbilical endothelial cells	Hackl et al. (2010) Aging Cell 9:291–6, Mogilyansky and Rigoutsos (2013) Cell Death Differentiation 20:1603–14; Grillari et al. (2010) Biogerontology 11:501–6

(continued)

Table 1 (continued)

miRNA	Organism/cell	Function and effect on aging	Reference
miR-22, miR-101a, miR-720 miR-721	*Mus musculus*	Target F1-Fo-ATPase and components of the mitochondrial electron transport chain, miR-22 regulates smooth muscle differentiation, elevated levels during brain aging	Li et al. (2011c) Neurobiol Aging 32:944–55, Xu et al. (2011) J Cell Biol 193:409–24
miR-30e, miR-34a miR-181a	*Mus musculus*	Targets BCL-2, a known regulator of apoptosis, under calorie restriction showed age-dependent decrease in brain	Meza-Sosa et al. (2014) Front Cell Neurosci 8:175 doi:10. 3389/fncel.2014. 00175, Khanna et al. (2011) Aging 3: 223–36
miR-7, miR-468, miR-542 miR-698	*Mus musculus*	Involved in muscle differentiation, epithelial-mesenchymal transition and many cancers, elevated expression in aging muscles	Hamrick et al. (2010) Biochem Biophys Res Commun 400:379–83
miR-124a, miR-181a, miR-221, miR-382, miR-434 and miR-455	*Mus musculus*	Muscle differentiation, decreased expression levels during aging	Smith-Vikos and Slack (2012) J Cell Sci 125:7–17, Hamrick et al. (2010) Biochem Biophys Res Commun 400:379–83
let-7	*Homo sapiens, Drosophila melanogaster*	Regulates cyclin-dependent kinase 6 (CDK6), CDC25A and CDC34, PAX7; targets molecules of Insulin/IGF-1/mTOR pathway via binding with *Lin28*, downregulated with increasing age	Toledano et al. (2012) Nature 485:605–10, Drummond et al. (2011) Physiol Genomics 43, 595–03
miR-15b miR-24 miR-25 miR-141	*Homo sapiens diploid fibroblast cells*	Targets MAP2K4, component of MAPK pathway, miR-141 regulates HOTAIR lncRNA expression in human cancer cells, upregulated levels in senescent fibroblasts	Chiyomaru et al. (2014) J Biol Chem 289:12550–65, Noren Hooten et al. (2010) PLoS ONE 5:e10724, Marasa et al. (2009) Sci Signal 2:ra69
miR-22	*Homo sapiens*	Regulates bone formation by targeting mimecan/osteoglycine (OGN) and SIRT1, induced with increasing life span, causes senescence and migration of cardiac fibroblasts	Jazbutyte et al. (2013) Age 35:747–62, Zhao et al. (2015) Arterioscler Thromb Vasc Biol 35:918–29, Zheng et al. (2014) Cell Physiol Biochem 34:1547–55

(continued)

Table 1 (continued)

miRNA	Organism/cell	Function and effect on aging	Reference
miR-146a and b	*Diploid fibroblast cells*	Regulates IRAK expression in heart by targeting inflammatory mediators, IL-6 and IL8, upregulated expression during aging in skeletal muscle and in senescent aortic endothelial cells	Vasa-Nicotera et al. (2011) Athero-sclerosis 217:326–30, Chassin et al. (2012) EMBO Mol Med 4:1308–19, Gao et al. (2015) J Immunol 195:672–82
miR-144	*Homo sapiens, Macau maculate, chimp*	Involved in spinocerebellar ataxia type 1 by targeting ataxin 1, induced expression during aging	Persengiev et al. (2011) Neurobiol Aging 32:2316. e17–27
miR-470, miR-669b miR-681	*Ames dwarf mice*	Regulates insulin pathway by suppressing IGF1R levels and AKT phosphorylation, highly upregulated in aging hippocampus	Liang et al. (2011) Aging Cell 10:1080–88
miR-27a	*Ames dwarf mice*	Target genes involved in glutathione metabolism, urea cycle and polyamine synthesis, increased expression is responsible for 70 % increase in longevity compared to normal mouse	Bates et al. (2010) Aging Cell 9:1–18, Dellago et al. (2013) Aging Cell 12:446–58
miR-1	*Mouse model of Hutchinson Gilford progeria syndrome*	Targets IGF-1, higher levels in liver leads to characteristic accelerated aging of the syndrome	Mariño et al. (2010) Proc Natl Acad Sci USA 107:16268–73
miR-107	*Mus musculus, Homo sapiens*	Targets Progranulin, decreased expression in Alzheimer's and frontotemporal dementia	Wang et al. (2010) Am J Pathol 177:334–45
mir-29	*Mouse model of Hutchinson Gilford progeria syndrome, Homo sapiens*	Regulates smooth muscle cell differentiation, stabilizes p53 by suppressing phosphatase PPMID, miR-29 targets type IV collagen genes, BACE1, ARP2/3 actin nucleation complex, increased levels in normal aging and Hutchinson-Gilford progeria mouse model, causes Angiotensin II-induced aneurysms in aged mice, loss of function causes sporadic Alzheimer's	Cushing et al. (2015) PLoS Genet 11:e1005238, Ugalde et al. (2011) EMBO J 30:2219–32, Merk et al. (2012), Martinez et al. (2011) Proc Natl Acad Sci USA 108:522–27

Table 2 Examples of long noncoding RNAs associated with aging and age-related diseases

LncRNA	Organism/cell/tissue	Function and effect during aging	Reference
H19	Human, Rat, Mouse, other vertebrates	Imprinting of IGF2 gene, cell growth and proliferation. Highly expressed during aging	Monnier et al. (2013) Proc Natl Acad Sci USA 110:20693–98, Ratajczak (2012) Folia Histochem Cytobiol 50:171–79
Kcnq1ot1 (KCNQ1-overlapping transcript 1)	Human, Mouse	Gene imprinting, loss of expression in type -2 diabetes and various cancers Lower expression could be implicated in age-related diseases	Redrup et al. (2009) Development 136:525–30, Korostowski et al. (2012) PLoS Genet 8: e1002956
HOTAIR (HOX transcript antisense RNA)	Human, Rat, Mouse, Chimpanzee and Rhesus Monkey	Embryo body plan, imprinting, loss of expression is linked to cancer development and metastasis, upregulated in senescent cells	Gupta et al. (2010) Nature 464:1071–76, Zhang et al. (2015) Oncotarget 6:537–46
ANRIL (antisense non-coding RNA in the INK4 locus)	Human	Antisense to tumour suppressor genes-p16^{INK4A}, p14^{ARF}and p15^{INK4B}, linked with various diseases and cancer, highly expressed in senescent fibroblasts and implicated in aging through regulation of cell growth and proliferation	Bochenek et al. (2013) Human Mol Genet 22:4516–27, Yap et al. (2010) Mol Cell 38:662–74
AIR (antisense Igf2r RNA)	Human, Mouse	Embryo development, gene imprinting of Igf2r, Slc22a2 and Slc22a3, implicated in aging process by regulating Igf2r	Latos et al. (2012) Science 338:1469–72, Santoro F et al. (2013) Development 140:1184–95
ANASSF1 (Antisense to RASSF1)	Human	Antisense to RASSF1, tumour suppressor gene, cell cycle regulation and apoptosis, implicated in aging by regulating RASSF1 expression	Beckedorff et al. (2013) PLoS Genet 9: e1003705
MALAT1 (metastasis associated lung adenocarcinoma transcript 1)	Human, Mouse	Alternative splicing, cell cycle regulation, downregulated during senescence	Abdelmohsen et al. (2013) Aging Cell 12:890–900, Wang et al. (2015) J Biol Chem 290:3925–35

(continued)

Table 2 (continued)

LncRNA	Organism/cell/tissue	Function and effect during aging	Reference
MIAT (Myocardial Infarction Associat-ed Transcript)	Human, Mouse, Chicken	Involved in Brain development, differentiation and alternative splicing, downregulated during senescence	Barry et al. (2014) Mol Psychiatry 19:486–94, Abdelmohsen et al. (2013) Aging Cell 12:890–900
XIST (X inactive specific transcript)	Human, Rat, Mouse, Chimpanzee	X chromosome inactivation, downregulated during senescence	Fukuda et al. (2014) Nat Commun 5:5464, Abdelmohsen et al. (2013) Aging Cell 12:890–900
TERC (Telomere RNA component)	Human, Mouse and other mammals	Telomere maintenance (RNA component of telomerase)., downregulated with aging	Bernardes de Jesus et al. (2012) EMBO Mol Med 4:691–04, Samper et al. (2001) EMBO Rep 2:800–07
TERRA (telomeric repeat containing RNA)	Human, Mouse, and other mammals	Involved in organization and maintenance of telomere, expression increases with increasing age	Wang et al. (2015) Int J Biol Sci 11:316–23, Maicher (2012) Nucleic Acids Res 40:6649–59
ecCEBP [extra coding CEBP (CAAT enhancer-binding protein)]	Human, Mouse	Adipogenesis and cell cycle regulations by regulating CEBP expression, CEBP expression is highly reduced in aged tissues, implicated in aging process by regulating CEBP	Hong et al. (2014) J Biol Chem 289:1106–18, Huggins et al. (2013) Mol Cell Biol 33:3242–58
pRNA	Human, Mouse, Xenopus	Regulates ribosomal RNA (rRNA) expression and protein translation, could be involved in the onset of aging by regulating rRNA expression (nucleolar stress)	Wehner et al. (2014) RNA Biol 11:3–9, Schmitz et al. (2010) Genes Dev 24:2264–69, Machwe et al. (2000) FASEB J 14:1715–24
PTENpg1	Human	Negatively regulates PTEN expression, implicated in aging by regulating cell growth and proliferation	Johnsson et al. (2013) Nat Struct Mol Biol 20:440–46
MEG3 (Maternally expressed gene 3)	Human, Sheep, Mouse	Brain development, differentiation, apoptosis and autophagy, downregulated in Huntington's Disease, mediate senescence by activating p53	Zhang et al. (2013) J Cell Biochem 116:542–50, Yin et al. (2015) Tumour Biol 36:4851–9

(continued)

Table 2 (continued)

LncRNA	Organism/cell/tissue	Function and effect during aging	Reference
PANDA (p21 associated ncRNA DNA damage activated)	Human	Bidirectional lncRNA from CDKN1A promoter, upon DNA damage regulates cell cycle arrest, regulates cellular senescence entry and exit	Puvvula et al. (2014) Nat Commun 5:5323
PINT (p53-induced non-coding transcript)	Mouse	Cell cycle regulation, mediate cellular senescence by regulating p53 signalling	Marin-Bejar et al. (2013) Genome Biol 14: R104
TUG1 (Taurine Up-Regulated 1)	Human, Rat, Mouse, Dog, Cattle	Retina development, linked to various cancers, high levels in aged human brain subependymal zone (SEZ), downregulated in Huntington's disorder	Barry et al. (2015) Front Neurol 6:45, Johnson et al. (2012) Neurobiol Dis 46:245–54
LincRNA-p21 (long intergenic non-coding RNA p21)	Human, Mouse	Cell cycle arrest, causes heterochromatin formation, upon DNA damage induces apoptosis and cellular senescence	Dimitrova et al. (2014) Mol Cell 54:777–90, Bao et al. (2015) Cell Research 25:80–92
7SL	Human	Regulates p53 expression and ER mediated transport, low levels in senescence cells, causes p53 translation and cell cycle arrest	Abdelmohsen et al. (2014) Nucleic Acids Res 42:10099–111
LincROR (lincRNA regulator of reprogramming)	Human	Reprogramming and pluripotency, upon DNA damage prevents p53 translation, could cause senescence by regulating p53 levels	Cheng et al. (2013) Dev Cell 25:1–2, Zhang et al. (2013) Cell Res 23:340–50
BACE1-AS (β site amyloid precursor protein cleavage enzyme-AS)	Human, Mouse	Upregulates *BACE1* expression by increasing its mRNA stability, elevated in Alzheimer's Disease	Modarresi et al. (2011) Int J Alzheimers Dis 2011:929042. doi:10.4061/2011/929042
BC200 (brain cytoplasmic RNA 1)	Human and other Primates	Regulates translation and synaptic plasticity, decreased levels in aged brain but higher in AD brains	Tiedge et al. (1993) J Neurosci 13:2382–90, Mus et al. (2011) Proc Natl Acad Sci USA 104:10679–84

(continued)

Table 2 (continued)

LncRNA	Organism/cell/tissue	Function and effect during aging	Reference
naPINK (PTEN Induced Putative Kinase 1)	Human, Drosophila	Antisense to PINK1gene, regulates mitochondrial function, dopamine release and motor function, increased levels in Parkinson's Disease (PD)	Scheele et al. (2007) BMC Genomics 8:74
HTTAS_V1(HTT-antisense transcript exon VI)	Human	Negatively regulates Huntingtin gene (HTT) expression, extremely lower levels in HD brain cortex	Chung et al. (2011) Human Mol Genet 20:3467–477
CTNI-NATs	Human, Rat	Regulates cardiac troponin I expression, required for normal cardiac function, age-dependent differential antisense to sense expression	Podlowski et al. (2002) J Cell Biochem 85:198–07
17A	Human	Regulates GABA receptor mediated signalling pathways, increased expression in AD brains following inflammation	Massone et al. (2011) Neurbiol Dis 41:308–17
Lethe	Mouse	Induced by NF-κB and negatively regulates its expression and associated with inflammation, downregulated with aging	Atianand et al. (2014) Trends Mol Med 20:623–31
THRIL	Human	Causes inflammation by inducing TNF-α levels via binding with hnRNP L, causes age-related inflammatory response	Li et al. (2015) Proc Natl Acad Sci USA 111:1002–7
LINC-RBE	Rat	Expression decreases in the old brain; cortex, hippocampus and cerebellum show decreased expression during aging	Kour and Rath (2015) Int J Dev Neurosci 46:55–66
LINC-RSAS	Rat	Expression decreases in the old brain; cortex, hippocampus and cerebellum show decreased expression during aging	Kour and Rath (2016) Mol Neurobiol (Jan 11 Epub ahead of print)

RNA-protein/RNA-chromatin interactions, this provides them with sensory, guiding, scaffolding and allosteric capacities (Flintoft 2013; Gonzalez-Buendia et al. 2015; Mortimer et al. 2014; Somarowthu et al. 2015). Their biological functions include sex-dependent dosage compensation, gene imprinting (McHugh et al. 2015; Mercer and Mattick 2013; Rinn 2014), chromatin modification and heterochromatinization (Bao et al. 2015; Khalil et al. 2009; Marchese and Huarte 2014; Mercer and Mattick 2013; Wang et al. 2015), epigenetic regulation of gene expression, post-transcriptional modulation of gene expression, via regulation of alternative splicing, RNA editing, RNA transport (Singh 2012), maintenance of nuclear structure, translational regulation by modulating RNA stability and degradation (Legnini et al. 2014) (Huarte et al. 2010), nuclear-cytoplasmic protein trafficking (Hu et al. 2014; Willingham et al. 2005) and many others (Fig. 1). The altered expression of lncRNAs, therefore, has been linked to the progression and prognosis of various cardiovascular and neurological diseases and disorders along with development of various types of cancers (Akula et al. 2014; Sigdel et al. 2015; Xue et al. 2015; Yang et al. 2014). Recently, many lncRNAs have been found to regulate the onset of cellular senescence or various genes/gene networks/processes directly associated with the progression of aging and age-related disorders (Table 2), however, direct implication of repeat sequence containing lncRNA in the aging process has not yet been fully explored. We have recently reviewed the possible link between various lncRNAs, associated with chromatin modulation, telomeric maintenance, p53-mediated cell cycle regulation, with the onset and pathophysiology of aging and age-related neurological, cardiovascular and immunological diseases/disorders (Kour and Rath 2016c).

In mammals, the higher order sensory-regulatory, cognitive and behavioral functions are performed by the complex, dynamic and intricate networks of neurons and glial cells in the brain or central nervous system (CNS). Brain is structurally and functionally heterogeneous in nature but with high levels of coordination among its different functional regions. This has been attributed to various cell types and developmental stage-specific gene expression patterns and their epigenetic regulatory mechanisms such as chromatin remodelling via differential histone modifications and DNA-methylation marks that drive, control and coordinate the gene expression patterns (Graff and Mansuy 2008; Weichenhan and Plass 2013). In a recent study, many cell/tissue/region-specific and spatio-temporally expressed long noncoding RNAs, which bind and recruit epigenetic regulatory enzyme complexes and other transcriptional factors, have been characterized to be essential for plethora of brain functions (Goff et al. 2015; Guennewig and Cooper 2014; Khalil et al. 2009; Lasalle et al. 2013; Mercer et al. 2008) such as brain development (Feng et al. 2006; Lin et al. 2014a; Lv et al. 2013), differentiation (Lin et al. 2014a; Mercer et al. 2010; Ramos et al. 2015), myelination (Lin et al. 2014b), synaptic transmission, strength and plasticity (Bernard et al. 2010), learning and memory, neurogenesis (Aprea et al. 2013; Ng et al. 2013). They are also dysregulated in many neurological diseases (Clark and Blackshaw 2014; Johnson 2012; Roberts et al. 2014b; Ziats and Rennert 2013). However, involvement of lncRNAs in brain aging and age-associated neurological diseases needs further investigation.

In our laboratory, a novel 1339 bp long, repeat sequence containing cDNA, named as *LINC-RBE* (long intergenic noncoding RNA-rat brain expressed; Accession no. GQ463152) has been isolated by screening of a rat testis λgt11 cDNA library by using a 227 bp rat genomic simple repeat DNA (Accession No. X97459) as a probe (Bajaj 2002; Dey 2000; Dey and Rath 2005). Bioinformatically, *LINC-RBE* (cDNA) was characterized as a trans-spliced non-protein coding transcript from the rat chromosome 5 and 3, with ~11.7 % of different types of repeat sequences (GA, CA and SINE B2/B4) (Mishra 2009). Through expression analysis by RT-PCR and Northern blotting, *LINC-RBE* was found to be strongly expressed in multiple rat tissues and when used as a probe, detected various large cellular RNAs in the size range of 10–0.2 kb as well as small RNAs of 20–30 nt., which is in accordance with the in silico based sequence homology of *LINC-RBE* (cDNA) with eight rat-specific piRNAs, therefore, suggesting their possible processing from the larger transcript(s) and playing role as precursor transcripts in the generation of the small regulatory RNAs. Thus, *LINC-RBE* represented a class of repeat sequence containing lncRNAs from the rat genome with unknown function. In the present study, analysis of *LINC-RBE* expression in different neurobiologically distinct compartments of the brain and their age-dependent functional significance was studied in 4 weeks (young), 16 weeks (adult) and 70 weeks (old) rats. Through RT-PCR and RNA in situ hybridization, *LINC-RBE* was reported to be expressed distinctly and differentially as a sense transcript in the cortex, hippocampus and cerebellum regions of the rat brain. *LINC-RBE* expression increased from young to adult and decreased from adult to old. Therefore, it suggests that repeat-rich lincRNAs could be crucially linked to various complex functions of the brain during maturation from young to adult phases of life, and they are also profoundly altered in the brain during aging from adulthood to old age (Kour 2015).

2 Materials and Methods

2.1 Bioinformatic Analyses

The genomic organization of *LINC-RBE* cDNA, its chromosomal location, sequence conservation among different vertebrates and homology with other RNA sequences was analyzed by using ESEMBL, UCSC genome browser (http://genome.ucsc.edu/cgibin/hg Blat) and BLAST (http://www.Ncbi.nlm.nih.gov/BLAST) search engines. Its coding potential was found through ORF finder (www.ncbi.nlm.nih.gov/gorf/gorf.htm) and Codon Potential Calculator (CPU) (http://cpc.cbi.pku.edu.cn/docs/terms.jsp), respectively. The functional regulatory elements (putative promoter region, transcriptional factor binding sites, polyA signal, presence of untranslated region or UTR element and possible miRNA target site) in the 2000 bp up/downstream genomic region of *LINC-RBE* was scanned by using PromoScan (www.bimas.cit.nih.gov/molbio/proscan), GENSCAN (http://genes.mit.edu/GENSCAN.html),

RegRNA2.0 (http://regrna2.mbc.nctu.edu.tw/), and Promo3 (http://alggen.lsi.upc.es/cgi-bin/promo_v3), respectively.

2.2 *Semi-quantitative Reverse Transcription Polymerase Chain Reaction (RT-PCR)*

Most of the molecular biology methods were performed as per the instructions given in the Sambrook's molecular cloning laboratory manual (Sambrook 2001). The 4 weeks (young), 16 weeks (adult) and 70 weeks (old) male rats were anesthetized with 80 mg/kg Ketamine and 20 mg/kg Xylazine injection (as approved by the Institutional Animal Ethics Committee) and the whole brains were collected. Total RNA was isolated from the brains of 4 weeks (young), 16 weeks (adult) and 70 weeks (old) Wistar rats (*Rattus norvegicus*) by following the Trizol-method as recommended by the manufacturer (Sigma-Aldrich). Prior to cDNA synthesis, 10 μg of total RNA was treated with RNase-free DNase I (Sigma-Aldrich) to remove the DNA contamination and then purified by phenol-chloroform extraction method. The cDNA was synthesized by reverse transcription from 1.0 μg of DNase-treated RNA in a 25 μl reaction mixture containing 500 ng of oligo-(dT) primer, 0.5 mM dNTPs, 1× M-MLV RT-reaction buffer, 20 U RNasin and 100 U of M-MLV Reverse Transcriptase at 37 °C for 1 h. For the amplification of *LINC-RBE* and GAPDH (a positive internal control) cDNAs, PCR was carried out by using 2.5–5.0 μl of cDNA in a 25 μl PCR-reaction mixture containing 1× Taq-buffer with 2 mM $MgCl_2$, dNTPs (0.2 mM), Taq DNA polymerase (1 U) and 25 pmol of primer pair specific for *LINC-RBE* (RBE Fwd: 5′CCCAAAATGAGCAAGTAAGGAA3′ and RBE Rev: 5′TGTCAACAGAAGCCCTTTTTCA3′) and GAPDH mRNA (GAPDHFwd: 5′ ACCACAGTCCATGCCATCAC3′; GAPDH Rev: 5′ TCCACCACCCTGTTGCT GTA3′). The amplification conditions used were as follows: initial denaturation at 95 °C for 4 min; followed by 35 cycles of denaturation at 95 °C for 45 s, annealing at 54 °C (*LINC-RBE*) or 60 °C (GAPDH) for 45 s and extension at 72 °C for 45 s and a final extension at 72 °C for 10 min.

2.3 *Strand-Specific Expression by RT-PCR*

For strand-specific RT-PCR, cDNA was synthesized from 1 μg DNase I-treated total RNA at 50 °C for 1 h by using 0.7–0.9 μM of either sense- or antisense strand-specific primer in a 25 μl reaction mixture containing dNTPs (1 mM), RNase-inhibitor (40 U), 1× M-MLV RT-buffer, M-MLV reverse transcriptase (200 U). The synthesized cDNA was PCR-amplified by using *LINC-RBE* specific primer pairs for total of 35 cycles under conditions as follows: denaturation at 94 °C for 45 s, primer annealing at 54 °C for 45 s and extension at 72 °C for 45 s. The amplification from the oligo-(dT) primer synthesized cDNA was used as a positive control.

2.4 *RNA In Situ Hybridization in Rat Brain Tissue*

2.4.1 Gelatin-Coated Slide Preparation

The glass slides, pre-cleaned with 0.2 % HCl and DEPC-treated H_2O, were incubated with gelatin solution (0.5 % gelatin and 0.05 % chromium potassium sulphate) in a coupling jar for 1–2 min. The excess gelatin was drained out and the slides were dried at 42 °C for 2–3 h and stored at RT for future use.

2.4.2 Tissue Perfusion, Paraffin Wax Embedding and Tissue Sectioning

The brain tissues from 4 weeks (young), 16 weeks (adult) and 70 weeks (old) rats were incubated in 4 % formaldehyde solution for overnight for further fixation and later on dehydrated with a series of increasing ethanol concentrations: 30 % for 30 min., 50 % for 30 min, 70 % for overnight, 95 % for 1 h and absolute ethanol for 1 h at 4 °C. The dehydrated tissue was incubated in xylene and paraffin wax (pre-melted at 60 °C) solution (1:1) at 60 °C for 30 min and subsequently immersed in pre-melted paraffin wax for 3 h at 60 °C. Then ~6–10 μm thick sagittal sections of the wax-moulded tissue were cut by using microtome and the sections were mounted on 0.5 % gelatine-coated slides.

2.4.3 Preparation of RNA Probe by In Vitro Transcription

The cDNA for the noncoding RNA was subcloned with either T_7 or T_3 RNA polymerase promoter region flanking at the 5′-end or 3′-end. The pBluescript KS(+) construct (1 ng) containing such DNA template was PCR-amplified in a 25 μl PCR-reaction consisting of 1 × Taq buffer, $MgCl_2$ (2 mM), dNTPs (0.2 mM), Taq DNA polymerase (1U) and 25 pmol of strand-specific primer pair for *LINC-RBE* (sense primer pair: *RBE*-T7Fwd: 5′TAATACGACTCACTATAGGCGGCCCAAA ATGAG3′; *RBE*-T7Rev: 5′ATGCAATTCT TTGTGTT 3′) and antisense primer pair (*RBE*-T3Fwd: 5′AATTAACCCTCACTAAAGGA TGCAATTCTTTGTGTT3′; *RBE*-T3Rev: 5′CGGCCCAAAATGAG3′), with forward primer containing either T_7 or T_3 promoter sequence at their 5′-end, to amplify the cDNA from both orientations. The PCR conditions used were: 30 cycles of denaturation at 94 °C for 45 s, annealing at 55 °C for 45 s and extension at 72 °C for 1 min, 30 s. Finally, 500 ng of purified PCR-amplified DNA was used for in vitro transcription of the RNA probes at 37 °C for 1 h in a reaction mixture (50 μl) containing T_3/T_7 transcription buffer (1×), DTT (2 mM), rNTP mixture (1 mM each of ATP, GTP, CTP; 0.65 mM of UTP and 0.35 mM of UTP-Digoxygenin (dig), RNase inhibitor (40 U) and 40 U of T_7/T_3 RNA polymerase. The integrity of the RNA product was checked by electrophoresis in a 1.2 % TBE-agarose gel.

2.4.4 Hydrolysis of RNA Probe

The optimal length (~500 bp) of the dig-labelled RNA probes was obtained by hydrolysis method as given by Cox et al. (1984) with few modifications (Cox et al. 1984). Briefly, the in vitro synthesized RNA probe was incubated with two volumes of carbonate buffer (pH 10.2) at 65 °C for optimum time period 't', calculated using the formula, $t = (L_o - L_f)/k\ (L_o)(L_f)$, where t = time in minutes; L_f = final probe size; L_o = initial probe size and k = 0.11. The hydrolysis reaction was stopped by adding equal volume of neutralizing solution (sodium acetate, pH 5.0), the RNA probe was precipitated with 2.5 volumes of ethanol and one tenth volume of 3 M sodium acetate (pH 5.2) and diluted into a 50 % deionized formamide and 4× SSC solution.

2.4.5 RNA In Situ Hybridization

For RNA in situ hybridization, paraffin-embedded tissue sections were dewaxed, hydrated and permeabilized with 5 μg/ml RNase-free Proteinase K solution in 1× PBS-Triton X-100 for 30 min. at 37 °C. After stopping the reaction with 100 mM glycine solution, the sections were post-fixed in 4 % paraformaldehyde for 10 min. at 4 °C, rinsed twice with DEPC-treated 1× PBS and acetylated twice with 0.1 M tri-ethanolamine (TEA) buffer (pH 8.0) containing 0.25 % (v/v) acetic anhydride for 5 min each. The sections were pre-hybridized in a buffer containing 50 % deionized paraformaldehyde, 4× SSC, 1× Denhardt's solution at 50 °C and after 2 h, the solution was replaced with hybridization buffer (50 % deionized paraformaldehyde, 4× SSC, 1× Denhardt's solution, 1 mg/ml denatured herring sperm DNA, 10 % dextran sulphate) containing 30 ng of either sense or antisense digoxygenin-labelled RNA probes and 1 mg/ml tRNA, which was pre-incubated at 65 °C for 5 min. The sections were incubated at 50 °C overnight in a moist chamber. After hybridization, the sections were successively washed in 2× SSC, 1× SSC and 0.1× SSC, treated with 20 μg/ml RNase A to remove any single stranded RNA probe and incubated with blocking solution (100 mM Tris.HCl, pH 7.5, 100 mM NaCl, 2 % BSA and 0.1 % Triton X-100) for 30 min at RT. The dig-labelled hybridized probe was detected by incubating the sections in a blocking solution containing 1:1000 dilution of sheep anti-digoxigenin-alkaline phosphatase-Fab fragment antibody (Roche) for 2 h in a humid chamber followed by development of color due to the alkaline phosphatase activity with the NBT/BCIP substrate solution. The sections were stored at RT for overnight and images were captured under bright field microscope (Nikon-TiS) at 50× and 600× optical magnifications.

2.5 Quantification and Statistical Analysis

For comparison of the RNA expression patterns in the brains of 4 weeks (young), 16 weeks (adult) and 70 weeks (old) rats, the number of in situ hybridization (ISH)-positive cells were counted and normalized with the number of haematoxylin and eosin-stained cells counted from total of nine different areas of the cortex, three from cerebellum and one each from CA1, CA2, CA3 and dentate gyrus of hippocampus. Similarly, by keeping the base value of black and white to 0 and 1, the mean intensity of ISH-positive cells were quantified by using the Nikon NIS-AR software and the percentage mean intensity of expression relative to its expression in the adult was calculated. The significance of RNA expression was determined by using ANOVA at 5 % significance level and the comparison of the significance between different age-groups was carried out with Turkey test. Each experiment was repeated three times.

3 Results

In order to explore biological function of repetitive DNA present in mammalian genome, we had earlier isolated a novel genomic simple repeat DNA sequence (Accession No. X97459) from the Wistar rat (*Rattus norvegicus*) genome, which showed triplex (H-DNA) structure in vitro; RNA-homology with many eukaryotic mRNAs mostly in the 5′-untranslated region (5′-UTR) and 3′-UTR due to presence of simple repeat sequences in the mRNAs and expression of endogenous, large-sized, complementary RNA in the rat brain suggesting constitutive expression of long noncoding RNAs from the repetitive DNA of the rat genome under in vivo conditions (Dey 2000; Dey and Rath 2005). This was one of the initial examples of long noncoding RNA expression in mammalian tissue. In order to clone RNAs with repeat sequences, the X97459 genomic simple repeat DNA was used as a probe to screen cDNA library from the rat and human testis and several repeat sequence containing cDNAs were isolated (Bajaj 2002; Dey 2000). Upon DNA sequencing, bioinformatic analyses and RNA expression studies, these cDNAs represented some long noncoding RNAs from the rat genome (Mishra 2009). *LINC-RBE* is one such repeat sequence containing lncRNA and was originally deposited in the NCBI Genbank as pRT5.5 (Accession no. GQ463152). This was later renamed as *LINC-RBE*. Here we describe sequence analysis and differential expression patterns of *LINC-RBE* in the rat brain by RT-PCR and in different parts of the rat brain by in situ RNA hybridization in the young (4 weeks), adult (16 weeks) and old (70 weeks) rats, which has been recently reported (Kour and Rath 2015). The results are discussed in the light of an example of the lincRNAs, differentially expressed in the cortex, hippocampus (subregions: CA1, CA2, CA3 and dentate gyrus) and cerebellum regions of the mammalian brain in an age-dependent manner and its possible functional significance during maturation and aging of the brain.

This example argues in favour of the possible role of lncRNAs in maturation and aging of mammalian brain.

3.1 Sequence Characterization of LINC-RBE *in the Rat Genome*

The sequence analysis of any given transcript in relation to its genomic location, homology to other functional RNAs, evolution and conservation over time among different species, presence of putative regulatory elements in its up/downstream genomic region provide sufficient basic information necessary for analyzing its biological function in an organism. Similarly, we tried to find out the functional sequence characteristics of *LINC-RBE* in rat genome.

By using rat Mar. 2012 (RGSC 5.0/rn5) assembly on UCSC Genome browser, we found that the 5′1–1213 nt. (~92 %) region of *LINC-RBE* (cDNA) sequence is homologous to the intergenic region of rat chromosome 5 (Chr.5 q33) and the remaining 1213–1334 3′ nt. (8 %) region is homologous to the intergenic region of rat chromosome 3 (Chr.3 q36), therefore, suggesting it to be a trans-spliced intergenic transcript. Also, Chr.5 q33 region, upstream of *LINC-RBE* (genomic DNA), contains transcription sites for various piRNAs (Girard et al. 2006; Lau et al. 2006), eight of which (DQ73550, DQ628056, DQ752533, DQ747205, DQ619691, DQ751276, DQ748367 and DQ613867) were found to be generated from within the *LINC-RBE* (cDNA) sequence (Fig. 2a; Table 3). Although the sequence comparison study of 5′1–1213 nt. region of *LINC-RBE* (cDNA) in UCSC genome browser showed *LINC-RBE* to be less evolutionarily constrained and moderately conserved among 13 different vertebrate species (Chodroff et al. 2010; Johnsson et al. 2014; Ponjavic and Ponting 2007), the presence of various tandem conserved regions in its upstream region at chromosome 5 and its highly conserved syntenic similarity with the piRNA clusters in the mouse chromosome 4 intergenic region argues for its possible function as a piRNA-precursor long noncoding RNA (Fig. 2b) (Kour and Rath 2015). One of these highly conserved regions, lod = 19, showed homology to four mouse (*Mus musculus*) piRNAs (DQ687281, DQ725370, DQ699388 and DQ726299) and one rat (*Rattus norvegicus*) (DQ618908) piRNA (Fig. 2a).

Furthermore, the scanning of 2000 bp up/downstream sequence of *LINC-RBE* (cDNA) on the chromosome 5 by using RegRNA2.0 and Genomatix Genome analyzer found binding sites for many important transcriptional factors involved in cell growth, proliferation and development, such as Activator Protein-1 (AP-1), SRY (sex determining region Y)-box 9 (Sox9), Retinoic Acid Receptor (RAR-alpha: RXR-alpha), Glucocorticoid Receptor (GR) and many others (Fig. 3b); 14 binding sites of the RNA binding protein, *Musashi*, which is a characteristic translational activator of certain temporally expressed mRNAs in *Xenopus* oocyte and a known context-dependent translational regulator in proliferating mammalian cells through its modulation of cytoplasmic polyadenylation (Rutledge et al. 2014; Sutherland et al. 2013) (Fig. 3c); two polyadenylation signals (Fig. 3b), one within (5′ **AAATAA**ATCCAAACTCCAAATTGCCTTT3′) and one ~903 bp downstream of *LINC-RBE* (genomic DNA) sequence (5′ **AATAA**TCATCTGATGGTTTCATGT TACCTTTGT TT TC3′); and one target site of the miRNA, rno-mir466-b within *LINC-RBE* (genomic DNA) (Fig. 3d). Thus, it suggests the multi-factorial regulation of *LINC-RBE* expression in the rat genome. Besides, a putative promoter region (250 bp) containing the regulatory binding sites for the transcription factors: Nuclear Factor kappa B (NF-κB), Serum Response Factor (SRF), Octamer-binding transcription factor (Oct factors) and Transcription Factor IID (TFIID) at 1250 bp downstream of *LINC-RBE* (genomic DNA) and −26 to −276 bp from the putative transcriptional start site (TSS) (+1) at chromosome 5 was also found by using PromoterScan (Fig. 3a). However, through Codon Potential Calculator (CPC) (Grote et al. 2005) and NCBI Open Reading Frame (ORF) finder tools, we found that *LINC-RBE* has a characteristic feature of a non-protein coding transcript with a low coding potential score of −1.11689 (lod score 31.92) and ORF coverage of 21.73 % (lod score value = 31.92) (Fig. 2c). The predicted small ORFs (128, 292, 132, 138 codon size) showed considerably small sequence homology with any known peptides in the protein database, thus, falling short of the criteria for a polypeptide to be considered as a functional protein (Fig. 2d). Overall, bioinformatic studies have confirmed that *LINC-RBE* is an intergenic lncRNA, which despite being less evolutionarily conserved, could possibly be regulated by various factors and might act as a precursor RNA for various piRNAs in rat cells and tissues.

3.2 *Transcriptional Strand-Specificity of* LINC-RBE *in Adult Rat Brain*

The different modes of transcription of various lncRNAs, such as sense, antisense, overlapping and bidirectional, in the mammalian genome could provide information about their biological role. Therefore, we assessed the transcriptional strand-specificity of *LINC-RBE* in the adult rat brain to gain insight into their possible role by using strand-specific sense or antisense primer during the reverse transcription step of RT-PCR (Fig. 4a). We found that in case of *LINC-RBE,* an amplicon of 545 bp was obtained from the reaction containing cDNA synthesized by using the antisense *LINC-RBE* specific-primer (*RBE*-Fwd), whereas no amplification was obtained in case of the sense-specific primer (*RBE*-Rev) (Kour and Rath 2015). Therefore, based on the complementarity of *RBE-Fwd* primer to the reverse (minus) strand of chromosome 5, we inferred that *LINC-RBE* is transcribed as a sense-transcript from the intergenic region of rat chromosome 5. The amplification from the oligo-(dT) primer synthesized cDNA suggested that *LINC-RBE* is a polyadenylated lincRNA.

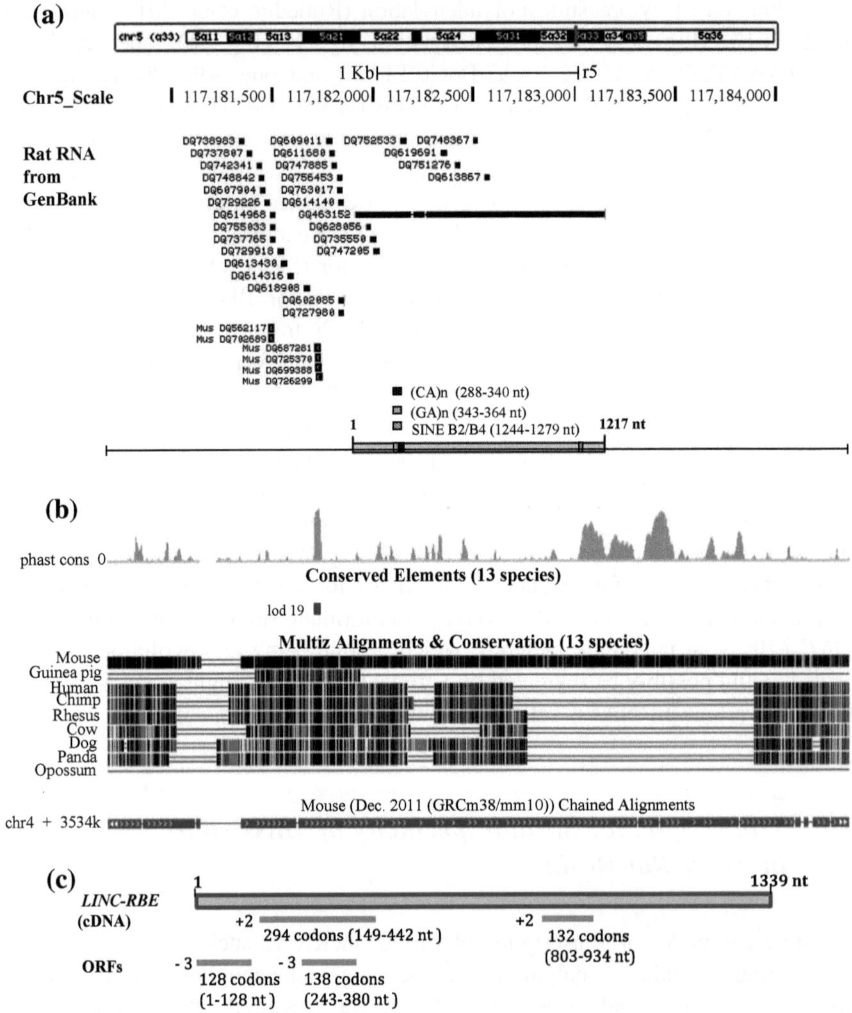

(d)

ID	C/NC	Coding Potential score	ORF coverage	Log Odds score
LINC-RBE (cDNA)	noncoding	-1.11689	21.7%	31.92

Fig. 2 Sequence analysis of *LINC-RBE* (cDNA). **a** Sequence homology of 1339 bp *LINC-RBE* (cDNA) (5′-1217 bp region) to the rat chromosome 5 (q33.3) and various rat and mouse piRNAs (already described in the rat genome database) transcribed from this region. **b** Conservation studies using UCSC Genome browser showed *LINC-RBE* (cDNA) to be moderately conserved among vertebrates with one highly conserved syntenic region in the intergenic region of mouse chromosome 4. **c** The coding potential of *LINC-RBE* (cDNA), analyzed by using ORF Finder tool, showed putative ORF of less than 300 codons but the predicted peptides did not show any homology with known peptides in the protein database. **d** Similarly, the coding potential of *LINC-RBE* (cDNA) by using codon potential calculator showed very low coding score and ORF coverage. Thus, *LINC-RBE* is a long noncoding transcript [from Fig. 1a and 1b of Kour and Rath (2015) with permission]

Table 3 Rat piRNAs homologous to *LINC-RBE* sequence

piRNA	Reference Id	Sequence (5′-3′)
piR-71600	DQ628056	tccaattggtgtttgagccagatatttcagt
piR-150872	DQ735550.1	aggacttcaaggtatatggaatggttccaa
piR-162527	DQ747205.1	tgggatgtgaatttacaggacttcaaggt
piR-167855	DQ752533.1	atatagttttcacatggcatagacaaaagaga
piR-86803	DQ619691.1	ttcaacctatcgaataccagtgtccaaaga
piR-163689	DQ748367	tattctgtcctaccttgaaactc
piR-81979	DQ613867	tggctcgttaggtggatatcttgtgccct
piR-166598	DQ751276	tatgtgttcttttaaggaggtttaaact

3.3 *Analysis of* LINC-RBE *Expression by RT-PCR in Rat Brain During Aging*

In mammals, through extensive high-throughput RNA sequence analysis, it has been shown that the brain has the highest transcriptome complexity in comparison to all other organs (Soumillon et al. 2013). Diverse groups of lncRNAs with different neurobiological functions such as brain development, differentiation, myelination etc. have been described in literature, however, association of lncRNAs with the process of aging and age-related brain diseases and disorders such as Alzheimer, Parkinson etc. have recently become evident. Since *LINC-RBE* was strongly expressed in the adult rat brain, we, therefore, investigated its likely involvement during maturation and aging of the brain by studying its expression pattern in the brains of 4 weeks (young), 16 weeks (adult), and 70 weeks (old) rats by RT-PCR (Fig. 4b, c). The expression of GAPDH mRNA was taken as an internal control and used for normalization of *LINC-RBE* expression. We found that the expression of *LINC-RBE*, measured as integrated density value (IDV), significantly ($\sim 2\times$ fold) increased from young to adult ($p = 0.0075$) and then decreased ($\sim 1.7\times$ fold) with aging from adult to old ($p = 0.017$) in the rat brain (Fig. 4c) (Kour and Rath 2015). Thus, the age-dependent differential expression patterns of *LINC-RBE* in the rat brain suggest its possible involvement in functions related to maturation and aging of the brain.

3.4 *Expression and Localization of* LINC-RBE *by RNA In Situ Hybridization in Rat Brain During Aging*

Recently, numerous studies have shown that many functionally characterized lncRNAs follow distinct cell type-, tissue- and developmental stage-specific expression patterns in mammalian brain, which define and relate to their biological roles. The age-dependent expression of *LINC-RBE* in rat brain has, therefore, led us to further elaborate on its possible function based on its cell type and sub-cellular

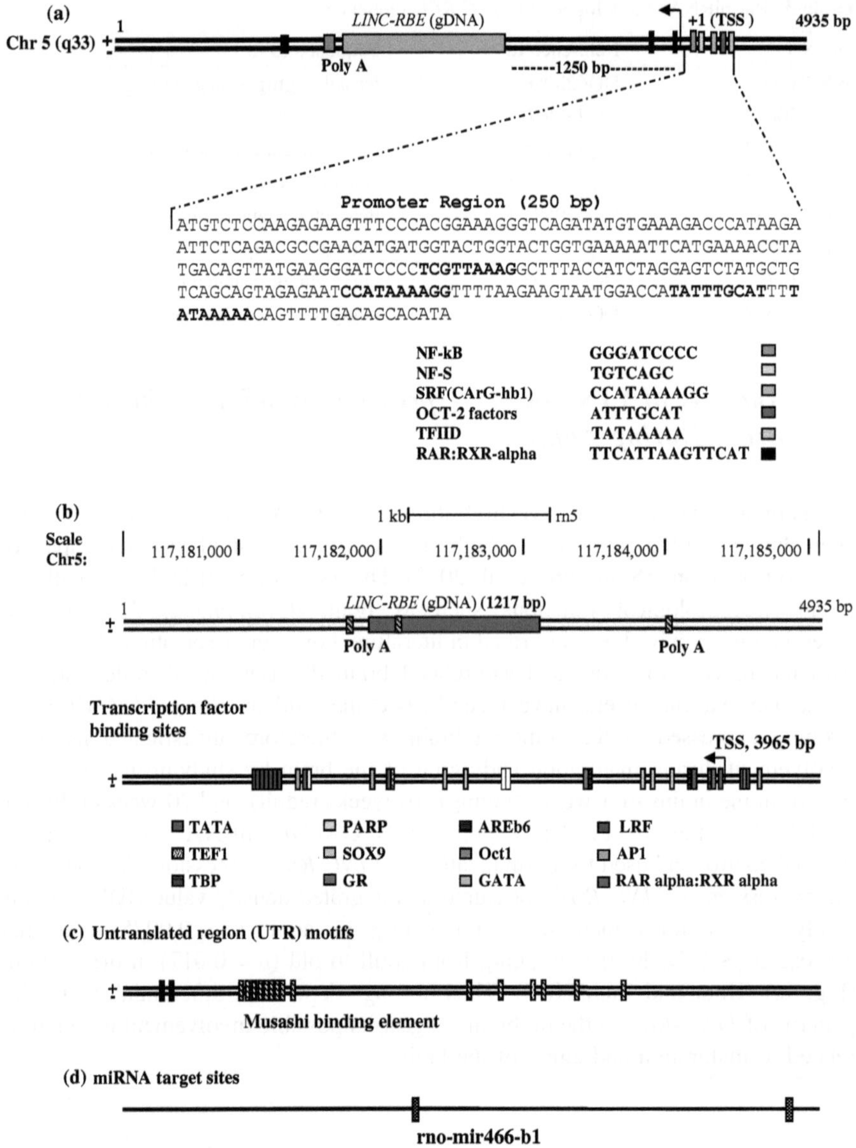

Fig. 3 Presence of various regulatory elements in the up/downstream genomic region of *LINC-RBE*. **a** Schematic representation showing presence of the putative promoter region, consisting of binding sites for the transcription factors: NF-κB, Oct-2, NF-S, TFIID, at 1250 bp downstream of *LINC-RBE* (genomic DNA) in the rat chromosome 5. **b–d** Schematic representation of presence of various regulatory elements such as polyadenylation signals (**b**), transcriptional factor binding sites (**b**), UTR-elements (**c**), and possible miRNA target sites (**d**) in the 2000 bp up/downstream region of *LINC-RBE* (genomic DNA) in the rat chromosome 5 were analyzed by using RegRNA 2.0, Genomatix Genome analyzer and Promo 3.0, respectively [from Fig. 1c of Kour and Rath (2015) with permission]

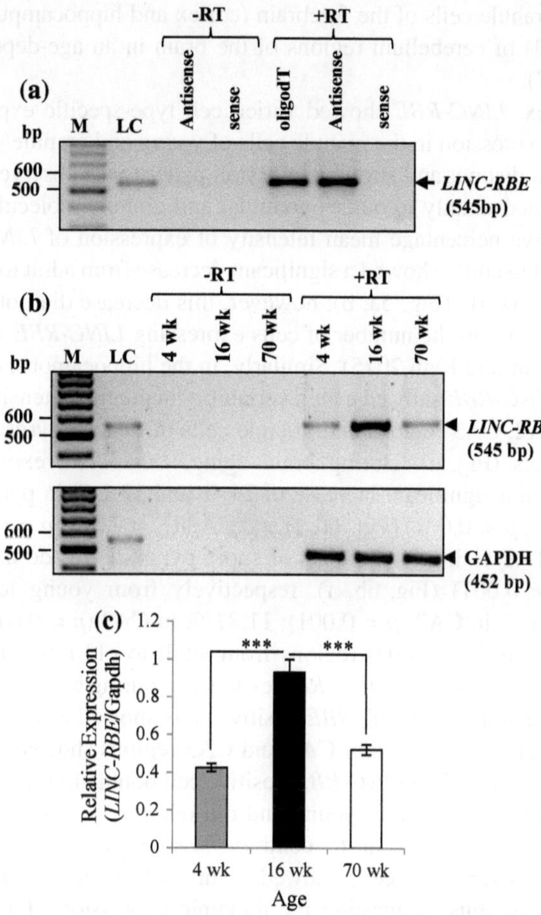

Fig. 4 Age-dependant expression of *LINC-RBE* in the rat brain by RT-PCR. **a** Strand-specificity of transcription of *LINC-RBE* in the adult (16 weeks) rat brain was determined by using strand-specific sense (*RBE*-Fwd) or antisense (*RBE*-Rev) primer during cDNA synthesis step of RT-PCR. **b** Expression of *LINC-RBE* in the brain of 4 weeks (young), 16 weeks (adult) and 70 weeks (old) rats was studied by RT-PCR. GAPDH mRNA was used as an internal control. **c, d** *LINC-RBE* expression was normalized to that of GAPDH mRNA. *−RT* Negative control; *M* 100 bp DNA-ladder; *LC* loading control (545 bp DNA-amplicon); *** $p < 0.001$ (n = 4). [from Fig. 2 of Kour and Rath (2015) with permission]

expression pattern in different functionally specialized neuroanatomical regions of rat brain during aging. This was performed by RNA in situ hybridization in paraffin-embedded brain sections of immature (4 weeks), adult (16 weeks) and old (70 weeks) rats by using digoxygenin-labelled sense- and antisense *LINC-RBE*-specific riboprobes. The *LINC-RBE* was found to be differentially expressed with respect to cell types, cell number and intensity of expression, specifically in the pyramidal and granule cells of the forebrain (cortex and hippocampus) and granule and Purkinje cells of cerebellum regions of the brain in an age-dependent manner (Figs. 5, 6 and 7).

In brain cortex, *LINC-RBE* showed varied cell type-specific expression profile with a punctate expression in the granule cells of young and granule/pyramidal cells of old rats, but, a distinct and strong expression pattern was observed in adult rats, which was confined mainly to outer pyramidal and granule molecular cortical cell layers. The relative percentage mean intensity of expression of *LINC-RBE*, calculated with respect to adult, showed a significant decrease from adult to young and old ($F = 93.844$, $p < 0.001$) (Fig. 5a, b), however, this decrease did not correlate with any possible decrease in the number of cells expressing *LINC-RBE* with increasing age (Fig. 5c) (Kour and Rath 2015). Similarly, in the hippocampus, expression and localization of *LINC-RBE* showed a high variability in relative intensity in pyramidal cells of CA1, CA2, CA3 regions and granule cells of dentate gyrus among young, adult and old rats (Fig. 6). During brain aging, *LINC-RBE* expression in hippocampus showed a significant increase of 24 % and 13.1 % in pyramidal cells of CA2 ($F = 38.105, p < 0.001$) (Fig. 6d, f) and CA3 ($F = 7.318, p = 0.025$) (Fig. 6e, f) regions, and 18.2 % in granule cells of supra-pyramidal blade of dentate gyrus ($F = 28.462$, $p < 0.001$) (Fig. 6b, f), respectively from young to adult, and a decrease of 19.41 % in CA2 ($p = 0.001$); 11.82 % in CA3 ($p = 0.046$) and 14.5 % in the dentate gyrus ($p = 0.003$) regions from adult to old rats. The CA1 region showed no profound change in *LINC-RBE* expression during brain aging (Fig. 6c, f). Besides this, the number of *LINC-RBE* positive cells showed a significant variation in CA2 and dentate gyrus, whereas CA1 and CA3 regions showed no change with increasing age (Fig. 6g). The *LINC-RBE* positive cell population was ~30–35 % in the hippocampal CA2 regions of young and old relative to ~45 % of cells in the adult ($F = 8.173, p = 0.019$). On the contrary, in suprapyramidal blade of dentate gyrus, *LINC-RBE* expressing cells showed ~threefold reduction from young and adult to old brains, thus, suggesting the dynamic expression of *LINC-RBE* with respect to maturation and aging in the rat hippocampal sub-regions (Fig. 6g). Furthermore, *LINC-RBE* expression was also observed in the Purkinje and granule cells of the cerebellum in young, adult and old rats. Similar to cortical region, the expression of *LINC-RBE* initially showed a significant increase of ~21 % from young to adult and then a decrease of ~8 % from adult to old in granule cells ($F = 158.151$, $p < 0.001$) (Fig. 7a, b). There were ~45 % of *LINC-RBE* positive cells in adult cerebellum as compared to ~35 % in young and old rats during aging ($F = 55.787, p < 0.001$) (Fig. 7a, c) (Kour and Rath 2015). Thus, the differential cell type and region-specific expression pattern of *LINC-RBE* in the three transcriptomically and functionally complex regions of the rat brain, i.e., cortex, hippocampus and cerebellum, with increasing age might suggest its potential functional significance in age-related cognitive processes and neurological diseases.

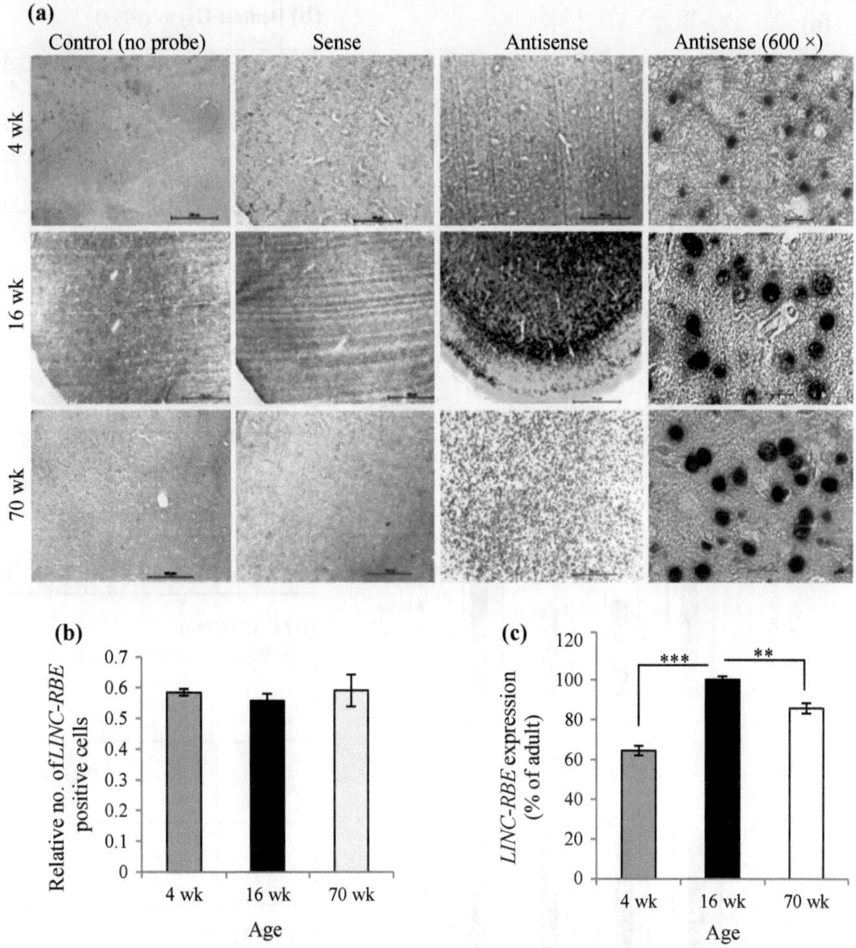

Fig. 5 Age-dependent expression of *LINC-RBE* in the cortex of the rat brain by RNA in situ hybridization. **a** Differential expression pattern of *LINC-RBE* in paraffin-embedded sections (10 μm thick) of the cortex of the brain of 4 weeks (young), 16 weeks (adult) and 70 weeks (old) rats by in situ RNA hybridization by using digoxygenin-labelled sense- and antisense strand-specific RNA probes as well as no probe negative control. Age-dependent differential expression pattern of *LINC-RBE* in the pyramidal and granule cells of different cortical layers of the rat brain with respect to the three age-groups are shown. Bar Scale, 100 and 5 μm. **b** Number of cells positive for *LINC-RBE* expression normalized to the number of cells stained by Haematoxylin/Eosin in the cortex of the brain. **c** Alterations in the intensity of expression of *LINC-RBE* in the cortex of the brain of young and old relative to that of the adult rats. *** $p < 0.001$; ** $p < 0.01$ (n = 3). [from Fig. 3 of Kour and Rath (2015) with permission]

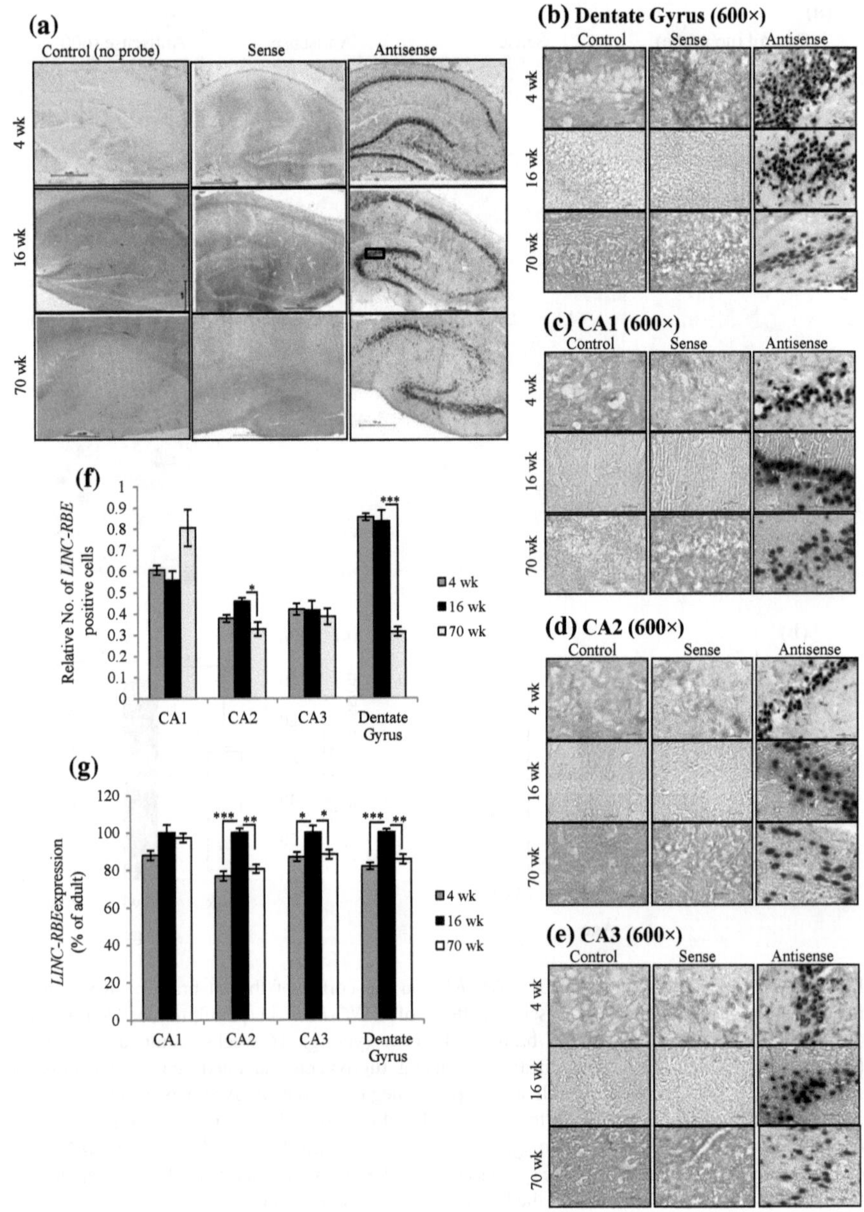
(a)
Control (no probe)
Sense
Antisense
4 wk
16 wk
70 wk
(b) Dentate Gyrus (600×)
Control
Sense
Antisense
(c) CA1 (600×)
(d) CA2 (600×)
(e) CA3 (600×)
(f)
Relative No. of LINC-RBE positive cells
4 wk
16 wk
70 wk
CA1
CA2
CA3
Dentate Gyrus
(g)
LINC-RBEexpression (% of adult)

◀ **Fig. 6** Age-dependent expression of *LINC-RBE* in the hippocampus of the rat brain by RNA in situ hybridization. **a** Differential expression pattern of *LINC-RBE* in paraffin-embedded sections (10 μm thick) of the hippocampus of the brain of 4 weeks (young), 16 weeks (adult) and 70 weeks (old) rats by in situ RNA hybridization by using digoxygenin-labelled sense- and antisense strand-specific RNA probes as well as no probe negative control. Bar Scale, 100 μm. **b–e** Expression of *LINC-RBE* in the granule cells of the dentate gyrus (**b**), and pyramidal cells of CA1 (**c**), CA2 (**d**), CA3 (**e**) sub-regions of the hippocampus. Bar Scale, 5 μm. (**f**) Number of cells positive for *LINC-RBE* expression normalized to the number of cells stained by Haematoxylin/Eosin in the hippocampal sub-regions of the brain. (**g**) Alterations in the intensity of *LINC-RBE* expression in the hippocampal sub-regions of the brain of young and old relative to that of the adult rats. *** $p < 0.001$; ** $p < 0.01$; $p < 0.05$ (n = 3). [from Fig. 4 of Kour and Rath (2015) with permission]

4 Discussion

Evolutionary conservation of newly discovered protein-coding genes are generally based on their nucleotide or amino acid sequences, which provide invaluable means to understand and evaluate their relatedness, functional significance and probable lineage-specific phenotypes. However, although lncRNAs are abundantly expressed and functionally diverse, these criteria do not explain their selection process during evolution and functional relevance among various species in depth (Chodroff et al. 2010; Guo et al. 2014; Johnsson et al. 2014). Refraining from the conventional concept of linking conservation of sequence to function, there are examples of many functionally characterized mammalian lncRNAs such as Myocardial Infarction Associated Transcript (*MIAT*), HOX transcript antisense RNA (*HOTAIR),* X inactive-specific transcript (*XIST*), Embryonic ventral forebrain-2 *(Evf-2)* and Antisense *Igf2*r RNA (*Air*) with poor sequence conservation among species (Diederichs 2014; Johnsson et al. 2014; Roberts et al. 2014a; Wood et al. 2013). Further, despite showing rapid evolution with accumulating mutation and lacking sequence orthólouge, many lncRNAs show syntenic locus conservation among different species with highly conserved regulatory regions and show specific sub-cellular, cell/tissue and developmental expression patterns in mammals, which as a whole, argues for their function (Diederichs 2014; Mercer et al. 2008; Ponting et al. 2009).

Similarly, by homology studies we found modest sequence conservation of *LINC-RBE* among 13 different vertebrates with the presence of a highly conserved syntenic locus in mouse chromosomes 4. The presence of various small conserved regions in the up/downstream region of *LINC-RBE* in chromosome 5 are the sites for synthesis for different piRNAs in the rat and mouse genome, further asserting its function as a precursor lncRNA that may be processed by RNA processing mechanisms to produce many small noncoding regulatory RNAs (piRNAs). In accordance with presence of up/downstream conserved spots, through sequence analysis, we found presence of binding sites for various cell growth/differentiation-specific regulatory transcriptional factors such as RAR-alpha:RXR-alpha, Oct-1, PARP, GR

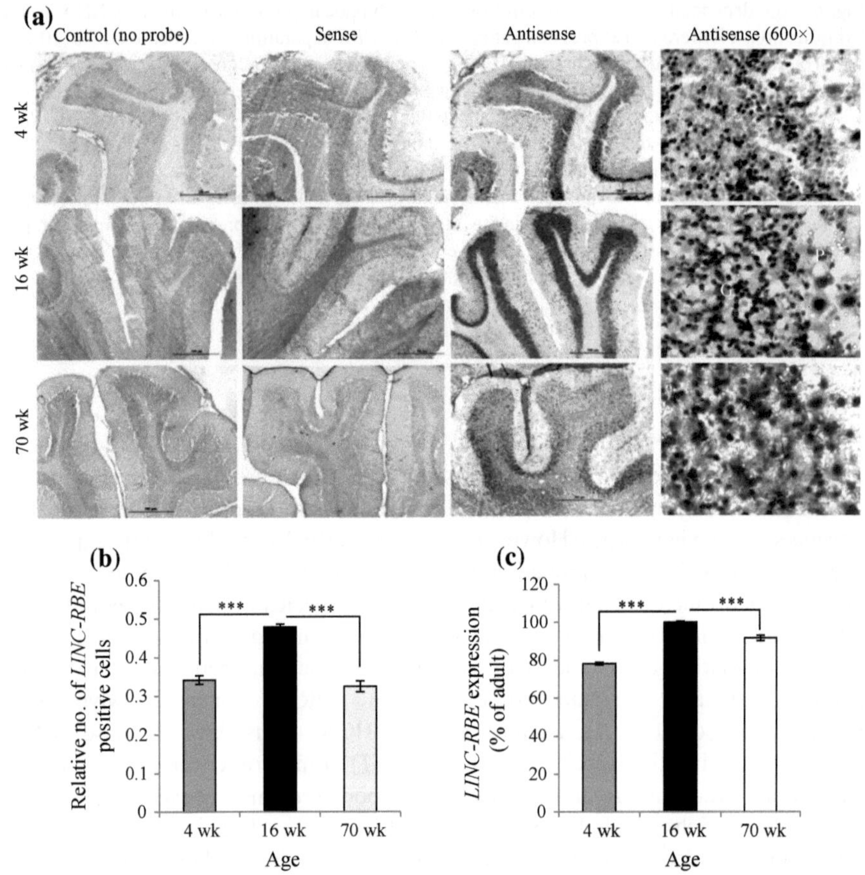

Fig. 7 Age-dependent expression of *LINC-RBE* in the cerebellum of the rat brain by RNA in situ hybridization. **a** Differential expression pattern of *LINC-RBE* in the granule (G) and Purkinje (P) cells of the cerebellum in paraffin-embedded sections (10 μm thick) of the brain of 4 weeks (young), 16 weeks (adult) and 70 weeks (old) rats by in situ RNA hybridization by using digoxigenin-labelled sense- and antisense strand-specific RNA probes as well as no probe negative control. Bar Scale, 100 and 5 μm. **b** Number of cells positive for *LINC-RBE* expression normalized to the number of cells stained by Haematoxylin/Eosin in the cerebellum of the brain. **c** Alterations in the expression of *LINC-RBE* in the cerebellum of the young and old relative to the adult rats. *** $p < 0.001$ ($n = 3$). [from Fig. 5 of Kour and Rath (2015) with permission]

and AP-1, NF-κB and SOX9 in the 2000 bp up/downstream region of *LINC-RBE* (genomic DNA) in rat chromosome 5. Taken together, our results may suggest the multifactorial regulation of expression and functional relevance of *LINC-RBE* in the rat genome. Involvement of *LINC-RBE* in regulation of aging process or its related diseases is emphasized due to the presence of binding sites for major age-related transcriptional factors, GR and NF-κB, (Adler et al. 2007; Rapicavoli et al. 2013), in the up/downstream sequence of *LINC-RBE* (genomic DNA).

Mammalian genome expresses a huge number of different lncRNAs that are transcribed in sense/antisense orientation with respect to protein-coding genes (Wood et al. 2013). One such example is the long intergenic noncoding RNA repeat-rich sense-antisense transcript (*LINC-RSAS*) from the chromosome 17, which is expressed in both sense and antisense orientations in the rat brain (Kour and Rath 2016a). Similarly, transcriptional orientation studied for *LINC-RBE* from the rat chromosome 5 showed its expression in sense orientation in the brain. In mammals, many lncRNAs with variable biological roles showed distinct tissue/organ and developmental stage-specific expression, which specified their tissue-specific function. *LINC-RBE* was found to be expressed in multiple rat tissues (Mishra 2009) and its expression was high in the adult rat brain. We, therefore, studied its expression in the rat brain during aging.

Aging is defined as a global intrinsic biological phenomenon of continuous and cumulative deterioration of neurological, immunological and physiological as well as cellular and molecular functions in an organism with increasing age (Robert et al. 2010). Alteration of many biological pathways/processes/mechanisms as well as cellular protein levels are proposed to be involved in its onset and progression. Aging is considered to be a major risk factor in the patho-physiology of various diseases and disorders (Robert et al. 2010; Sinha et al. 2014). Recently, through advanced RNA sequencing techniques, vast number of functionally diverse regulatory RNAs, sncRNAs and lncRNAs, have been reported, which could provide another way to explore and discourse the complexity of aging process and its related diseases (Kour and Rath 2016c).

Involvement of small noncoding RNAs in aging and age-related diseases such as cardiac malfunctions (*mir-18, mir-19, mir-241, mir-214, mir-217, mir-146*) (van Almen et al. 2011; van Balkom et al. 2013), neurological disorders (*miR-34a, mir-29, mir-144, mir-107*) have been studied extensively. However, function and age-related expression of lncRNAs are much less explored. Till date, only a few reports of modulation of long noncoding RNAs during cellular senescence are known, e.g., senescence-associated known long noncoding RNAs such as Metastasis-associated lung adenocarcinoma transcript 1 (*MALAT1*), *MIAT,* Taurine up-regulated gene 1 (*TUG1*) and novel lncRNAs such as XLOC025918, XLOC025931, XLOC023166 mediate proliferation defects and induce senescence phenotype in cultured human fibroblasts upon their down-regulation (Abdelmohsen et al. 2013). Many lncRNA such as *H19,* Antisense noncoding RNA in the INK4 locus (*ANRIL*), *Air,* Telomeric repeat-containing RNAs (*TERRA*), *HOTAIR,* Long intergenic non coding RN-p21 (*lincRNA-p21*) could possibly be implicated in aging process based on their regulatory role in heterochromatin and telomere maintenance, p53-mediated cell proliferation and apoptosis, a paramount factor during progression of aging and its related diseases (Kour and Rath 2016c). Similarly, *LINC-RSAS*, a repeat-rich 942 bp intergenic lncRNA has been found to be differentially expressed in sense-antisense orientations and age-dependent manner in the rat brain (Kour and Rath 2016a). However, still much is unclear regarding their direct implication and mechanism through which they regulate aging process. In our study, by using RT-PCR and in situ RNA hybridization, we found strong and differential expression of a novel lincRNA, *LINC-RBE* in the rat brain during aging. Its expression in the brain showed a marked increase from the young (4 weeks) to adult (16 weeks) followed by a subsequent and significant decrease from the adult to old (70 weeks) rats. This suggested the possible involvement of *LINC-RBE* in functions related to maturation and aging of the brain.

Mammalian brain is a heterogeneous, complex functional system, primarily consisting of various arrangements of interconnected neurons and glial cells in transcriptomically diverse compartments but with elaborate networks. The cortex has many functional areas associated with visual, auditory, motor functions and cognitive functions such as memory, language, emotions, creativity and judgement. Cortex contains neurons of various shapes, sizes, density and different arrangements of neural fibres organized into multiple molecular layers with variable functions. The layer V and VI consisting mainly of pyramidal neurons are highly developed in the cortical regions involved in motor functions and interconnect cortex to subcortical regions by giving rise to efferent cortical projections to brain stem, spinal cord, basal ganglia and thalamus. The hippocampus consists of varied arrangements of extremely larger pyramidal neurons into CA1, CA2 and CA3 regions, and smaller granule neurons in the dentate gyrus region. The changes in dendritic complexity, synapse number, transmission, plasticity and formation of new neurons or synapses via neurogenesis in the hippocampus throughout life plays an important role in learning, generation and storage of memory processes. The cerebellum accounts for 50 % of total neurons in the brain. Anatomically, cerebellar cortex comprises of innermost granular layer, middle Purkinje cell layer and outermost molecular layer consisting of axons of granule cells and the dendrites of Purkinje cells. Cerebellum is primarily involved in motor functions (voluntary movements, balance and posture) and in cognitive functions such as language. The cortex, hippocampus and cerebellum regions of brain function together in controlling various cognitive functions such as spatial learning and generation of episodic memory.

High-throughput sequencing and microarray techniques to evaluate organ level transcriptome along with inputs from Allen Brain Atlas have reported the cell type- and developmental stage-specific dynamic expression of vast number of sense and antisense transcripts from specific regions/compartments of mammalian brain, which via their innumerable associations with various proteins/complexes are involved in complex brain functions such as learning, memory formation, synaptic variations and plasticity, myelination, neuron development and differentiation (Carrieri et al. 2015; Goff et al. 2015; Kadakkuzha et al. 2015; Mercer et al. 2008). Of late, analysis of RNAseq dataset of mouse brain has reported region-specific expression of 2759 lncRNAs in the hippocampus and 2561 lncRNAs in the prefrontal cortex region, the two main regions involved in various cognitive functions and neuropsychiatric disorders, and together they account for 70 % of the total annotated lncRNAs in the mouse genome (Kadakkuzha et al. 2015).

Similarly, through RNA in situ hybridization based expression studies in brain of young, adult and old rats, we found dynamic, age-dependent expression profile of *LINC-RBE* in different neuroanatomical regions of the brain such as cortex; CA1, CA2, CA3, and dentate gyrus subregions of hippocampus and cerebellum, involved primarily in generation of episodic and spatial learning and memory. The strong, dynamic expression profile of *LINC-RBE,* in terms of both cell number and intensity, in adult compared to young and old rat brains in the complex brain regions involved in cognitive functions (learning and memory), i.e., cortex, hippocampus and cerebellum, might suggest the possible involvement of *LINC-RBE* in modulation of various brain functions including neurogenesis during maturation and lack of it during aging. Furthermore, progression of many neurological disorders and diseases with impaired cognitive functions (memory and learning) such as Alzheimer's disease, autism, Huntington disease have been associated with onset of brain aging (Carrieri et al. 2015). Therefore, further evaluation of *LINC-RBE* function(s) in the hippocampal and cortical regions of the brain during aging or onset and prognosis of many age-related neurological diseases/disorders would highlight a way to study its relevance in various brain functions and neuro-pathological conditions. Since the varied transcriptome based functional compartmentalization of mammalian brain is the result of differential regional- and cell-specific, epigenetically chromatin-modulated gene expression patterns, transcription of *LINC-RBE* in the cortex, hippocampus and cerebellum regions might suggest either its chromatin association or chromatin-based regulation of gene expression in the brain during maturation and aging. It may also be involved in various RNA processing pathways in the brain during maturation and aging. In another study, the relevance of *LINC-RBE* in the function of hippocampus was investigated in primary hippocampal neurons from the adult rat in the presence of a vitamin A derivative, all-trans retinoic acid (atRA), a known regulator of brain development, adult neurogenesis, synaptic plasticity and memory formation. AtRA was found to significantly upregulate expression of *LINC-RBE* in the nucleus and cytoplasm of the neurons along with the dendrites in a time- and dose-dependent manner (Kour and Rath 2016b). The atRA-mediated induction of *LINC-RBE* expression was found to be inhibited by actinomycin D, hence, it was regulated at transcriptional level. The possible binding of atRA along with the retinoic acid receptor (RAR: RXR) to the two binding sites present in the putative promoter of *LINC-RBE* was proposed. Altogether, the decrease in *LINC-RBE* expression in the hippocampus during brain aging and its transcriptional induction by atRA in the primary hippocampal neurons, could possibly argue for its significance in cognitive functions such as synaptic regulation, learning and memory formation as well as their decline during aging.

5 Conclusion

Transcription, gene expression and RNA processing are complex and heterogeneous in the brain. LncRNAs have emerged as major regulators of these processes. This study is a conclusive representation of possible role of lincRNAs in mammalian brain during maturation from young to adulthood and aging from adulthood to old age. Since the onset of many neurological diseases such as dementia, Alzheimer, Parkinson, Huntington occur late in life, study of molecular aspects of brain functions such as impairment of cognitive functions like short and long term memory formation, thinking capability, decision making etc. with increasing age would provide an important basis for understanding details of such age-related diseases and, therefore, would pave new avenues for development of successful therapeutics and treatments. LncRNAs in coordination with various factor(s)/complex(es) have been found to modulate almost each and every process of the flow of genetic information as depicted in the revised "central dogma of molecular biology", i.e., chromatin organization, gene expression, RNA processing, translation and RNA/protein trafficking, thus, regulating nearly every aspect of biological processes required for life. The immense transcription and functional implication of lncRNAs in the mammalian brain, thus, would provide a foremost way to explore the biological process of aging and its related diseases extensively. The age-dependent and differential expression of *LINC-RBE* in specific cell types/regions of the brain, e.g., cortex, hippocampus and cerebellum, the three interconnected regions involved in memory processing and other cognitive functions, may further contribute to more detailed study of brain maturation, aging and age-related diseases at molecular level. *LINC-RBE* may be used as a biomarker for brain aging.

Acknowledgments We thank financial support from the University Grants Commission (UGC)-Research Networking Resource Centre, Department of Science and Technology (DST)-FIST & PURSE to the School of Life Sciences (SLS), Jawaharlal Nehru University (JNU) & PCR and the UGC-Junior & Senior Research Fellowship to SK for Ph.D.

References

Abdelmohsen K, Panda A, Kang MJ, Xu J, Selimyan R, Yoon JH, Martindale JL, De S, Wood WH 3rd, Becker KG et al (2013) Senescence-associated lncRNAs: senescence-associated long noncoding RNAs. Aging Cell 12:890–900

Adler AS, Sinha S, Kawahara TL, Zhang JY, Segal E, Chang HY (2007) Motif module map reveals enforcement of aging by continual NF-kappaB activity. Genes Dev 21:3244–3257

Akula N, Barb J, Jiang X, Wendland JR, Choi KH, Sen SK, Hou L, Chen DT, Laje G, Johnson K et al (2014) RNA-sequencing of the brain transcriptome implicates dysregulation of neuroplasticity, circadian rhythms and GTPase binding in bipolar disorder. Mol. Psychiatry 19:1179–1185

Amiel JL (1965) The Nobel prize for Medicine 1965 is awarded to professors Lwoff, Monod and Jacob. Rev Fr Etud Clin Biol 10:1015–1017

Ananda G, Takemon Y, Hinerfeld D, Korstanje R (2014). Whole-genome sequence of the C57L/J mouse inbred strain. G3 (Bethesda) 4:1689–1692

Aprea J, Prenninger S, Dori M, Ghosh T, Monasor LS, Wessendorf E, Zocher S, Massalini S, Alexopoulou D, Lesche M et al (2013) Transcriptome sequencing during mouse brain development identifies long non-coding RNAs functionally involved in neurogenic commitment. EMBO J 32:3145–3160

Arning L, Ocklenburg S, Schulz S, Ness V, Gerding WM, Hengstler JG, Falkenstein M, Epplen JT, Gunturkun O, Beste C (2015) Handedness and the X chromosome: the role of androgen receptor CAG-repeat length. Sci Rep 5:8325. doi:10.1038/srep08325

Bajaj G (2002) Molecular cloning and characterization of human cDNAs by a simple repeat DNA probe, identification of novel candidate genes, Ph.D thesis. Jawaharlal Nehru University, New Delhi

Ballarino M, Cazzella V, D'Andrea D, Grassi L, Bisceglie L, Cipriano A, Santini T, Pinnaro C, Morlando M, Tramontano A et al (2015) Novel long noncoding RNAs (lncRNAs) in myogenesis: a miR-31 overlapping lncRNA transcript controls myoblast differentiation. Mol Cell Biol 35:728–736

Bao X, Wu H, Zhu X, Guo X, Hutchins AP, Luo Z, Song H, Chen Y, Lai K, Yin M et al (2015) The p53-induced lincRNA-p21 derails somatic cell reprogramming by sustaining H3K9me3 and CpG methylation at pluripotency gene promoters. Cell Res 25:80–92

Bates DJ, Li N, Liang R, Sarojini H, An J, Masternak MM, Bartke A, Wang E (2010) MicroRNA regulation in Ames dwarf mouse liver may contribute to delayed aging. Aging Cell 9:1–18

Bernard D, Prasanth KV, Tripathi V, Colasse S, Nakamura T, Xuan Z, Zhang MQ, Sedel F, Jourdren L, Coulpier F et al (2010) A long nuclear-retained non-coding RNA regulates synaptogenesis by modulating gene expression. EMBO J 29:3082–3093

Birney E, Stamatoyannopoulos JA, Dutta A, Guigo R, Gingeras TR, Margulies EH, Weng Z, Snyder M, Dermitzakis ET, Thurman RE et al (2007) Identification and analysis of functional elements in 1% of the human genome by the ENCODE pilot project. Nature 447:799–816

Boehm M, Slack F (2005) A developmental timing microRNA and its target regulate life span in C. elegans. Science 310:1954–1957

Boon RA, Iekushi K, Lechner S, Seeger T, Fischer A, Heydt S, Kaluza D, Tréguer K, Carmona G, Bonauer A et al (2013) MicroRNA-34a regulates cardiac ageing and function. Nature 495:107–110

Cajigas I, Leib DE, Cochrane J, Luo H, Swyter K, Chen S, Clark BS, Thompson J, Yates JR 3rd, Kingston RE et al (2015) Evf2 lncRNA/BRG1/DLX1 interactions reveal RNA-dependent chromatin remodeling inhibition. Development 142:2641–2652

Capobianco E (2014) RNA-Seq data: a complexity journey. Comput Struct Biotechnol J 11:123–130

Carninci P, Kasukawa T, Katayama S, Gough J, Frith MC, Maeda N, Oyama R, Ravasi T, Lenhard B, Wells C et al (2005) The transcriptional landscape of the mammalian genome. Science 309:1559–1563

Carrieri C, Forrest AR, Santoro C, Persichetti F, Carninci P, Zucchelli S, Gustincich S (2015) Expression analysis of the long non-coding RNA antisense to Uchl1 (AS Uchl1) during dopaminergic cells' differentiation in vitro and in neurochemical models of Parkinson's disease. Front Cell Neurosci 9:114. doi:10.3389/fncel.2015.00114

Chassin C, Hempel C, Stockinger S, Dupont A, Kübler JF, Wedemeyer J, Vandewalle A, Hornef MW (2012) MicroRNA-146a-mediated downregulation of IRAK1 protects mouse and human small intestine against ischemia/ reperfusion injury. EMBO Mol Med 4:1308–1319

Chen D, Fu LY, Zhang Z, Li G, Zhang H, Jiang L, Harrison AP, Shanahan HP, Klukas C, Zhang HY et al (2014) Dissecting the chromatin interactome of microRNA genes. Nucleic Acids Res 42:3028–3043

Chiyomaru T, Fukuhara S, Saini S, Majid S, Deng G, Shahryari V, Chang I, Tanaka Y, Enokida H, Nakagawa M et al (2014). Long non-coding RNA HOTAIR is targeted and regulated by miR-141 in human cancer cells. J Biol Chem 289:12550–12565

Chodroff RA, Goodstadt L, Sirey TM, Oliver PL, Davies KE, Green ED, Molnar Z, Ponting CP (2010) Long noncoding RNA genes: conservation of sequence and brain expression among diverse amniotes. Genome Biol 11:R72. doi:10.1186/gb-2010-11-7-r72
Chow JC, Ciaudo C, Fazzari MJ, Mise N, Servant N, Glass JL, Attreed M, Avner P, Wutz A, Barillot E et al (2010) LINE-1 activity in facultative heterochromatin formation during X chromosome inactivation. Cell 141:956–969
Clark BS, Blackshaw S (2014) Long non-coding RNA-dependent transcriptional regulation in neuronal development and disease. Front Genet 5:164. doi:10.3389/fgene.2014.00164
Cox KH, DeLeon DV, Angerer LM, Angerer RC (1984) Detection of mrnas in sea urchin embryos by in situ hybridization using asymmetric RNA probes. Dev Biol 101:485–502
Cushing L, Costinean S, Xu W, Jiang Z, Madden L, Kuang P, Huang J, Weisman A, Hata A, Croce CM (2015) Disruption of miR-29 Leads to Aberrant Differentiation of Smooth Muscle Cells Selectively Associated with Distal Lung Vasculature. PLoS Genet 11:e1005238
de Lencastre A, Pincus Z, Zhou K, Kato M, Lee SS, Slack FJ (2010) MicroRNAs both promote and antagonize longevity in C. elegans. Curr Biol 20:2159–2168
Dellago H, Preschitz-Kammerhofer B, Terlecki-Zaniewicz L, Schreiner C, Fortschegger K, Chang MW, Hackl M, Monteforte R, Kühnel H, Schosserer M et al (2013) High levels of oncomiR-21 contribute to the senescence-induced growth arrest in normal human cells and its knock-down increases the replicative lifespan. Aging Cell 12:446–458
Denas O, Sandstrom R, Cheng Y, Beal K, Herrero J, Hardison RC, Taylor J (2015) Genome-wide comparative analysis reveals human-mouse regulatory landscape and evolution. BMC Genom 16:87. doi:10.1186/s12864-015-1245-6
Deng S, Zhu S, Wang B, Li X, Liu Y, Qin Q, Gong Q, Niu Y, Xiang C, Chen J et al (2014) Chronic pancreatitis and pancreatic cancer demonstrate active epithelial-mesenchymal transition profile, regulated by miR-217-SIRT1 pathway. Cancer Lett 355:184–191
Dey I (2000) A study of chromatin structure and transcription using LINE DNA and simple repeat DNA probes. PhD thesis. School of Life Sciences, Jawaharlal Nehru University, New Delhi
Dey I, Rath PC (2005) A novel rat genomic simple repeat DNA with RNA-homology shows triplex (H-DNA)-like structure and tissue-specific RNA expression. Biochem Biophys Res Commun 327:276–286
Diederichs S (2014) The four dimensions of noncoding RNA conservation. Trends Genet 30:121–123
Dimmeler S, Nicotera P (2013) MicroRNAs in age-related diseases. EMBO Mol Med 5:180–190
Drummond MJ, McCarthy JJ, Sinha M, Spratt HM, Volpi E, Esser KA, Rasmussen BB (2011) Aging and microRNA expression in human skeletal muscle: a microarray and bioinformatics analysis. Physiol Genomics 43:595–603
Faulkner GJ, Kimura Y, Daub CO, Wani S, Plessy C, Irvine KM, Schroder K, Cloonan N, Steptoe AL, Lassmann T et al (2009) The regulated retrotransposon transcriptome of mammalian cells. Nat Genet 41:563–571
Feng J, Bi C, Clark BS, Mady R, Shah P, Kohtz JD (2006) The Evf-2 noncoding RNA is transcribed from the Dlx-5/6 ultraconserved region and functions as a Dlx-2 transcriptional coactivator. Genes Dev 20:1470–1484
Flintoft L (2013) Non-coding RNA: structure and function for lncRNAs. Nat Rev Genet 14:598. doi:10.1038/nrg3561
Flynn RA, Chang HY (2014) Long noncoding RNAs in cell-fate programming and reprogramming. Cell Stem Cell 14:752–761
Gao M, Wang X, Zhang X, Ha T, Ma H, Liu L, Kalbfleisch JH, Gao X, Kao RL, Williams DL (2015) Attenuation of Cardiac Dysfunction in Polymicrobial Sepsis by MicroRNA-146a Is Mediated via Targeting of IRAK1 and TRAF6 Expression. J Immunol 195:672–682
Geiduschek EP, Haselkorn R (1969) Messenger RNA. Annu Rev Biochem 38:647–676
Gibbs RA, Weinstock GM, Metzker ML, Muzny DM, Sodergren EJ, Scherer S, Scott G, Steffen D, Worley KC, Burch PE et al (2004) Genome sequence of the Brown Norway rat yields insights into mammalian evolution. Nature 428:493–521

Girard A, Sachidanandam R, Hannon GJ, Carmell MA (2006) A germline-specific class of small RNAs binds mammalian Piwi proteins. Nature 442:199–202

Goff LA, Groff AF, Sauvageau M, Trayes-Gibson Z, Sanchez-Gomez DB, Morse M, Martin RD, Elcavage LE, Liapis SC, Gonzalez-Celeiro M et al (2015) Spatiotemporal expression and transcriptional perturbations by long noncoding RNAs in the mouse brain. Proc. Natl. Acad. Sci. U. S. A. 112:6855–6862

Gong C, Li Z, Ramanujan K, Clay I, Zhang Y, Lemire-Brachat S, Glass DJ (2015) A long non-coding RNA, LncMyoD, regulates skeletal muscle differentiation by blocking IMP2-mediated mRNA translation. Dev Cell 34:181–191

Gonzalez-Buendia E, Saldana-Meyer R, Meier K, Recillas-Targa F (2015) Transcriptome-wide identification of in vivo interactions between RNAs and RNA-binding proteins by RIP and PAR-CLIP assays. Methods Mol Biol 1288:413–428

Graff J, Mansuy IM (2008) Epigenetic codes in cognition and behaviour. Behav Brain Res 192:70–87

Grillari J, Hackl M, Grillari-Voglauer R (2010) miR-17-92 cluster: ups and downs in cancer and aging. Biogerontology 11:501–506

Grote A, Hiller K, Scheer M, Munch R, Nortemann B, Hempel DC, Jahn D (2005) JCat: a novel tool to adapt codon usage of a target gene to its potential expression host. Nucleic Acids Res 33:W526–W531

Guennewig B, Cooper AA (2014) The central role of noncoding RNA in the brain. Int Rev Neurobiol 116:153–194

Guo L, Zhao Y, Yang S, Zhang H, Wu Q, Chen F (2014) An integrated evolutionary analysis of miRNA-lncRNA in mammals. Mol Biol Rep 41:201–207

Guttman M, Amit I, Garber M, French C, Lin MF, Feldser D, Huarte M, Zuk O, Carey BW, Cassady JP et al (2009) Chromatin signature reveals over a thousand highly conserved large non-coding RNAs in mammals. Nature 458:223–227

Hackl M, Brunner S, Fortschegger K, Schreiner C, Micutkova L, Mück C, Laschober GT, Lepperdinger G, Sampson N, Berger P (2010) miR-17, miR-19b, miR-20a, and miR-106a are down-regulated in human aging. Aging Cell 9:291–296

Hamrick MW, Herberg S, Arounleut P, He HZ, Shiver A, Qi RQ, Zhou L, Isales CM, Mi QS (2010) The adipokine leptin increases skeletal muscle mass and significantly alters skeletal muscle miRNA expression profile in aged mice. Biochem Biophys Res Commun 400:379–383

Hoffman Y, Pilpel Y, Oren M (2014) microRNAs and Alu elements in the p53-Mdm2-Mdm4 regulatory network. J Mol Cell Biol 6:192–197

Hu G, Lou Z, Gupta M (2014) The long non-coding RNA GAS5 cooperates with the eukaryotic translation initiation factor 4E to regulate c-Myc translation. PLoS ONE 9:e107016

Hu J, Chen Z, Xia D, Wu J, Xu H, Ye ZQ (2012) Promoter-associated small double-stranded RNA interacts with heterogeneous nuclear ribonucleoprotein A2/B1 to induce transcriptional activation. Biochem J 447:407–416

Huarte M, Guttman M, Feldser D, Garber M, Koziol MJ, Kenzelmann-Broz D, Khalil AM, Zuk O, Amit I, Rabani M et al (2010) A large intergenic noncoding RNA induced by p53 mediates global gene repression in the p53 response. Cell 142:409–419

Iyer MK, Niknafs YS, Malik R, Singhal U, Sahu A, Hosono Y, Barrette TR, Prensner JR, Evans JR, Zhao S et al (2015) The landscape of long noncoding RNAs in the human transcriptome. Nat Genet 47:199–208

Jazbutyte V, Fiedler J, Kneitz S, Galuppo P, Just A, Holzmann A, Bauersachs J, Thum T (2013) MicroRNA-22 increases senescence and activates cardiac fibroblasts in the aging heart. Age 35:747–762

Jiang L, Wang C, Lei F, Zhang L, Zhang X, Liu A, Wu G, Zhu J, Song L (2015) miR-93 promotes cell proliferation in gliomas through activation of PI3K/Akt signalling pathway. Oncotarget 6:8286–8299

Johnson R (2012) Long non-coding RNAs in Huntington's disease neurodegeneration. Neurobiol Dis 46:245–254

Johnsson P, Lipovich L, Grander D, Morris KV (2014) Evolutionary conservation of long non-coding RNAs; sequence, structure, function. Biochim Biophys Acta 1840:1063–1071
Kadakkuzha BM, Liu XA, McCrate J, Shankar G, Rizzo V, Afinogenova A, Young B, Fallahi M, Carvalloza AC, Raveendra B et al (2015) Transcriptome analyses of adult mouse brain reveal enrichment of lncRNAs in specific brain regions and neuronal populations. Front Cell Neurosci 9:63. doi:10.3389/fncel.2015.00063
Kapusta A, Kronenberg Z, Lynch VJ, Zhuo X, Ramsay L, Bourque G, Yandell M, Feschotte C (2013) Transposable elements are major contributors to the origin, diversification, and regulation of vertebrate long noncoding RNAs. PLoS Genet 9:e1003470
Karp X, Hammell M, Ow MC, Ambros V (2011) Effect of life history on microRNAs expression during C. elegans development. RNA 17:639–651
Kato M, Chen X, Inukai S, Zhao H, Slack FJ (2011) Age-associated changes in expression of small, noncoding RNAs, including microRNAs, in C. elegans. RNA 17:1804–1820
Kawaji H, Lizio M, Itoh M, Kanamori-Katayama M, Kaiho A, Nishiyori-Sueki H, Shin JW, Kojima-Ishiyama M, Kawano M, Murata M et al (2014) Comparison of CAGE and RNA-seq transcriptome profiling using clonally amplified and single-molecule next-generation sequencing. Genome Res 24:708–717
Khalil AM, Guttman M, Huarte M, Garber M, Raj A, Rivea Morales D, Thomas K, Presser A, Bernstein BE, van Oudenaarden A et al (2009) Many human large intergenic noncoding RNAs associate with chromatin-modifying complexes and affect gene expression. Proc Natl Acad Sci USA 106:11667–11672
Khanna A, Muthusamy S, Liang R, Sarojini H, Wang E (2011) Gain of survival signalling by down-regulation of three key miRNAs in brain of calorie-restricted mice. Aging 3:223–236
Kim K, Vinayagam A, Perrimon N (2014) A rapid genome-wide microRNA screen identifies miR-14 as a modulator of Hedgehog signaling. Cell Rep 7:2066–2077
Kim DH, Marinov GK, Pepke S, Singer ZS, He P, Williams B, Schroth GP, Elowitz MB, Wold BJ (2015) Single-cell transcriptome analysis reveals dynamic changes in lncRNA expression during reprogramming. Cell Stem Cell 16:88–101
Kour S (2015) Functional characterization of the noncoding RNAs from the repetitive DNA of mammalian genome. Jawaharlal Nehru University, New Delhi, PhD
Kour S, Rath PC (2015) Age-dependent differential expression profile of a novel intergenic long noncoding RNA in rat brain. Int J Dev Neurosci 47:286–297
Kour S, Rath PC (2016a) Age-related expression of a repeat-rich intergenic long noncoding RNA in the rat brain. Mol Neurobiol [Epub ahead of print]
Kour S, Rath PC (2016b) All-trans retinoic acid induces expression of a novel intergenic long noncoding RNA in adult rat primary hippocampal neurons. J Mol Neurosci 58:266–276
Kour S, Rath PC (2016c) Long noncoding RNAs in aging and age-related diseases. Ageing Res Rev 26:1–21
Kurokawa R (2011) Promoter-associated long noncoding RNAs repress transcription through a RNA binding protein TLS. Adv Exp Med Biol 722:196–208
Hyun S, Lee JH, Jin H, Nam J, Namkoong B, Lee G, Chung J, Kim VN (2009) Conserved MicroRNA miR-8/miR-200 and its target USH/FOG2 control growth by regulating PI3K. Cell 139:1096–1108
Lasalle JM, Powell WT, Yasui DH (2013) Epigenetic layers and players underlying neurodevelopment. Trends Neurosci 36:460–470
Lau NC, Seto AG, Kim J, Kuramochi-Miyagawa S, Nakano T, Bartel DP, Kingston RE (2006) Characterization of the piRNA complex from rat testes. Science 313:363–367
Lee JT (2011) Gracefully ageing at 50, X-chromosome inactivation becomes a paradigm for RNA and chromatin control. Nat Rev Mol Cell Biol 12:815–826
Legnini I, Morlando M, Mangiavacchi A, Fatica A, Bozzoni I (2014) A feedforward regulatory loop between HuR and the long noncoding RNA linc-MD1 controls early phases of myogenesis. Mol Cell 53:506–514

Li N, Muthusamy S, Liang R, Sarojini H, Wang E (2011a) Increased expression of miR-34a and miR-93 in rat liver during aging, and their impact on the expression of Mgst1 and Sirt1. Mech Ageing Dev 132:75–85

Li X, Khanna A, Li N, Wang E (2011b) Circulatory miR34a as an RNA based, noninvasive biomarker for brain aging. Aging 3:985–1002

Li N, Bates DJ, An J, Terry DA, Wang E (2011c) Up-regulation of key microRNAs, and inverse down-regulation of their predicted oxidative phosphorylation target genes, during aging in mouse brain. Neurobiol Aging 32:944–955

Liang R, Khanna A, Muthusamy S, Li N, Sarojini H, Kopchick JJ, Masternak MM, Bartke A, Wang E (2011) Post-transcriptional regulation of IGF1R by key microRNAs in long-lived mutant mice. Aging Cell 10:1080–1088

Liang C, Forrest AR, Wagner GP (2015) The statistical geometry of transcriptome divergence in cell-type evolution and cancer. Nat Commun 6:6066. doi:10.1038/ncomms7066

Lin N, Chang KY, Li Z, Gates K, Rana ZA, Dang J, Zhang D, Han T, Yang CS, Cunningham TJ et al (2014a) An evolutionarily conserved long noncoding RNA TUNA controls pluripotency and neural lineage commitment. Mol Cell 53:1005–1019

Lin ST, Heng MY, Ptacek LJ, Fu YH (2014b) Regulation of Myelination in the Central Nervous System by Nuclear Lamin B1 and Non-coding RNAs. Transl Neurodegener 3:4. doi:10.1186/2047-9158-3-4

Liu N, Landreh M, Cao K, Abe M, Hendriks GJ, Kennerdell JR, Zhu Y, Wang LS, Bonini NM (2012) The microRNA miR-34 modulates ageing and neurodegeneration in Drosophila. Nature 482:519–523

Lizio M, Harshbarger J, Shimoji H, Severin J, Kasukawa T, Sahin S, Abugessaisa I, Fukuda S, Hori F, Ishikawa-Kato S et al (2015) Gateways to the FANTOM5 promoter level mammalian expression atlas. Genome Biol 16:22. doi:10.1186/s13059-014-0560-6

Loewer S, Cabili MN, Guttman M, Loh YH, Thomas K, Park IH, Garber M, Curran M, Onder T, Agarwal S et al (2010) Large intergenic non-coding RNA-RoR modulates reprogramming of human induced pluripotent stem cells. Nat Genet 42:1113–1117

Lucanic M, Graham J, Scott G, Bhaumik D, Benz CC, Hubbard A, Lithgow GJ, Melov S (2013) Age-related micro-RNA abundance in individual C. elegans. Aging 5:394–411

Lv J, Liu H, Huang Z, Su J, He H, Xiu Y, Zhang Y, Wu Q (2013) Long non-coding RNA identification over mouse brain development by integrative modeling of chromatin and genomic features. Nucleic Acids Res 41:10044–10061

Maes OC, An J, Sarojini H, Wang E (2008) Murine microRNAs implicated in liver functions and aging process. Mech Ageing Dev 129:534–541

Marasa BS, Srikantan S, Masuda K, Abdelmohsen K, Kuwano Y, Yang X, Martindale JL, Rinker-Schaeffer CW, Gorospe M (2009). Increased MKK4 abundance with replicative senescence is linked to the joint reduction of multiple microRNAs. Sci Signal 2:ra69

Marchese FP, Huarte M (2014) Long non-coding RNAs and chromatin modifiers: their place in the epigenetic code. Epigenetics 9:21–26

Mariño G, Ugalde AP, Fernández AF, Osorio FG, Fueyo A, Freije JM, López-Otín C (2010) Insulin-like growth factor 1 treatment extends longevity in a mouse model of human premature aging by restoring somatotroph axis function. Proc Natl Acad Sci USA 107:16268–16273

Martinez I, Cazalla D, Almstead LL, Steitz JA, DiMaio D (2011) miR-29 and miR-30 regulate B-Myb expression during cellular senescence. Proc Natl Acad Sci USA 108:522–527

Marques AC, Ponting CP (2014) Intergenic lncRNAs and the evolution of gene expression. Curr Opin Genet Dev 27:48–53

Mattick JS, Rinn JL (2015) Discovery and annotation of long noncoding RNAs. Nat Struct Mol Biol 22:5–7

McHugh CA, Chen CK, Chow A, Surka CF, Tran C, McDonel P, Pandya-Jones A, Blanco M, Burghard C, Moradian A et al (2015) The Xist lncRNA interacts directly with SHARP to silence transcription through HDAC3. Nature. doi:10.1038/nature14443

Menghini R, Casagrande V, Cardellini M, Martelli E, Terrinoni A, Amati F, Vasa-Nicotera M, Ippoliti A, Novelli G, Melino G et al (2009) MicroRNA 217 modulates endothelial cell senescence via silent information regulator 1. Circulation 120:1524–1532

Mercer TR, Dinger ME, Sunkin SM, Mehler MF, Mattick JS (2008) Specific expression of long noncoding RNAs in the mouse brain. Proc Natl Acad Sci USA 105:716–721

Mercer TR, Mattick JS (2013) Structure and function of long noncoding RNAs in epigenetic regulation. Nat Struct Mol Biol 20:300–307

Mercer TR, Qureshi IA, Gokhan S, Dinger ME, Li G, Mattick JS, Mehler MF (2010) Long noncoding RNAs in neuronal-glial fate specification and oligodendrocyte lineage maturation. BMC Neurosci 11:14. doi:10.1186/1471-2202-11-14

Meza-Sosa KF, Pedraza-Alva G, Pérez-Martínez L (2014) MicroRNAs: key triggers of neuronal cell fate. Front Cell Neurosci 8:175

Milligan MJ, Lipovich L (2014) Pseudogene-derived lncRNAs: emerging regulators of gene expression. Front Genet 5:476. doi:10.3389/fgene.2014.00476

Mishra RR (2009) Expression and functional characterization of candidate cDNAs isolated by a simple repeat DNA probe. Ph.D thesis. School of Life Sciences, Jawaharlal Nehru University, New Delhi

Mogilyansky E, Rigoutsos I (2013) The miR-17/92 cluster: a comprehensive update on its genomics, genetics, functions and increasingly important and numerous roles in health and disease. Cell Death Differ 20:1603–1614

Mortimer SA, Kidwell MA, Doudna JA (2014) Insights into RNA structure and function from genome-wide studies. Nat Rev Genet 15:469–479

Nelson C, Ambros V, Baehrecke EH (2014) miR-14 regulates autophagy during developmental cell death by targeting ip3-kinase 2. Mol Cell 56:376–388

Ng SY, Bogu GK, Soh BS, Stanton LW (2013) The long noncoding RNA RMST interacts with SOX2 to regulate neurogenesis. Mol Cell 51:349–359

Noren Hooten N, Abdelmohsen K, Gorospe M, Ejiogu N, Zonderman AB, Evans MK (2010) MicroRNA expression patterns reveal differential expression of target genes with age. PLoS One 5:e10724

Olive V, Minella AC, He L (2015) Outside the coding genome, mammalian microRNAs confer structural and functional complexity. Sci Signal 8, re2. doi:10.1126/scisignal.2005813

Persengiev S, Kondova I, Otting N, Koeppen AH, Bontrop RE (2011) Genome-wide analysis of miRNA expression reveals a potential role for miR-144 in brain aging and spinocerebellar ataxia pathogenesis. Neurobiol Aging 32:2316.e17–27

Philippe JV, Ayadi L, Branlant C, Behm-Ansmant I (2015) Probing small non-coding RNAs structures. Methods Mol Biol 1296:119–136

Pincus Z, Smith-Vikos T, Slack FJ (2011) MicroRNA predictors of longevity in Caenorhabditis elegans. PLoS Genet 7(9):e1002306

Pnueli L, Rudnizky S, Yosefzon Y, Melamed P (2015) RNA transcribed from a distal enhancer is required for activating the chromatin at the promoter of the gonadotropin alpha-subunit gene. Proc Natl Acad Sci USA 112:4369–4374

Ponjavic J, Ponting CP (2007) The long and the short of RNA maps. BioEssays 29:1077–1080

Ponting CP, Oliver PL, Reik W (2009) Evolution and functions of long noncoding RNAs. Cell 136:629–641

Porro A, Feuerhahn S, Lingner J (2014) TERRA-reinforced association of LSD1 with MRE11 promotes processing of uncapped telomeres. Cell Rep 6:765–776

Quek XC, Thomson DW, Maag JL, Bartonicek N, Signal B, Clark MB, Gloss BS, Dinger ME (2015) lncRNAdb v2.0: expanding the reference database for functional long noncoding RNAs. Nucleic Acids Res 43:D168–D173

Ramos AD, Andersen RE, Liu SJ, Nowakowski TJ, Hong SJ, Gertz CC, Salinas RD, Zarabi H, Kriegstein AR, Lim DA (2015) The long noncoding RNA Pnky regulates neuronal differentiation of embryonic and postnatal neural stem cells. Cell Stem Cell 16:439–447

Rapicavoli NA, Qu K, Zhang J, Mikhail M, Laberge RM, Chang HY (2013) A mammalian pseudogene lncRNA at the interface of inflammation and anti-inflammatory therapeutics. Elife 2:e00762

Redon S, Zemp I, Lingner J (2013) A three-state model for the regulation of telomerase by TERRA and hnRNPA1. Nucleic Acids Res 41:9117–9128

Rinn JL (2014) lncRNAs: linking RNA to chromatin. Cold Spring Harb Perspect Biol 6. doi:10.1101/cshperspect.a018614

Robert L, Labat-Robert J, Robert AM (2010) Genetic, epigenetic and posttranslational mechanisms of aging. Biogerontology 11:387–399

Roberts TC, Morris KV, Weinberg MS (2014a) Perspectives on the mechanism of transcriptional regulation by long non-coding RNAs. Epigenetics 9:13–20

Roberts TC, Morris KV, Wood MJ (2014b) The role of long non-coding RNAs in neurodevelopment, brain function and neurological disease. Philos Trans R Soc Lond B Biol Sci 369. doi:10.1098/rstb.2013.0507

Rokavec M, Li H, Jiang L, Hermeking H (2014) The p53/miR-34 axis in development and disease. J Mol Cell Biol 6:214–230

Rutledge CE, Lau HT, Mangan H, Hardy LL, Sunnotel O, Guo F, MacNicol AM, Walsh CP, Lees-Murdock DJ (2014) Efficient translation of Dnmt1 requires cytoplasmic polyadenylation and Musashi binding elements. PLoS ONE 9:e88385

Saksouk N, Barth TK, Ziegler-Birling C, Olova N, Nowak A, Rey E, Mateos-Langerak J, Urbach S, Reik W, Torres-Padilla ME et al (2014) Redundant mechanisms to form silent chromatin at pericentromeric regions rely on BEND3 and DNA methylation. Mol Cell 56:580–594

Sambrook JRD (2001) Molecular Cloning. Cold Spring Harbour Press, New York, A laboratory manual

Serna E, Gambini J, Borras C, Abdelaziz KM, Belenguer A, Sanchis P, Avellana JA, Rodriguez-Manas L, Vina J (2012) Centenarians, but not octogenarians, up-regulate the expression of microRNAs. Sci Rep 2:961. doi:10.1038/srep00961

Shahryari A, Jazi MS, Samaei NM, Mowla SJ (2015) Long non-coding RNA SOX2OT: expression signature, splicing patterns, and emerging roles in pluripotency and tumorigenesis. Front Genet 6:196. doi:10.3389/fgene.2015.00196

Shi Z, Hayes G, Ruvkun G (2013) Dual regulation of the lin-14 target mRNA by the lin-4 miRNA. PLoS One 8:e75475

Sigdel KR, Cheng A, Wang Y, Duan L, Zhang Y (2015) The emerging functions of long noncoding RNA in immune cells: autoimmune diseases. J Immunol Res 2015:848790. doi:10.1155/2015/848790

Siggens L, Ekwall K (2014) Epigenetics, chromatin and genome organization: recent advances from the ENCODE project. J Intern Med 276:201–214

Singh DK, Rath PC (2012) Long interspersed nuclear elements (LINEs) show tissue-specific, mosaic genome and methylation-unrestricted, widespread expression of noncoding RNAs in somatic tissues of the rat. RNA Biol 9:1380–1396

Singh M (2012) Dysregulated A to I RNA editing and non-coding RNAs in neurodegeneration. Front Genet 3:326. doi:10.3389/fgene.2012.00326

Sinha JK, Ghosh S, Swain U, Giridharan NV, Raghunath M (2014) Increased macromolecular damage due to oxidative stress in the neocortex and hippocampus of WNIN/Ob, a novel rat model of premature aging. Neuroscience 269:256–264

Smith-Vikos T, Slack FJ (2012) MicroRNAs and their roles in aging. J Cell Sci 125:7–17

Somarowthu S, Legiewicz M, Chillon I, Marcia M, Liu F, Pyle AM (2015) HOTAIR forms an intricate and modular secondary structure. Mol Cell 58:353–361

Song R, Ro S, Michaels JD, Park C, McCarrey JR, Yan W (2009) Many X-linked microRNAs escape meiotic sex chromosome inactivation. Nat Genet 41:488–493

Soumillon M, Necsulea A, Weier M, Brawand D, Zhang X, Gu H, Barthes P, Kokkinaki M, Nef S, Gnirke A et al (2013) Cellular source and mechanisms of high transcriptome complexity in the mammalian testis. Cell Rep 3:2179–2190

St Laurent G, Wahlestedt C, Kapranov P (2015) The Landscape of long noncoding RNA classification. Trends Genet 31:239–251

Sutherland JM, McLaughlin EA, Hime GR, Siddall NA (2013) The Musashi family of RNA binding proteins: master regulators of multiple stem cell populations. Adv Exp Med Biol 786:233–245

Takahashi M, Eda A, Fukushima T, Hohjoh H (2012) Reduction of type IV collagen by upregulated miR-29 in normal elderly mouse and klotho-deficient, senescence-model mouse. PLoS One 711:e48974

Trümbach D, Prakash N (2015) The conserved miR-8/miR-200 microRNA family and their role in invertebrate and vertebrate neurogenesis. Cell Tissue Res 359:161–177

Toledano H, D'Alterio C, Czech B, Levine E, Jones DL (2012) The let-7-Imp axis regulates ageing of the Drosophila testis stem-cell niche. Nature 485:605–610

Turner MJ, Jiao AL, Slack FJ (2014) Autoregulation of lin-4 microRNA transcription by RNA activation (RNAa) in C. elegans. Cell Cycle 13:772–781

Ugalde AP, Ramsay AJ, de la Rosa J, Varela I, Mariño G, Cadiñanos J, Lu J, Freije JM, López-Otín C (2011) Aging and chronic DNA damage response activate a regulatory pathway involving miR-29 and p53. EMBO J 30:2219–2232

van Almen GC, Verhesen W, van Leeuwen RE, van de Vrie M, Eurlings C, Schellings MW, Swinnen M, Cleutjens JP, van Zandvoort MA, Heymans S et al (2011) MicroRNA-18 and microRNA-19 regulate CTGF and TSP-1 expression in age-related heart failure. Aging Cell 10:769–779

van Balkom BW, de Jong OG, Smits M, Brummelman J, den Ouden K, de Bree PM, van Eijndhoven MA, Pegtel DM, Stoorvogel W, Wurdinger T, et al (2013) Endothelial cells require miR-214 to secrete exosomes that suppress senescence and induce angiogenesis in human and mouse endothelial cells. Blood 121:3997–4006, S3991–3915

Vasa-Nicotera M, Chen H, Tucci P, Yang AL, Saintigny G, Menghini R, Mahè C, Agostini M, Knight RA, Melino G et al (2011) miR-146a is modulated in human endothelial cell with aging. Atherosclerosis 217:326–330

Venter JC, Adams MD, Myers EW, Li PW, Mural RJ, Sutton GG, Smith HO, Yandell M, Evans CA, Holt RA et al (2001) The sequence of the human genome. Science 291:1304–1351

Vucicevic D, Corradin O, Ntini E, Scacheri PC, Orom UA (2015) Long ncRNA expression associates with tissue-specific enhancers. Cell Cycle 14:253–260

Wahlestedt C (2013) Targeting long non-coding RNA to therapeutically upregulate gene expression. Nat Rev Drug Discov 12:433–446

Wan DY, Zhang Z, Yang HH (2015) Cardioprotective effect of miR-214 in myocardial ischemic postconditioning by down-regulation of hypoxia inducible factor 1, alpha subunit inhibitor. Cell Mol Biol 61:1–6

Wang WX, Wilfred BR, Madathil SK, Tang G, Hu Y, Dimayuga J, Stromberg AJ, Huang Q, Saatman KE, Nelson PT (2010) miR-107 regulates granulin/progranulin with Implications for traumatic brain injury and neurodegenerative disease. Am J Pathol 177:334–345

Wang KC, Chang HY (2011) Molecular mechanisms of long noncoding RNAs. Mol Cell 43:904–914

Wang L, Zhao Y, Bao X, Zhu X, Kwok YK, Sun K, Chen X, Huang Y, Jauch R, Esteban MA et al (2015) LncRNA Dum interacts with Dnmts to regulate Dppa2 expression during myogenic differentiation and muscle regeneration. Cell Res 25:335–350

Waterston RH, Lindblad-Toh K, Birney E, Rogers J, Abril JF, Agarwal P, Agarwala R, Ainscough R, Alexandersson M, An P et al (2002) Initial sequencing and comparative analysis of the mouse genome. Nature 420:520–562

Watson JD, Crick FH (1953) Molecular structure of nucleic acids; a structure for deoxyribose nucleic acid. Nature 171:737–738

Weichenhan D, Plass C (2013) The Evolving Epigenome. Hum Mol Genet 22:R1–R6

Willingham AT, Orth AP, Batalov S, Peters EC, Wen BG, Aza-Blanc P, Hogenesch JB, Schultz PG (2005) A strategy for probing the function of noncoding RNAs finds a repressor of NFAT. Science 309:1570–1573

Wood EJ, Chin-Inmanu K, Jia H, Lipovich L (2013) Sense-antisense gene pairs: sequence, transcription, and structure are not conserved between human and mouse. Front Genet 4:183. doi:10.3389/fgene.2013.00183

Xu D, Takeshita F, Hino Y, Fukunaga S, Kudo Y, Tamaki A, Matsunaga J, Takahashi RU, Takata T, Shimamoto A et al (2011) miR-22 represses cancer progression by inducing cellular senescence. J Cell Biol 193:409–424

Xue Y, Gu D, Ma G, Zhu L, Hua Q, Chu H, Tong N, Chen J, Zhang Z, Wang M (2015) Genetic variants in lncRNA HOTAIR are associated with risk of colorectal cancer. Mutagenesis 30:303–310

Yang G, Lu X, Yuan L (2014) LncRNA: a link between RNA and cancer. Biochim Biophys Acta 1839:1097–1109

Yu X, Zhang L, Wen G, Zhao H, Luong LA, Chen Q, Huang Y, Zhu J, Ye S, Xu Q, Wang W, Xiao Q (2015) Upregulated sirtuin 1 by miRNA-34a is required for smooth muscle cell differentiation from pluripotent stem cells. Cell Death Differ 22:1170–1180

Zaramela LS, Vencio RZ, ten-Caten F, Baliga NS, Koide T, 2014. Transcription start site associated RNAs (TSSaRNAs) are ubiquitous in all domains of life. PLoS One 9, e107680

Zhang X, Zabinsky R, Teng Y, Cui M, Han M (2011) MicroRNAs play critical roles in the survival and recovery of Caenorhabditis elegans from starvation-induced L1 diapause. Proc Natl Acad Sci USA 108:17997–18002

Zhao H, Wen G, Huang Y, Yu X, Chen Q, Afzal TA, Luong le A, Zhu J, Ye S, Zhang L et al (2015) MicroRNA-22 regulates smooth muscle cell differentiation from stem cells by targeting methyl CpG-binding protein 2. Arterioscler Thromb Vasc Biol 35:918–929

Zheng Y, Xu Z (2014) MicroRNA-22 induces endothelial progenitor cell senescence by targeting AKT3. Cell Physiol Biochem 34:1547–1555

Ziats MN, Rennert OM (2013) Aberrant expression of long noncoding RNAs in autistic brain. J Mol Neurosci 49:589–593

Chromatin and Aging

Jogeswar Satchidananda Purohit and Madan Mohan Chaturvedi

Abstract Recent studies from a number of model organisms have indicated chromatin structure and its remodeling as a major contributory agent for aging. Few recent experiments also demonstrate that modulation in the chromatin modifying agents also affect the life span of an organism and even in some cases the change is inherited epigenetically to subsequent generations. Hence, in the present report we discuss the chromatin organization and its changes during aging.

1 Introduction

Aging is a complex phenomenon and is a result of many biochemical and molecular changes in the cellular level. The major aging mediated changes in a cell include: reduction in number of mitotic cells; alteration in permeability of membrane; decreased hormone and enzyme secretion; decreased immunity; decreased antioxidant defence with concomitant increase in free radical generation; increased DNA damage (Ashok and Ali 2003). There are many theories to explain aging, such as the gene regulation theory. Recently, change in chromatin structure and epigenetic alterations have also been correlated to aging (Gravina and Vijg 2010). The change in the chromatin structure during aging can be traced back to the observation that chromatin became compact in aged rats with concomitant decline in transcription (Chaturvedi and Kanungo 1985). However, progress in the area of chromatin as a mediator of aging has been relatively slow and was only dependent on few sporadic observations. Few observations of the last decade including role of histone

J.S. Purohit
Cluster Innovation Centre, University Stadium, Delhi University, Sports Complex, G.C. Narang Marg, North Campus, Delhi 110007, India

M.M. Chaturvedi (✉)
Laboratory for Chromatin Biology, Department of Zoology, University of Delhi, Room no. 317, 2nd Floor, North Campus, Delhi 110007, India
e-mail: mchaturvedi@zoology.du.ac.in; director@cic.du.ac.in

© Springer Science+Business Media Singapore 2017

P.C. Rath et al. (eds.), *Topics in Biomedical Gerontology*,
DOI 10.1007/978-981-10-2155-8_11

deacetylases in extension of life span in few model organisms has reinforced the link of chromatin to aging.

Here, we describe age dependent changes in the chromatin structure and function. In the first section we give a brief account of the chromatin structure and organization. Then we describe different modes of chromatin remodeling such as ATP-dependent and ATP-independent (post-translational modifications of histones) remodeling. Then we describe different model organisms used for aging study. Finally we describe different changes in the chromatin organization that takes place during aging.

2 The Chromatin Organization

The amount of DNA in a eukaryotic cell is enormous (approximately two meters in case of a diploid human nucleus). Hence, in order to be packaged inside the tiny nucleus, which is approximately 10 μm in diameter, a huge level of folding is needed. To facilitate this, the eukaryotic DNA forms a large macromolecular complex called as chromatin, by associating with almost equal amount of histone and non-histone proteins (Kornberg 1974; Kornberg and Thomas 1974). The fundamental structural unit of chromatin is called nucleosome. Each nucleosome is a complex of a core histone protein octamer, around which 146/147 base pairs of DNA wraps around ~1.7 turns and ~54 bp (10–90 bp) remains as linker DNA. Arrays of nucleosomes appear as beads on a string structure of approximately 11 nm diameter and 5.5 nm in height (Kornberg 1974; Luger et al. 2012). These arrays of nucleosomes form the primary structure of chromatin and popularly named as 10 nm fibre.

The core histone octamer consists of histone H2A, H2B, H3 and H4, each in two copies, which are highly conserved among eukaryotes. These are lysine and arginine rich small basic proteins of molecular weight 11–22 kDa (Luger et al. 1997). The C-terminal regions of each core histone contain five α helical regions. Out of which, the central histone fold motif, comprise of α1 (9–14 a. a.), α2 (~29 a. a.) and α3 (9–14 a. a.). This central histone fold motif is flanked on either side by two other α-helices. These α-helices are connected to each other by short β-loops. The histone fold domains interact with each other to form H2A–H2B and H3–H4 heterodimers. In physiological salt concentrations, two such H3–H4 heterodimers interact with each other by a H3:H3 interaction, to form a tetramer. Similarly a H2A–H2B dimer associates on either side of the H3–H4 tetramer via a H2B:H4 interaction (Luger et al. 1997).

Binding of the core histones to DNA is mostly achieved by weak interactions, such as electrostatic interactions, between the positively charged histone and the DNA phosphate backbone (Luger et al. 1997). The minor groove of the DNA faces towards the histones. There are about 14 such contacts per nucleosome.

The loops of adjacent histones come together and a pair of such loops forms a β-bridge, containing two arginine residues, which interact with the minor groove of

DNA. The linker DNA and the nucleosome entry and exit points are bound by H1. The nucleosome containing H1 is called as "chromatosome" which spans ~168 bp (Fletcher and Hansen 1996).

The N-terminal tails of all the core histones, and the C-terminal tails of histone H2A are highly flexible and do not contribute much in the nucleosome formation. They emanate out of nucleosome core, are unstructured and are subject to diverse post-translational modifications (described later). These post-translational modifications act as epigenetic marks for the transcriptional activation and repression. In addition, these tails are responsible for making contact with other nucleosomes and subsequently form higher-order chromatin structure (Li and Reinberg 2011).

These 10 nm fibres undergo next level of compaction to form the 30 nm fibres, in which another six fold condensation of DNA takes place to form a secondary structure. Subsequent formation of the tertiary structure requires interaction between long-distance chromatin fibres by the involvement of histone N-terminal tails (Hansen 2002). The stabilization of higher order chromatin is achieved by the incorporation of linker histone H1 and its variants to the chromatin (Carruthers et al. 1998).

The exact structure of the 30 nm chromatin fibre still remains elusive (van Holde and Zlatanova 1995). However, to explain the conformation several models are proposed (see below). In low ionic strength, the chromatin appears as a beaded fibre, in which beads represent the nucleosomes (Olins and Wright 1973). The 10 nm fibres are wrapped in a left-handed fashion with six nucleosomes per turn to form the 30 nm chromatin structure. Two different models have been proposed to explain the 30 nm chromatin fibre structure; namely the solenoid model and the zigzag model (Robinson et al. 2006). According to solenoid model, the nucleosomes are folded to a simple helix, where the adjacent nucleosomes remain present next to each other in a chromatin fibre. In the zigzag model, the linker DNA is present in a straight path between nucleosomes (Woodcock et al. 1993). Here, two rows of nucleosomes are arranged in a zigzag fashion, due to interaction between alternative nucleosomes. This type of interaction produces a double helical structure (Bednar et al. 1998). Protein-protein cross-linking analysis revealed that from a single stack of 12 nucleosomes, two rows of six nucleosomes were produced as predicted for the zigzag model (Robinson et al. 2006).

The 300 nm fibre, which is the tertiary structure of the chromatin, is formed by folding of the 30 nm fibres. To achieve this folding, the 30 nm fibre assembles into loops, which are also associated with non-histone proteins. These loops are organized being part of scaffold proteins, which maintain its higher order structure (Woodcock and Ghosh 2010). They remain firmly attached to eukaryotic DNA even after its extraction under harsh conditions like high salt, strong detergents and polyelectrolytes. The scaffold is mainly comprised of two proteins i.e. scaffold protein 1 and 2, of M.W. 170 and 135 kDa respectively. The stretches of DNA loops that are associated with the protein scaffold are known as scaffold attachment regions (SARs). The 30 nm chromatin fibre loops are attached to the nuclear matrix via matrix attachment regions (MAR).

The 300 nm chromatin fibres further compact to form 700 nm chromatin fibre during the division phase of cell cycle. The mechanism of 700 nm fibre formation and its exact structure is still unknown. However, the metaphase chromosome is 1400 nm in diameter having two side-by-side 700 nm fibres. Altogether, it can be stated that the structure of 10 nm fibre is well understood and the structure of higher order chromatin structure still remains elusive (Woodcock and Ghosh 2010).

3 The Euchromatin and Heterochromatin: The Classical Domains of Chromatin

Historically, through staining methods, the chromatin was distinguished into a darkly stained heterochromatin and a light stained euchromatin regions. However, it is now established that these domains are differently folded (Grigoryev 2001). Some portions of the chromosomes have a dense, compact structure during interphase, in which genes are repressed, and can be stained with certain dyes. These regions are known as heterochromatin. The regions of chromatin that are relatively unfolded and permissive for gene expression are defined as euchromatin. The euchromatin and heterochromatin domains are interspersed along the chromosome and are demarcated by chromatin barriers known as boundary or insulator elements (Bi and Broach 2001). With recent studies, they have been shown to have differential pattern of post-translational modification in histones and DNA (described later).

4 Chromatin Remodeling

Chromatin becomes inhospitable to nuclear events, such as replication, transcription, recombination and repair, since the DNA remains occluded, being in the form of nucleosomes. This occlusion is reversed by chromatin remodeling. Chromatin remodeling refers to regulated alteration in the chromatin or in the nucleosome structure, making it transiently permissive/occlusive, so as to enhance/obscure binding of factors requiring DNA as template. ATP dependent chromatin remodeling factors and ATP-independent chromatin remodeling factors, with support of each other, mostly bring about the chromatin remodeling (Vaquero et al. 2003).

The ATP-dependent chromatin remodelers are multi-subunit protein complexes of >1 MDa. They utilize ATP and cause disruption in the nucleosome so that the DNA-histone contacts are loosened; mobilize the octamer in cis along a DNA sequence so that a particular DNA sequence becomes accessible; transfer the octamers in trans so that the promoters are freed to be accessed by transcription factors and RNA polymerase; cause looping in DNA; or incorporate histone variants to the nucleosomes. The above mentioned remodeling complexes mostly contain a highly conserved ATPase subunit of helicase superfamily. Depending on the presence of other flanking domains and accessory subunits, the remodeling complexes are further divided into SWI/SNF, ISWI, CHD and INO80 families (Clapier and Cairns 2009).

4.1 SWI/SNF (Switch Defective/Sucrose Non-fermenting) Family

The SWI/SNF group of remodelers (with an approximate size of ~2 MDa) contain an ATPases domain of the SNF2 subfamily and 8–15 other subunits (Vaquero et al. 2003). The C-terminal domain contains a bromodomain for binding to acetylated histones. The SWI/SNF complex was initially identified in yeast. Homologues of yeast *SWI2/SNF2* were subsequently identified and characterized from *Drosophila* (*brahma*), mouse (*mBrm* and *mBrg*) and human (*hBRM* and *BRG1*). Similar *BRG1* containing homologues have also been identified in rat, *Xenopus* and chicken (Panigrahi et al. 2003; Vignali et al. 2000). In yeast, there is another SWI/SNF like complex called as RSC (*R*emodels *S*tructure of *C*hromatin, size of ~1.5 MDa) (Cairns et al. 1996). The RSC contains four homologues of yeast SWI/SNF components.

The SWI/SNF and related complexes broadly function in reorganization and positioning of nucleosomes in the promoter region for facilitating transcription factor binding. They can cause random positioning of the nucleosomes in a chromatin domain using evenly positioned nucleosomes as substrates. They are also responsible for ejection of nucleosomes and exposing a target DNA site (Dechassa et al. 2010).

4.2 ISWI (Imitation Switch) Group of Chromatin-Remodeling Complexes

Identified first in *Drosophila,* ISWI-group of remodeling factors has a helicase-ATPase domain, but they lack the bromodomain, unlike the *SWI2/SNF2*. Instead their C-terminal domain contains a SANT or a SLIDE domain (Clapier and Cairns 2009). Later the ISWI protein was found to be a principal constituent of three ATP-dependent chromatin-remodeling complexes in *Drosophila,* namely NURF (*nu*cleosome *r*emodelling *f*actor), CHRAC (*chr*omatin *a*ccessibility *c*omplex) and ACF (*A*TP-utilizing *c*hromatin assembly and remodeling *f*actor) (Vignali et al. 2000). NURF was purified from *Drosophila* embryonic extracts as an activity that stimulated transcription factor binding to chromatin promoter. Similarly, CHRAC was purified as an activity that made chromatin accessible to nucleases. ACF, on the other hand, was responsible for nucleosome assembly with the help of histone chaperons (Vignali et al. 2000). ISWI-related complexes were also identified from yeast, *Xenopus* and human (Dirscherl and Krebs 2004). There is a considerable functional heterogeneity with the ISWI containing complexes. Except NURF, no other ISWI-complex has been shown to disrupt nucleosomes. However, they slide nucleosomes along DNA (cis-transfer) (Dirscherl and Krebs 2004).

4.3 Mi-2/CHD (Chromodomain, Helicase, DNA Binding) Chromatin-Remodeling Factors

These protein complexes were initially identified in mouse (Delmas et al. 1993). They broadly contain an N-terminal chromodomain, usually an essential domain for proteins that interact with and modify chromatin, an ATPase domain (a classical SNF2 group of helicases/ATPases), and a plant homeodomain fingers (a DNA-binding domain). Similar complexes were also identified in yeast, *Drosophila*, *Xenopus*, mouse and human (Woodage et al. 1997). In human cells *nu*cleosome *r*emodeling and *d*eacetylase complex, NuRD was identified (Xue et al. 1998). Apart from the regular components of CHD, it also includes *h*istone *deac*etylase *1 & II*. NuRD plays very significant role in cellular processes converging histone deacetylation, DNA methylation and cell-cycle regulation (Fugita et al. 2004).

4.4 INO80 (Inositol Requiring) Factors

These factors are characterized by a split ATPase domain due to a long insertion in the middle. Homologous factors of INO80 exist in yeast (INO80 and SWRI) and human (INO80, SRCAP and TIP60) (Bao and Shen 2007). They demonstrate different kinds of remodeling activities such as nucleosome repositioning, eviction, replacement and exchange of histone with histone variants, mediated by replication independent pathways. They have also role in double strand break repairs of DNA. The TIP60 complex has both histone acetylation and ATPase activity (Morrison et al. 2007).

5 ATP Independent Chromatin Remodeling: Covalent Modification of Histones

The ATP independent chromatin remodelers covalently modify the histone tails or the globular domains. Histones are one of the most conserved eukaryotic proteins and are subject to most diverse number of post-translational modifications such as:

acetylation, methylation, phosphorylation, ADP-ribosylation, ubiquitination, biotinylation, sumoylation, etc. (Vaquero et al. 2003). Altogether, they generate accessible or repressible chromosomal domains for further interaction of chromatin complexes.

5.1 Acetylation

Acetylation of histones involves transfer of an acetyl group to lysine residues, present mostly in the N-terminal tails and some core residues of histones. Acetylation of histones is a reversible process. Histone acetyl transferases (HATs) add acetyl groups while histone deacetylases (HDACs) remove it. The acetylated histones become less positive, resulting in decreased affinity for the negatively charged DNA that in turn weakens DNA binding (Hong et al. 1993). This brings about a permissive structure in the chromatin, for binding of different proteins factors to DNA (Lee et al. 1993). Transcriptional activation has been shown mostly synergistic to histone acetylation (Struhl 1998). Further, acetylated histone tails can directly recruit other chromatin interacting machineries, such as bromodomains containing factors. Hence, they act as docking sites for downstream events (Berger 2002; Zeng and Zhou 2002).

In case of vertebrates, all core histones are acetylated at multiple sites. The sites are: H2A (K5, K9); H2B (K5, K12, K15 and K20); H3 (K9, K14, K18 and K23); H4 (K5, K8, K12 and K16), except in yeast, where acetylation sites of H2A, H2B are little different from vertebrates (Roth et al. 2001). There are different HATs, with differential histone substrate specificities. The substrate specificity is accomplished by the interacting partners. The HATs can be broadly divided into type A and type B (Roth et al. 2001). The type B HATs are mostly cytoplasmic and acetylate free histones at sites which are crucial for deposition into nucleosomes. The type A HATs acetylate nucleosome bound histones and broadly belong to three families. The H3 and H4 acetylation is carried by HAT family member Gcn5, while, H4 and H2A are acetylated by the nucleosome acetyl transferase Esa1 (Suka et al. 2001). In vitro all four core histones are acetylated by p300 (Kuo et al. 1996; Ogryzko et al. 1996; Yang et al. 1996).

Histone deacetylation is carried out by histone deacetylases and it results in gene repression. All four core histones are deacetylated by Rpd3 deacetylase (Roth et al. 2001), while, H3 and H2B are deacetylated by Hda1 (Histone deacetylase 1) (Xu et al. 1998). Rpd3 and Hda1 cause deacetylation of larger regions of euchromatin (Vogelauer et al. 2000). Histone deacetylases are also responsible for deacetylation of non-histone proteins, such as p53, tubulin and various transcription factors (Jones and Baylin 2002; Timmermann et al. 2001). Depending on the functional motifs and domains, mammalian histone deacetylases are classified into Class I HDACs (1, 2, 3 and 8), Class II HDACs (4, 5, 6, 7, 9, 10) and class-III HDACs (NAD+ dependent Sir2 family) (de Ruijter et al. 2003; Grozinger et al. 2001). HDAC 11 is shared by both class-I and class-II HDACs. Histone deacetylases have role in turnover of histones, initiation of transcription and repression of active genes.

5.2 *Phosphorylation*

Phosphorylation of histone occurs in Ser, Thr and Tyr residues. It is also a reversible modification being regulated by kinases and phosphatases (Bannister and Kouzarides 2015). Phosphorylation of H1 and H3 has been correlated with chromosome condensation during mitosis (Hanks et al. 1983; Hirano 2000). Phosphorylation also adds a negative charge to histones. The kinases responsible for phosphorylation of histones can be divided broadly into two categories: (i) those dependent on cyclic nucleotide monophosphates for activation and (ii) those whose activity is independent of cyclic nucleotide monophosphates. However, many other functions of histone phosphorylation still remain elusive.

5.3 *ADP Ribosylation*

ADP-ribose molecules are transferred either to glutamic acids or arginine residues by poly (ADP-ribosyl) polymerase (Jacobson and Jacobson 1999). Though, all histones are subject to ADP-ribosylation; H2B and H1 are the most preferred substrates (Kappus et al. 1993). ADP-ribosyl transferase transfers single ADP-ribose to arginine or lysine residues of free histones in the cytoplasm. Subsequently, poly-(ADP-ribosyl) polymerase (PARP), carry out the elongation process in nucleus. PARP recognizes and binds to single or double strand DNA breaks and then exposes the DNA strand breaks to proteins involved in repair process (Le Rhun et al. 1998). The ADP-ribosylation is reversed by poly-ADP-ribose-glycohydrolase. Poly ADP-ribosylated histones are part of the relaxed chromatin. Further, it is known that by ADP-ribosylation, the H3K4me3 demethylase is excluded from the chromatin, making the chromatin domain transcriptionally active (Li et al. 2014).

5.4 *Ubiquitination*

Ubiquitin is a 76-amino acid protein (Conaway et al. 2002). Ubiquitination is another reversible modification, which occurs at lysines residues of histone H2A and H2B during cell cycle. H3 and H1 are also reported to be modified by the same mechanism (Belz et al. 2002). Monoubiquitination is a mark of both transcriptional activation and repression depending on the position and histone subtype. Monoubiquitination of H2BK123 in yeast is crucial for methylation of H3K4, a mark of transcriptionally active chromatin (Sun and Allis 2002). Similarly, monoubiquitination of H2AK119 is a mark if transcriptionally inactive chromatin. In vitro, polyubiquitination of histone H2A, H2B and H3 are targets for degradation (Jentsch and Schlenker 1995).

5.5 *Sumoylation*

Small Ubiquitin related modifiers (SUMO) is also involved in posttranslational modification of histones on lysine residues (Melchior 2000; Nathan et al. 2003; Robzyk et al. 2000; Shiio and Eisenman 2003). SUMO shares only 18 % identity with Ubiquitin group. Out of the four histones, histone H4 is efficiently modified by SUMO-1 or SUMO-3, both in vitro as well as in vivo. Acetylated histone H4 usually gets sumoylated. It leads to decreased acetylation of histone H3 due to recruitment of HDACs and increase in HP1 binding. Hence, sumoylation has a role in repressing transcribing chromatin (Girdwood et al. 2003). It has been hypothesized that probably sumoylation initiates attenuation followed by gene repression.

5.6 *Lysine Propionylation and Butyrylation*

Histones are subject to propionylation at K5, K8 and K12 of H4 and lysine butyrylation at K5 and K12 of histone H4 (Nathan et al. 2003). These sites are also known to be acetylated (Tanaka et al. 2004). Probably these two new types of modification create novel docking sites for downstream recruitment of other complexes which are yet to be elucidated.

5.7 *Deimination*

In this type of modification, an arginine or monomethyl arginine residue of the histone is converted to citruline (Cuthbert et al. 2004). In mammals, this conversion is catalysed by peptidyl deiminase 4 (PADI4) and this enzyme is known as arginine demethylase. The resultant change of the citruline instead of arginine reduces in the histones affects the positive charge of the histones and it also abolishes the respective docking site.

5.8 *Methylation and Demethylation*

Histone H3, H4 and H1 can be methylated on their arginine (R) and lysine (K) residues (DeLange et al. 1969; Gershey et al. 1969; Murray 1964; Patterson and Davies 1969). Depending on the number of methyl groups attached to ε amino group, lysines can be mono, di or tri methylated. Similarly, arginines can be mono or dimethylated. Lysine methylation can have mixed effects. It depends on the position of lysine residue which is methylated and the levels of methylation (mono, di or trimethylated). Methylation at K4, K36 and K79 of histone H3 are marks of transcriptionally active chromatin, while methylation at K9, K27 of H3 and K20 of H4 are marks of transcriptionally inactive chromatin (Cao et al. 2002; Peters et al. 2003; Rice et al. 2003; Schotta et al. 2004). The lysine methylation is reversibly regulated by histone methyl transferases and demethylases. Histone methyl transferases (HMTs) methylate histone lysine and arginine residues. They have been broadly classified into three protein families: the PRMT family methylates arginine residues (Bedford and Richard 2005; Zhang and Reinberg 2001), while the SET domain containing family and the Dot1/DOTL families methylate lysine residues (Bannister et al. 2002; Kouzarides 2002; Lachner et al. 2003). The SET1 group is specific for H3K4 (Ringrose and Paro 2004; Smith et al. 2004), the SUV 39 group for H3K9 (O'Carroll et al. 2000; Rea et al. 2000; Tachibana et al. 2001; Yang et al. 2002), the SET2 group for H3K36 (Peters et al. 2001) while ASH1 shows broad specificity and methylates H3K4, H3K9 and H4K20. The EZH family methylate K27 and K9 of histone H3 (Czermin et al. 2002; Muller et al. 2002), Dot1 containing proteins (Singer et al. 1998) methylates H3K79 of the core domain (Feng et al. 2002; Lacoste et al. 2002; Ng et al. 2002; van Leeuwen et al. 2002). However, in many of these methylases, cross-reactivity is also reported.

Unlike acetylation, in methylation, there is no major change in total charge on histones. Rather specific lysine residues, after being differentially methylated (mono, di or tri), act as docking sites for further recruitment of other interacting machineries. It finally results in generation of transcriptionally permissive/occlusive chromatin template (Strahl and Allis 2000; Turner 2000). For example, the chromodomain containing protein HP1 (heterochromatin protein 1) is recruited to methylated histone H3K9 and generates a heterochromatin domain (Bannister et al. 2001; Nakayama et al. 2001). The heterochromatinization spreads by self-association of many HP1 (Nielsen et al. 2002). Antagonistically, chd1, the chromodomain containing protein (a component of SAGA and HAT complex) is recruited to methylated H3 K4, and it further recruits SAGA and HAT complexes and ultimately generates a euchromatic chromatin domain (Pray-Grant et al. 2005).

5.9 Lysine Methylation in Transcriptional Silencing

Methylation of K9 and K27 of H3 and K20 of H4 are marks of transcriptionally silenced chromatin domains (Bannister and Kouzarides 2015). Methylation of H3K9 is a deterministic mark of constitutive heterochromatin and also marks chromatin domains for transcriptional silencing. The differential role is played by the fact that euchromatic genes are regulated being dimethylated at H3K9 and the heterochromatic silencing takes place by generating trimethylated H3K9. However, few deviations are also available (Vakoc et al. 2005). Similarly, H3K27 methylation is also a mark of transcriptional silencing and X-chromosome inactivation.

There are reports that methylation of H9 of H3 (H3K9me) is a prerequisite for DNA methylation (Jackson et al. 2002; Lehnertz et al. 2003; Tamaru and Selker 2001), as binding of HP1 to H3K9me recruits DNA methyl transferases. It has been further demonstrated that E2H2 complex, which methylates H3K27, also interacts with DNA methyl transferases (Vire et al. 2006). However, results supporting the opposite hypothesis have also been documented. There is significant reduction in H3K9 methylation marks in DNA methyl transferase deficient cells (Sarraf and Stancheva 2004).

H4K20 methylation also correlates with the condensed regions of chromatin. In vitro its methylation inhibits of H4K16 acetylation, which is a mark of transcriptionally active chromatin domain. The H4K20me is also an epigenetic mark of repression and generally goes synergistically to H3K9 methylation (Zhang and Reinberg 2001).

5.10 Lysine Methylation and Active Transcription

5.10.1 Methylation of H3K4

The complex associated with *setl* (COMPASS) in association with elongating RNA Polymerase II, methylates H3K4 (Briggs et al. 2001; Daser and Rabbitts 2004; Gerber and Shilatifard 2003; Krogan et al. 2002; Miller et al. 2001). The COMPASS of the MLL complex are classically known as generators of euchromatin marks (Roguev et al. 2001). Components of COMPASS generate trimethyl K4 of histone H3 (H3K4me3) marks (Schneider et al. 2005). Actively transcribing genes are enriched with both dimethyl and trimethyl H3K4 marks. While, the dimethyl modifications mostly restricted to the coding regions, the trimethyl modification localize specifically to the regulatory regions of genes (Bernstein et al. 2005). WDR5 and *Chd1* are recruited to the methylated H3K4, and they acetylate other lysine residues like H3K14 and K9. The K9 acetylation mark is antagonistic to K9 methylation mark and thus the respective chromatin domain remains in an activated state (Timmers and Tora 2005). Further, the WDR5 binds to H3K4me2 and is responsible for maintaining global trimethylated H3K4 condition (Wysocka et al. 2005).

5.10.2 Methylation of H3K36

Set2 was initially identified in yeast to have HMTase activity specific to H3K36 (Strahl et al. 2002). *Set2* was speculated to have role in transcriptional elongation as was reported to associate with phosphorylated RNA Polymerase II (Krogan et al. 2003). It is thus hypothesized that recruitment of *Set2* along with RNA Pol II is crucial for establishing K36 methylation on chromatin (Gerber and Shilatifard 2003, Hampsey and Reinberg 2003).

5.10.3 Methylation of H3K79

The nonset domain containing protein *dot1* (disrupter of telomeric silencing 1) methylates the core domain of nucleosomal histone H3. It is further seen that 90 % of the nucleosomal histone H3K79 remains methylated (van Leeuwen et al. 2002). The H3K79me mark prevents binding of silent domain Sir Proteins and maintains the chromatin in euchromatic state and prevents further heterochromatinization.

5.10.4 Histone Lysine Demethylation

Histone methylation was considered to be a relatively permanent modification since, the histone demethylases were not known. However, subsequently the histone demethylases have been identified and characterized, at least for quite a few methylated residues. Lysine specific demethylase-1 (LSD1) was shown to be specific for H3K4me2 (Shi et al. 2004). LSD1 is a component of Co-REST co-repressor complexes (Hakimi et al. 2002; Shi et al. 2003; You et al. 2001). Demethylation by LSD1 generates unmethylated histones, formaldehyde and H_2O_2 and in the process FAD^+ is reduced to $FADH_2$, which is then reoxidised by (O). The (O) is reduced to form H_2O_2. However, LSD1 shifts its specificity from H3K4me2 to H3K9me2, and acts as a H3K9 demethylase and transcriptional activator in the presence of Androgen receptor (AR) (Metzger et al. 2006).

Another group of proteins called Jumonji (JmjC) proteins are shown to have histone dimethylase activity (Tsukada et al. 2006). JHMD1/Fbx11 (JmjC domain containing histone demethylase1, alias Fax11) demethylates mono and di meH3K36. Similarly, JHDM2 demethylates H3K9me2 (Yamane et al. 2006). Recently JMJD 2A-2D groups are identified. They reverse trimethylation, specifically H3K9me3 and H3K36me3 (Whetstine et al. 2006). Since, H3K9me3 and H3K36me3 are total antagonists: as the former is a mark of transcriptionally inactive region and the later is a mark of region engaged in transcriptional elongation. The dual site specificity of the JMJD2A/C may suggest a coordinated role in regulating H3K9/K36 trimethylation for gene regulation (Carrozza et al. 2005).

5.10.5 Histone Arginine Methylation

Arginine methylation is mainly linked to active transcription. Protein arginine methyl transferases (PRMTs) transfers a methyl group to arginine residues (Gary and Clarke 1998). Arginine methylation occurs in R2, R17 and R26 of histone H3 and R3 of histone H4. Arginines can be mono- or dimethylated. Further, the dimethylation can be symmetric or asymmetric. The arginine methyl transferases (PRMT) which methylate Arg residues have been divided into two groups. The type-I enzyme catalyse the formation of -monomethyl and asymmetric dimethyl arginine residues, whereas the type-II enzyme catalyzes the formation of monomethyl and symmetric dimethyl arginines (Zhang and Reinberg 2001).

5.10.6 Reversal of Arginine Methylation

Monomethylated and unmethylated arginine can be converted to citrulline by arginine deiminase, PADI4 (Cuthbert et al. 2004). PADI4 can deiminate unmodified arginine and monomethyl (but not dimethyl) arginine. It is a Ca^{+2} dependent enzyme that acts as a repressor of transcription, in the *pS2* promoter. It demethylates H3 and H4 arginines (mono or unmethylate) by deimination, at positions R2, R8, R17 and R26 of histone H3 in vitro and R2, R8 and R17 in vivo. Similarly, JMJD6 demethylates H3R2me2 and H4R3me2 in vitro and in vivo (Chang et al. 2007).

5.10.7 DNA Methylation

Methylation also occurs in few nucleotides of DNA and is often synergistic or antagonistic to few histone methylations (described earlier). 5-methyl cytosine and subsequent guanine (5meCpG) dinucleotides are mostly marks of gene silencing and also confer epigenetic memory. DNA methylation is regulated by de novo DNA methyl transferases and maintenance methyl transferases (Marchal and Miotto 2015). The first group of methyl transferases, DNMT3a and DNMT3b, methylate one strand of the newly synthesized DNA and convert them to hemimethylated DNA. The hemimethylated DNA is converted to a methylated DNA by maintenance methylase, DNMT1. However, DNMT3a and DNMT3b have been shown to also act on hemimethylated DNA (Marchal and Miotto 2015). Few recent reports have also demonstrated methylation outside the CpG (Yang et al. 2013). The human genome has approximately 30 million CpG. Dense clustering of these CpG in some regions of the genome is observed and these regions are known as CpG islands. Presence of 5meCpG islands in the promoter region of a gene is a mark of silenced gene. 5meCpG in the coding region of a gene is also observed, however, its mode of regulation is yet to be properly understood (Shenker and Flanagan 2011). DNA demethylases were not known until recently and hence DNA methylation was believed to be a relatively permanent mark. Now, formation of 5-hydroxymethyl cytosine (5-hmeC) is believed to be an intermediate in DNA demethylation process (Pfaffeneder et al. 2011). Oxidation of 5-meC generates 5-hmeC and then to several intermediates which are observed (Jurkowski and Jeltsch 2011). These reactions are catalysed by TET (ten-eleven translocations) enzyme family. Embryonic stem cells have higher levels of 5-hmC and it decreases upon differentiation (Xu et al. 2011). However, few other methods are also proposed for demethylation of DNA. They include non-replacement of methylation marks in the daughter cells during division, removal of methyl group by DNA glycosylase and deamination of cytosine or oxidation.

5.10.8 The Histone Code and Cross-Talk in Histone Modification

Many of the histone modifications act as docking sites and cause further recruitment of protein complexes that bring about broadly two effects: a permissive chromatin domain for binding of factors; or an occlusive chromatin domain refractory for factor binding. Accordingly, a group of modifications coexist and work synergistically but are not compatible with others group of modifications (Jenuwein and Allis 2001). Say, for example, when H3K4 is methylated, there is synergistic acetylation of H3K14 and phosphorylation of H3S10. All are involved in transcriptional activation, and all of them are antagonistic to the inhibitory H3K9 methylation. Further, the complexity increases as lysines can be mono, di or tri methylated. Accordingly, they encode different signals for activation. Taken together, a language of histone modification can be created with the available information, which was initially framed as histone code (Strahl and Allis 2000). This code was then supported by yet another hypothesis that the simplest code can be binary (Jenuwein and Allis 2001). For example, H3K9 and H3K4 cannot be methylated simultaneously in the same chromatin domain. Likewise, H4K20 methylation and H4K16 acetylation are antagonistic. Distinct histone modifications, on one or more tails, act sequentially to form a "histone code", that is read by chromo and bromodomain containing proteins. This creates specific combinations of modifications on a local domain. All together they bring about formation of the euchromatic and heterochromatic domains of chromatin. A detailed account of the different post-translational modification and their cross-talk is represented in Fig. 1.

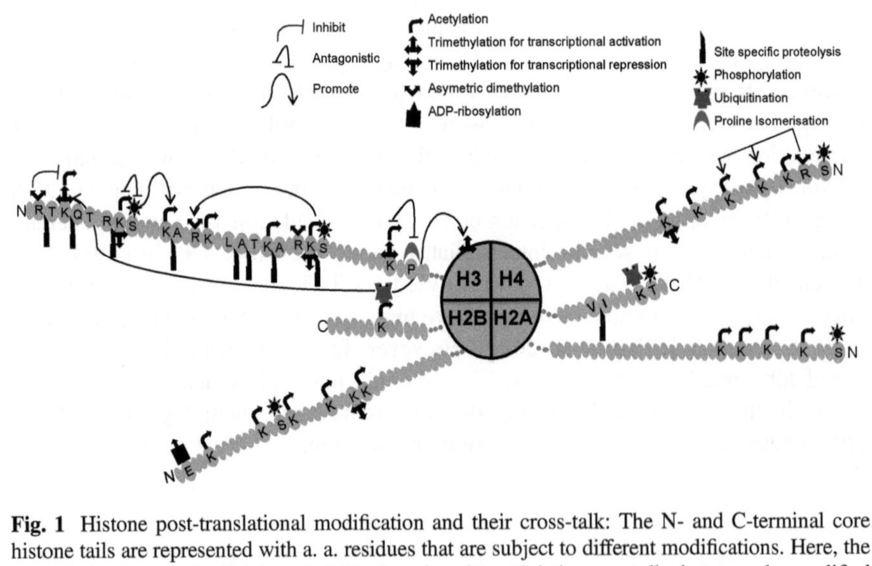

Fig. 1 Histone post-translational modification and their cross-talk: The N- and C-terminal core histone tails are represented with a. a. residues that are subject to different modifications. Here, the proteolytic cleavage is also included. Further, few deterministic cross-talks between the modified docking sites are also represented

5.10.9 The Proteolytic Processing of Histones

One permanent way of removal of the histone modification is by clipping the histone tails. Proteolysis of histones, though rare, is believed to be another mode of irreversible post-translational modification (Allis et al. 1980). The N-terminal/C-terminal tails of histones are more susceptible for the proteolysis, when histones exist in the form of chromatin. Proteolysis of histones emerged as a problem for chromatin isolators (Brandt et al. 1975). However, later the site specific histone proteolysis has been linked to aging and chromatin structure modulation (see below).

In calf thymus, chromatin-bound proteolytic activity which was specific to H2A was demonstrated (Eickbush et al. 1988). The cleaved product, which lacked 15 a. a. from the C-terminus, was named as cH2A. The cleavage site was between Val-114 and Leu-115. (Okawa et al. 2003). Two truncated forms of H2A have been purified from the nuclei of acute myeloid leukemia OCI/AML1a cells (Okawa et al. 2003). However, the enzyme catalysing this cleavage is not known. Proteolysis of histones has also been proposed during transformation of spermatids into spermatozoa (spermeogenesis), at the terminal stages of spermatogenesis (Marushige et al. 1976). During the process, chromosomal DNA becomes tightly packaged as a result of proteolytic replacement (Marushige and Marushige 1983) of somatic type histones, by protamines. The mechanism of histones displacement requires their proteolysis for removal. A protease activity associated with mouse testicular chromatin has also been observed, which might be involved in proteolytic removal of somatic histones during spermatogenesis (Faulkner and Bhatnagar 1987). Further, this specific proteolysis was absent in other tissues like thymus and liver suggesting tissue specific functional regulation.

Selective and site specific proteolysis of a specific histone subtype has also been observed during viral infection. There is histone H3 degradation in mammalian kidney cells (BHK cells) infected with Foot-and-Mouth Disease Virus (FMDV) (Grigera and Tisminetzky 1984). Further, it is demonstrated that the H3 degradation is catalysed by the FMDV 3C protease and cleavage site has been mapped between Leu-20 and Ala-21, from the N-terminal end (Falk et al. 1990). As the processed H3 lacks most of the N-terminal acetylation sites, it would shut off/reduce the host cellular transcription (Tesar and Marquardt 1990).

Proteolytic processing of histone H1 and H3 has been observed as a physiologically and developmentally regulated event in *Tetrahymena* nuclei (Allis et al. 1984). Histone H3 is cleaved at multiple site spanning between a.a. 21–27 by Cathepsin L during mouse embryonic stem cell differentiation (Duncan et al. 2008). The protease Cathepsin L is extensively characterized. The H3 cleavage by Cathepsin L is regulated by specific covalent modification code of histone H3. Recently, in chicken liver tissue, glutamate dehydrogenase was identified as a histone H3 N-terminus specific protease (Purohit et al. 2013). A yet to be identified histone H2A specific protease is also characterized from chicken liver (Panda et al. 2013). In summary, in recent years histone specific proteolytic cleavage and histone subtype non-specific degradation have also been observed simultaneously, suggesting that histone-specific and histone non-specific proteases exist in nature and they have evolved along with histones. Hence, it is recently opined to distinguish these proteases into broadly two groups: 1. histone proteases with broad spectrum specificity mostly function for turnover of histones. 2. Histone proteases with site specificity have role in modulating chromatin structure and function by modulating epigenetic marks (Dhaenens et al. 2015).

6 Change in Chromatin During Aging

With the preliminary understanding on the chromatin organization and function, in the following section we intend to describe the age dependent change in the chromatin organization and function. Since, aging is naturally a slow process to score, at least in a number of higher organisms; simple models are developed to study aging. So in the first part of the following section, we describe different suitable model systems which are presently being studied to understand the chromatin structure and aging. In the subsequent section we will describe different changes in the chromatin organization and function that take place during aging.

7 Organismic Models for Studying Chromatin and Aging

7.1 Yeast

Due to the ease in genetic manipulation, short life cycle and conservation of mechanisms of aging across the eukaryotes, the budding yeast *Saccharomyces cerevisiae* is a growing model for aging study (Wood and Helfand 2013). One of the measurement parameters for aging is the replicative life span in budding yeast or in other words the number of daughter cells a budding mother cell produces before it reaches non-dividing stage. It has been observed convincingly that there is loss of transcriptional silencing with concomitant decrease in chromatin compaction in budding yeast during the process of aging (Smeal et al. 1996). In budding yeast, three regions of the genome remain silent: viz. the mating type switching locus (MAT locus), the ribosomal DNA locus (rDNA) and the telomere proximal regions. In aged yeast, there is loss of silencing of the MAT locus leading to co-expression of both mating type genes, causing insensitivity to pheromones and increased sterility in aged yeast. Further, there is movement of silencing proteins such as HDAC (NAD^+ dependent histone deacetylase) and silencing information regulator 2 (Sir2) away from the telomeric region in the aged yeast (Kennedy et al. 1997). Also, in the aged yeast, there is loss in silencing of the rDNA region with concomitant rise in recombination and formation of extra ribosomal chromosomal circles and increase in genomic instability. These events can be monitored with limiting replicative life spans. It is further seen in yeast model that deletion of Sir2 reduces replicative life spans, while introduction of an extra copy of Sir2 gene increases the same (Etzioni et al. 1999). Hence, from these observations, it can be concluded from yeast model that chromatin of the silent regions remains in more relaxed state in aged yeast. Sir2 is known to deacetylate H4K16Ac (Imai et al. 2000). Further, there is an age dependent decrease in Sir2 proteins in yeast leading to increase in H4K16Ac mark in global level (Dang et al. 2009). The H4K16Ac is a unique epigenetic mark which inhibits formation of 30 nm fibres and subsequent folding of the chromatin into higher order. Thus, age dependent increase in H4K16Ac leads to formation of open chromatin structures globally (Shogren-Knaak et al. 2006).

There is also change in the gene expression pattern during aging in yeast (Wood and Helfand 2013). Some of the change in gene expression could be due to elevated stress responses and induction in repair genes during aging. However, a significant amount of change in the gene expression could also be due to altered chromatin structure during aging. It is observed that there is increased transcription of the core histones with aging. On the other hand, short-lived yeast mutants deficient in the ability to acetylate H3K56Ac have decreased level of core histone transcript in aged cells, which is in turn similar to that of telomerase deficient cells (Lesur and Campbell 2004). H3K56Ac is enriched in histone gene promoters for transcription. Conversely, elevation in the histone transcript levels in aged yeast does not correlate with increase in bulk histones in the aged yeast. In fact, there is decrease in bulk histones in aged yeast cells (Feser et al. 2010). This could be due to the feedback inhibition of the histone transcript translation. Further, by ChIP analysis it has been revealed that there is reduced histone occupancy in many regions of the chromatin including the telomere proximal region, the rDNA region and the MAT locus in the aged yeast (Wood and Helfand 2013). The loss of histone from the chromatin of aged yeast leads to the formation of a more open chromatin and in turn increases transcription of some genes. On the other hand, lower concentrations of the synthesized histones in the aged yeast leads to decreased pool of free histones and also alters epigenetic marks of the histones. In another line of evidences, it has been found that aged yeast cells have decreased polyamine synthesis (amines are required for cell proliferation). Yeast cells treated with spermidine showed hypoacetylation of H3K9, K14, K18 and H4K16 with concomitant decrease in HAT activity, specific for these sites and increase in replicating life span (Eisenberg et al. 2009). However, it was in contrast to the hyperacetylation of the H3K56 sites suggesting again the site specific role of diverse histone acetylations in modulation of aging in yeast model.

7.2 *Drosophila*

Similar to the change in the acetylation at multiple sites of histone H3 and H4, the marks of active chromatin, from aged yeast, a decline in few methylation sites (which are mark of active chromatin) have also been reported from aged

Drosophila models. There is decrease in H3K4me3 and H3K36me3 marks with age in *Drosophila* (Wood and Helfand 2013). However, there is also less enrichment in the H3K9me3, and Heterochromatin protein1, HP1 (mark of inactive chromatin) in the pericentric heterochromatin. These flies with decreased HP1 expression also exhibited increased levels of rRNA transcripts. Recently it has been reported that flies with decreased HP1 expression exhibited shorter lifespan whereas in flies where HP1 was overexpressed, there was increase in the lifespan (Larson et al. 2012).

7.3 *Caenorhabditis elegans*

In *C. elegans,* a number of histone modifications have been recently correlated with the life span of the organism. The methylation marks of K4, K9 and K27 of H3 are extensively studied in the worm (Padilla et al. 2014). It is found that disruption of the H3K4 methyl transferase complex (ASH-2), leads to increase in the lifespan of the worm, while disruption of the H3K4 demethylase (RBR-2) leads to decrease of the same. It has been further shown that this increase in life span is inherited up to certain generations (Greer et al. 2011). Similar observations are also reported for other modifications sites of histone H3. An aging dependent decline in the H3K27me3 is observed in the worm with concomitant increased activity of the respective histone demethylase (UTX-1). It has been further seen that disruption of UTX-1 results in increased levels of H3K27me3 marks in the worm genome and there is increase in the life span (Maures et al. 2011). Synergistically, disruption of the LSD1 complex that acts as a demethylase for H3K4me and H3K9me sites also results in increased life span in the worm (McColl et al. 2008). However, it is not clear from these observations, why there is an increase in life span on hypermethylation of activation marks such as H3K4 as well as repression marks such as H3K9 and H3K27. It can be suggested that probably gene specific transcription plays a more important role during aging than the global level change in the transcription of an organism which can also be tissue specific.

7.4 *Mammals*

Mouse, rats and few human diseases are also used as good models to study age related changes in chromatin. Analysis of the 30 nm fibres and *Micrococcal* nuclease (MNase) digestion of the nuclei have indicated that there is irregular positioning of the nucleosomes in the chromatin of aged fibroblasts. It suggests that there is loosening of the chromatin structure with age (Macieira-Coelho and Puvion-Dutilleul 1989). There is also decrease in heterochromatic regions like less efficient inactivation of the X-chromosomes in aged mice (Cattanach 1974).

Genetic disease such as progeria in mammals mimics the aging process and pose as a good model for studying aging (Sedivy et al. 2008). The disease is caused due to mutation in Lamin A gene which forms a part of the nuclear envelope (Oberdoerffer and Sinclair 2007). Mutation of the Lamin A gene by altered splicing, generates a truncated lamin A protein, termed as progerin (Feser and Tyler 2011). Accumulation of lamin A in the nuclear envelope causes a number of downstream effects. There is decline in methylation marks for heterochromatin, like H3K9me3 and H3K27me3 levels synergistic to decline in HP1 and EZH2 (H3K27 methyl transferase). It results in increased transcription from pericentric heterochromatin, shortening of telomeric regions and disruption of nuclear architecture. Progerin expresses at very basal level in young and old mammals. Overexpression of progerin results in all these change in the histone modifications status, described as above and disorganization of the nuclear architecture. Even, by blocking the cryptic splicing site and correcting the splicing defect, the abnormal cell can be reverted back to normal and delay aging process (Feser and Tyler 2011).

It has been shown that healthy aged cells have decreased levels of NURD complex (Meshorer and Gruenbaum 2009). Progeric cells also show decreased level of NURD complex and concomitant loss of HP1 and H3K9me3 along with severe DNA damage (Feser and Tyler 2011). Hence, these systems also pose a good model to study NURD mediated epigenetic signalling for aging.

8 Change in Chromatin Structure and Function During Aging

8.1 Change in Nuclear Architecture During Aging

As discussed earlier, in yeast, the heterochromatic regions mainly silence the mating type locus and stabilize the rRNA genes and telomere regions by preventing them from recombination. As the rDNA region is repetitive, it is also prone to recombination. In yeast, the rRNA region is stabilized and silenced by REST (regulator of nucleolar silencing and telomere exit) complex (Oberdoerffer and Sinclair 2007). This complex is a Sir2 containing complex. Sir2 is a major histone deacetylase responsible for deacetylating histone H4K16Ac. The Sir2 interacts with Sir3 and Sir4 not only in the rDNA region but also in the MAT and the telomeric regions and is regulated by them (Kaeberlein et al. 1999). In yeast, the highly repetitive rDNA region is prone to recombination. Recombination mediated excision in these regions generates extra-chromosomal circles. During subsequent cycles these extra-chromosomal circle keep on accumulating in the nucleolus and cause cell death. Hence, there is a need to silence the recombining rDNA region to increase genomic stability and delay aging, which is done by the Sir2 complex. A truncated Sir4 protein, no longer interacts with the telomeric region. Instead, more Sir2 and Sir3 are targeted to the nucleolus. It results in extension in life span of the yeast (Straight et al. 1999).

Similar observations are also obtained from human progeroid disease, Werner Syndrome (WS) (Oberdoerffer and Sinclair 2007). In this disease, there is a mutation in the DNA helicase gene. This mutation also causes genomic instability by extra-recombination in the rDNA region and results in early aging. In other similar diseases such as HGPS (Hutchinson-Gilford Progeria Syndrome), there is a mutation in the Lamin A gene which, encodes a protein of the nuclear membrane, as described earlier (Oberdoerffer and Sinclair 2007). It disturbs nuclear membrane architecture with concomitant loss of HP1 proteins. HP1 is very much essential for maintenance of the integrity of the constitutive heterochromatic region as it binds to methylated H3K9. Loss of HP1 leads to disturbance in the heterochromatic regions. Further findings have shown that a similar splice variant of lamin A also accumulates in normal aged individuals indicating similar events in the normal aging and disease induced aging processes (Scaffidi and Misteli 2006). Similarly, in the AT (ataxia telangiectasia) disease also there is mutation in the ATM gene which is a component of the DNA repair cascade important for telomere maintenance. Defective ATM products also disturb telomere and nuclear matrix interaction and enhance aging (Greenwell et al. 1995).

Age dependent changes in the heterochromatic regions, such as centromeric region are also reported. The pericentric-heterochromatin region around the centromere becomes transcriptionally more active in aged cardiac tissue (Gaubatz and Cutler 1990). There are further supporting observations hypothesizing that change in the perinuclear architecture contributes to aging in a number of organisms (Oberdoerffer and Sinclair 2007). Few recent observations have shown that there is a notable increase in the total facultative heterochromatin domains in the senescent cells (Adams 2007). By this process there is large scale reorganization of the heterochromatic regions in an age dependent manner.

8.2 Age Dependent Change in Core Histone Expression and Deposition into Nucleosomes

An age dependent change in expression of core histone is recently reported in many organisms such as yeast and mammals. There is reduction in the core histone expression during replicative aging in yeast with concomitant decrease in histone occupancy (Hu et al. 2014). It leads to aberrant up-regulation of associated genes and also causes genomic instability. Further, in yeast models, deletion of anti-silencing function-1 and CAF-1, (the central chaperone complex required for deposition and removal of histones from chromatin) causes drastic reduction in its life span (Feser et al. 2010). It has been seen that though there is significant increase in the histone transcripts during aging, there is global decrease in the chromatin bound core histones during aging. Supply of extra core histone or ectopic expression of histone H3 and H4 and not

H2A and H2B increases the replicative life span in yeast. Hence, it is concluded that expression of H3 and H4 dimer promotes nucleosome deposition and transforms again to a more uncompact form and thereby decrease genomic instability and increase life span. Though, to the best of our knowledge, similar experimental observations are limiting from mammalian cells, it can be argued that similar process can also be applicable to mammalian systems, as similar decline in histone expression is also observed in mammalian cells too.

8.3 Change in DNA Methylation

In vertebrates, transcriptionally silent regions are generally marked by the presence of 5-methyl cytosines in the CpG dinucleotide sites (described earlier). There is reduction in the 5meC in aged human cells. Further studies have indicated that there is decrease in the 5meC in the CpGs present outside the promoter regions in different human tissues (Benayoun et al. 2015). Conversely, there is an increase in 5meC in the promoter region of aged tissues of human and mice. It is further hypothesized that the methylation in the promoter regions of development related genes probably contribute to the aging mediated mis-regulation of gene expression. Recent reports have suggested a link between human age and 5meC status of DNA (Zou et al. 2014). The stem cells (embryonic and pluripotent stem cells) are estimated to be ageless in correlation to the DNA methylation age. Similarly, sperms and ovum have younger DNA methylation ages compared to respected differentiated tissues from the same individuals. Further, a comparison of the DNA methylation status of different tissues can also be useful for predicting the health of various tissues. For example, in obese individuals, there is increased DNA methylation age of liver than other tissues of the same individual. Similarly, individuals with Down's syndromes have increased methylation age of blood cells compared to normal individuals with similar biological age.

Similar observations are also seen in model organisms for aging study. In *Drosophila,* overexpression of dDnmt2 (DNA methyl transferase) gene increases longevity, whereas deletion of the gene generates short lived flies (Benayoun et al. 2015). Although, the mechanism of methylation mediated modulation of aging is still unclear, the results obtained from *Drosophila* and human samples pose a relationship between DNA methylation and aging and suggests that DNA methylation could be used as a general biomarker for aging.

8.4 Changes in Histone Methylation

As discussed earlier, methylation of core histones at specific sites, mark either the active or repressive chromatin. Further, histone methylation is reversibly regulated by histone methyl transferases and demethylases. With aging, there is global loss of heterochromatin domains with concomitant redistribution of heterochromatin silencing proteins (Tsurumi and Li 2012). Increase in H3K27me3 by decreasing expression of H3K27me3 demethylase results in increase in life span in *C. elegans* (Ni et al. 2012). However, the results are contradictory from different organisms. In *Drosophila*, reduction in expression of H3K27me3 methyl transferase with concomitant reduction in H3K27me3 increases life span (Ni et al. 2012). On the contrary, in muscle stem cells of old mice, there is increase in gross H3K27me3 (Baumgart et al. 2015; Liu et al. 2013). The contradictory increase or decrease in methylation status of H3K27 further suggests the fact that the methylation signals are tissue and organism dependent.

Age dependent redistribution of H3K4me (a mark of active chromatin) is also observed (Shah et al. 2013). Spreading and redistribution of H3K4me3 occurs in aging fibroblast of humans and hematopoietic stem cells of mice. Further, modulators of H3K4me also influence longevity. It is observed that knocking down the main component of the H3K4 methyl transferase complex increased lifespan in *C. elegans*. On the contrary, when the respective demethylase gene was knocked down, it resulted in shortened lifespan (Greer et al. 2011). Further, overexpression of the H3K4 demethylase in the germinal cells increased lifespan. Similar observations are also obtained from other histone H3 amino acid residues which act as site for methylation. Mutation of the H3K36 demethylase gene causes expansion in replicative age in yeast. Synergistically, a mutation the K36 residue in H3 of yeast, which makes it refractory to be methylated, shortens life span (Sen et al. 2015). Further, *in C. elegans*, knockdown of the H3K36me3 methyl transferase gene also shortens life span of the worm (Pu et al. 2015).

8.5 Changes in Histone Acetylation

The acetylation pattern of core histones has been reported to change during normal aging in a number of organisms (Benayoun et al. 2015). In yeast models, during replicative aging, there is bulk level decrease in H3K56Ac, while there is increase in H4K16Ac (Dang et al. 2009). On the contrary, there is age dependent decrease in H4K16Ac in mouse and progeroid models (Krishnan et al. 2011). In aged mice also there is a lack of transcription mediated upregulation in H4K12Ac, which is a mark of transcriptional elongation.

Concomitant to the age dependent increase in H4K16Ac, the enzyme specific for deacetylating the H4K16Ac, the Sir2 also modulates life span in yeast. Increasing dose of Sir2 (Sirtuin silent information regulator 2) or inducer of Sir2 such as Resveratrol, extends life span in yeast and other model systems (Kaeberlein et al. 1999). Similarly, orthologues of Sir2 such as SIRT6 in mice deacetylates H3K9Ac and H3K56Ac. Deficiency of SIRT6 generates progeroid like phenotypes in mice, hypothesizing that it has role in aging. On the contrary, overexpression of sirt6 increases longevity (Kanfi et al. 2012). Even senescence can be induced in these cells by exposing these cells to a HDAC inhibitor such as Trichostatin A. The exact mechanisms of modulation of aging by *Sirt* genes are unclear, but it is hypothesized that they recruit specific chromatin remodeling factors and promote genome stability. There is also decrease in HDAC-1 expression during serial passaging of human fibroblast cells.

8.6 Change in Other Histone Modifications

Though, not very well studied, histone modifications other than DNA methylation, histone methylation and histone acetylation also have contributory effects in modulating aging like phenomenon. For example histone H2A, H2B and H4 can be modified with an O-N-acetyl-glucosamine (OGNAc). These OGNAc attached histones are enriched in gene promoters important for aging and stress responses in *C. elegans* (Benayoun et al. 2015). Further, deletion of the OGNAc depositing gene shortens the lifespan of *C. elegans;* while, deletion of the gene that removes the OGNAc mark extends the life span. Age dependent changes in the oxidative modifications of histones such as carbonylation are also reported in rat liver (Sharma et al. 2006). While, the rate of carbonylation was higher for the above mentioned histones in young rat liver, the carbomylation was significantly lower in the liver histones of aged rats. Further, dietary restriction of older rats increased carbamylation rates. A detailed change in the histone modification is represented in Table 1.

8.7 Generation of Heterochromatic Foci

Senescent cells show discrete Senescence associated heterochromatic foci (SAHF) (Kosar et al. 2011). SAHF structures are readily stained by DAPI and show resistance to be digested by nucleases, suggesting a more compact structure. SAHF structures were first identified in aged human fibroblast cells (Glauche et al. 2011). SAHF contain chromatin domains that are resonant of the constitutive heterochromatin domains, such as enrichment of H3K9me3 and HP1and hypoacetylated histones and incorporation of macroH2A. However, these foci lack linker histone H1. It is further seen that each chromosome condenses into a single SAHF. Preliminary experiments have shown that disruption of the H3K9 methyl transferase promotes tumour formation in the lymphocytes of mice. In these cells there is prevention of senescence suggesting that SAHF complex formation is crucial for occurrence of senescence (Kosar et al. 2011).

8.8 Age Dependent Change in Chromatin Remodeling Factors

Age dependent changes in the expression of remodeling complexes are also observed, in a number of organisms. The components of NuRD (nucleosome

Table 1 A summary of aging dependent changes in chromatin organization and function: Here, '0' represents no change, '+' represents increase and '−' represents decrease in the specific modification

Epigenetic mark	Function	Age dependent change	Deviations if any
5-meC	Transcriptional repression	Global change: 0 Local changes: + Only local level increase in some region and decrease in the other	In cellular senescence model there is global decrease and local increase
Core histone expression	Form nucleosome and are generalized repressor	−	No change in histone H3 expression in Drosophila head tissue whereas decrease seen in other models such as *C. elegans*
Macro H2A	Component of heterochromatin	+	Nil
HP1	Component of heterochromatin	−	HP1β increases in human fibroblast senescence model.
H3K9me	Transcriptional repression mark of euchromatin	+ (Human fibroblast senescence model)	Nil
H3K9me2	Transcriptional repression mark of euchromatin	− (Human fibroblast senescence model and *Drosophila*)	Nil
H3K9me3	Enriched in heterochromatin	−	Increase in *Drosophila*
H3K27me3	Transcriptional repression mark of euchromatin	−	Increase in mouse brain, muscles and HSC
H4K20me2	Genomic instability and DNA repair	+ (Human fibroblast senescence model)	Nil
H4K20me3	Enriched in pericentric heterochromatin	+	Decrease in human fibroblast senescence model
H3K4me2	Enriched in transcriptionally active gene bodies	+	Nil
H3K4me3	Enriched in transcriptionally active gene promoters	0	Minor increase in HSC, muscle stem cells and neurons of human. Decrease in head tissue of *Drosophila*.
H3K56Ac	DNA damage, replication and nucleosome assembly	− (Yeast and cellular senescence model)	Nil

(continued)

Table 1 (continued)

Epigenetic mark	Function	Age dependent change	Deviations if any
H4K16Ac	Telomere silencing and chromatin compaction	–	Increase in yeast
H4K12Ac	Mark of transcriptional elongation Enriched in active gene body	–	Nil

remodeling and deacetylase) complex are found to be down regulated in aged healthy individuals as well as in HGPS patients as studied in fibroblasts (Pegoraro et al. 2009). Further, it is also demonstrated that in *Hela* cells, RNAi mediated down-regulation of NuRD results in loss of heterochromatin (Pegoraro et al. 2009). Similarly, in yeast and *C. elegans,* deletion of ISWI (chromatin remodeling imitation switch) complex increases life span (Benayoun et al. 2015). However, a functionally active SWI/SNF is required for promoting longevity in *C. elegans*. Few other observations also state that in human adrenal cortex carcinoma derived cell lines, senescence can be induced by *BRG1*. It is further seen that *BrmI* level increases in old mice liver correlating to age dependent repression complex formation. Further, when nuclear extract of young animals were incubated with *BrmI*, it resulted in the formation of similar complexes observed in aged animals (Benayoun et al. 2015).

8.9 Age Dependent Changes in Histone Variants and Histone Exchange

In non-replicating cells such as neurons, there is replication independent enrichment of histone H3.3, a histone H3 variant of the chromatin (Pina and Suau 1987). It has been further shown in chicken and mice model that during aging there is increase in H3.3 level (Urban and Zweidler 1983). This observation also correlates well with the hypothesis that aged cells have more open chromatin and hyperacetylated chromatin. In neurons, there is formation of an open chromatin due to incorporation of H3.3. In vitro also there is increased incorporation of H3.3 in human fibroblast cell lines entering aging. The H3.3 specific chaperons also increase in aged baboons (Jeyapalan et al. 2007). The histone H2A variant H2A.Z is also correlated with aging. Knockdown of H2A.Z variant or p400 (the H2A.Z exchanger) promotes aging in human fibroblasts (Chan et al. 2005).

8.10 Age Dependent Proteolysis of Histones

Two different forms of histone H3, namely a 'slower migrating' ($H3^S$) and a 'faster migrating' ($H3^F$) have been detected in *Tetrahymena thermophila* micronuclei. Using partial proteolytic peptide mapping it was suggested that the $H3^F$ was a proteolytically clipped product of $H3^S$. It was cleaved by six amino acids residues from its N-terminal end, and it was a physiologically regulated proteolytic processing event (Allis et al. 1984). It has been further shown that when macronuclei became senescent and transcription was inactivated, the N-terminal tails of the core histones were proteolytically removed (Lin et al. 1991). It has been further proposed that the macronuclei, which were transcriptionally active, did not own such a proteolytic event.

Few recent observations have demonstrated the proteolytic processing of histone H3 and histone H1 in old rat liver and chicken (Chaturvedi, M.M. group, unpublished observation) and in Japanese quail (Mahendra et al. 1999; Mahendra and Kanungo 2000; Mishra and Kanungo 1994). It has been further shown that progesterone induces the H3 cleavage in Japanese quail, suggesting that the protease specific to histone H3 may be regulated by progesterone. Shanti, 1995, observed an additional band in the histones prepared either from nuclei or purified nucleosome-core from liver of old rats (Chaturvedi, M.M. group, unpublished observation). This band migrated between histone H2A and H4. The appearance of this band had a strong correlation with decrease in the stoichiometry of the histone H3. It was speculated that this band might be a clipped product of H3, and hence named as ΔH3. However, the precursor-product relationship between H3 and ΔH3 was not established. The ΔH3 generation has also been demonstrated recently in case of old chicken liver (Purohit et al. 2012). Through N-terminal sequencing, it has been shown that the ΔH3 was an N-terminally clipped product of histone H3, being cleaved at 23 amino acids from the N-terminus. The other histone post-translational modification profile of the ΔH3 is presently being investigated by western blotting (Chaturvedi, M.M. group).

9 Conclusion and Future Prospects: Attaining Epigenetic Rejuvenation?

A number of increasing lines of evidence propose chromatin structure as a modulator of aging. There is aging dependent change in regulation of transcription and nuclear architecture. When organisms age, there is decline in maintenance of the cellular structure and function which brings about change in the chromatin structure and organization. However, the age dependent change in the chromatin structure is multifaceted and often antagonistic in different organisms. For instance, in most of the organisms, with aging there is decreased histone expression and decrease in heterochromatin abundance with concomitant formation of an open chromatin

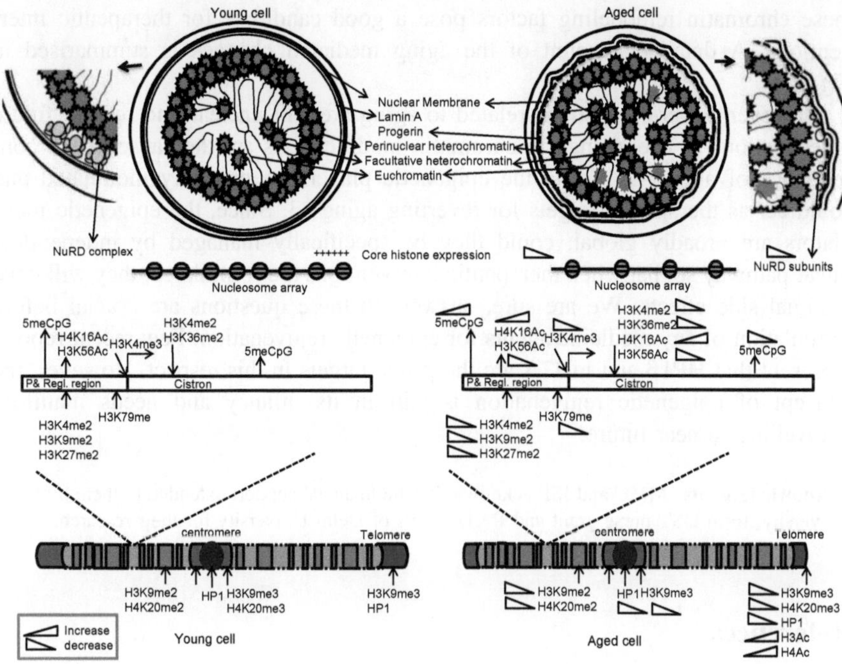

Fig. 2 Summary of the changes in the chromatin organization and function in aged cells. Normal cells have distinct perinuclear heterochromatin foci and few facultative heterochromatin loci. Aged cells on the contrary have disorganized perinuclear heterochromatin with concomitant increase in facultative heterochromatin loci. Also, in contrast to the Lamin A deposition in the nuclear membrane and interaction of NuRD complex to lamin A in young cells, there is an age dependent accumulation of progerin leading to disorganization of the nuclear architecture. In addition, there is decreased expression of NuRD complex in aged cells. Further, in aged cells, there is decreased core histone expression leading to reduced histone occupancy in the nucleosomal array. In addition to that there are a number of changes in the histone and DNA PTM, such as methylation and acetylation in the promoter and the coding region of the genes, which are indicated

structure. However, in the replicating senescence model there is age dependent accumulation of heterochromatic foci. The other examples also include the redistribution of SIR proteins in yeast during aging from the telomere and MAT locus to the rDNA regions promoting rDNA silencing and telomere shortening. At present restoring back to a youthful chromatin structure involves elevation in histone expression, reduction in acetylation of H3-N-terminal and H4K16 and inactivation of the HDAC Rpd3.

The epigenome is also influenced by environmental factors. Throughout the lifespan of an organism, the insults induced by the environment also contribute to the functioning of cells to organs. Thus it can also be proposed that the epigenomic instability induced by environment and life style also influence aging as represented by the methylation clock. It can be further stated that modulation of the epigenetic reprogramming by altering chromatin remodeling factors can revert aging. Hence, these chromatin remodeling factors pose a good candidate for therapeutic interventions. A detailed account of the aging mediated changes is summarised in Fig. 2.

However, several questions related to aging are still unclear and require future investigation. 1. The change in chromatin modifications are the cause or the consequence of aging? 2. Could the epigenetic players be precisely modulated and could act as therapeutic targets for reverting aging? 3. Since, the epigenetic modulators are broadly global; could they be specifically managed by independent linear pathway so that when therapeutic interventions are formulated, they will have minimal side effects. We are sure, answers to these questions are crucial before formulation of any medicine/agents for epigenetic rejuvenation. Few recent reports represent that HP1β and mH2A are the prime targets in this respect. However, the concept of epigenetic rejuvenation is still in its infancy and needs manifold unravelling in near future.

Acknowledgments MMC and JSP acknowledge the financial support extended to them by Delhi University, form DST purse grant and R&D grants of Delhi University for their research.

References

Adams PD (2007) Remodeling of chromatin structure in senescent cells and its potential impact on tumor suppression and aging. Gene 397:84–93

Allis CD, Allen RL, Wiggins JC, Chicoine LG, Richman R (1984) Proteolytic processing of h1-like histones in chromatin: a physiologically and developmentally regulated event in *Tetrahymena* micronuclei. J Cell Biol 99:1669–1677

Allis CD, Bowen JK, Abraham GN, Glover CV, Gorovsky MA (1980) Proteolytic processing of histone H3 in chromatin: a physiologically regulated event in *Tetrahymena* micronuclei. Cell 20:55–64

Ashok BT, Ali R (2003) Aging research in India. Exp Gerontol 38:597–603

Bannister AJ, Kouzarides T (2015) Regulation of chromatin by histone modifications. Cell Res 21:381–395

Bannister AJ, Schneider R, Kouzarides T (2002) Histone methylation: dynamic or static? Cell 109:801–806

Bannister AJ, Zegerman P, Partridge JF, Miska EA, Thomas JO, Allshire RC, Kouzarides T (2001) Selective recognition of methylated lysine 9 on histone H3 by the HP1 chromo domain. Nature 410:120–124

Bao Y, Shen X (2007) INO80 subfamily of chromatin remodeling complexes. Mutat Res 618:18–29

Baumgart M et al (2015) RNA-seq of the aging brain in the short-lived fish *N. furzeri*—conserved pathways and novel genes associated with neurogenesis. Aging Cell 13:965–974

Bedford MT, Richard S (2005) Arginine methylation an emerging regulator of protein function. Mol Cell 18:263–272

Bednar J, Horowitz RA, Grigoryev SA, Carruthers LM, Hansen JC, Koster AJ, Woodcock CL (1998) Nucleosomes, linker DNA, and linker histone form a unique structural motif that directs the higher-order folding and compaction of chromatin. Proc Natl Acad Sci USA 95:14173–14178

Belz T, Pham AD, Beisel C, Anders N, Bogin J, Kwozynski S, Sauer F (2002) In vitro assays to study protein ubiquitination in transcription. Methods 26:233–244

Benayoun BA, Pollina EA, Brunet A (2015) Epigenetic regulation of ageing: linking environmental inputs to genomic stability. Nat Rev Mol Cell Biol 16:593–610

Berger SL (2002) Histone modifications in transcriptional regulation. Curr Opin Genet Dev 12:142–148

Bernstein BE et al (2005) Genomic maps and comparative analysis of histone modifications in human and mouse. Cell 120:169–181

Bi X, Broach JR (2001) Chromosomal boundaries in S. cerevisiae. Curr Opin Genet Dev 11:199–204

Brandt WF, Bohm L, Von Holt C (1975) Proteolytic degradation of histones and site of cleavage in histone F2al and F3. FEBS Lett 51:88–93

Briggs SD, Bryk M, Strahl BD, Cheung WL, Davie JK, Dent SY, Winston F, Allis CD (2001) Histone H3 lysine 4 methylation is mediated by Set1 and required for cell growth and rDNA silencing in *Saccharomyces cerevisiae*. Genes Dev 15:3286–3295

Cairns BR, Lorch Y, Li Y, Zhang M, Lacomis L, Erdjument-Bromage H, Tempst P, Du J, Laurent B, Kornberg RD (1996) RSC, an essential, abundant chromatin-remodeling complex. Cell 87:1249–1260

Cao R, Wang L, Wang H, Xia L, Erdjument-Bromage H, Tempst P, Jones RS, Zhang Y (2002) Role of histone H3 lysine 27 methylation in Polycomb-group silencing. Science 298:1039–1043

Carrozza MJ, Florens L, Swanson SK, Shia WJ, Anderson S, Yates J, Washburn MP, Workman JL (2005) Stable incorporation of sequence specific repressors Ash1 and Ume6 into the Rpd3L complex. Biochim Biophys Acta 1731:77–87; discussion 75–76

Carruthers LM, Bednar J, Woodcock CL, Hansen JC (1998) Linker histones stabilize the intrinsic salt-dependent folding of nucleosomal arrays: mechanistic ramifications for higher-order chromatin folding. Biochemistry 37:14776–14787

Cattanach BM (1974) Position effect variegation in the mouse. Genet Res 23:291–306

Chan HM, Narita M, Lowe SW, Livingston DM (2005) The p400 E1A-associated protein is a novel component of the p53 –> p21 senescence pathway. Genes Dev 19:196–201

Chang B, Chen Y, Zhao Y, Bruick RK (2007) JMJD6 is a histone arginine demethylase. Science 318:444–447

Chaturvedi MM, Kanungo MS (1985) Analysis of conformation and function of the chromatin of the brain of young and old rats. Mol Biol Rep 10:215–219

Clapier CR, Cairns BR (2009) The biology of chromatin remodeling complexes. Annu Rev Biochem 78:273–304

Conaway RC, Brower CS, Conaway JW (2002) Emerging roles of ubiquitin in transcription regulation. Science 296:1254–1258

Cuthbert GL et al (2004) Histone deimination antagonizes arginine methylation. Cell 118:545–553

Czermin B, Melfi R, McCabe D, Seitz V, Imhof A, Pirrotta V (2002) Drosophila enhancer of Zeste/ESC complexes have a histone H3 methyltransferase activity that marks chromosomal Polycomb sites. Cell 111:185–196

Dang W, Steffen KK, Perry R, Dorsey JA, Johnson FB, Shilatifard A, Kaeberlein M, Kennedy BK, Berger SL (2009) Histone H4 lysine 16 acetylation regulates cellular lifespan. Nature 459:802–807

Daser A, Rabbitts TH (2004) Extending the repertoire of the mixed-lineage leukemia gene MLL in leukemogenesis. Genes Dev 18:965–974

de Ruijter AJ, van Gennip AH, Caron HN, Kemp S, van Kuilenburg AB (2003) Histone deacetylases (HDACs): characterization of the classical HDAC family. Biochem J 370:737–749

Dechassa ML, Sabri A, Pondugula S, Kassabov SR, Chatterjee N, Kladde MP, Bartholomew B (2010) SWI/SNF has intrinsic nucleosome disassembly activity that is dependent on adjacent nucleosomes. Mol Cell 38:590–602

DeLange RJ, Fambrough DM, Smith EL, Bonner J (1969) Calf and pea histone IV. II. The complete amino acid sequence of calf thymus histone IV; presence of epsilon-N-acetyllysine. J Biol Chem 244:319–334

Delmas V, Stokes DG, Perry RP (1993) A mammalian DNA-binding protein that contains a chromodomain and an SNF2/SWI2-like helicase domain. Proc Natl Acad Sci USA 90:2414–2418

Dhaenens M, Glibert P, Meert P, Vossaert L, Deforce D (2015) Histone proteolysis: a proposal for categorization into 'clipping' and 'degradation'. BioEssays 37:70–79

Dirscherl SS, Krebs JE (2004) Functional diversity of ISWI complexes. Biochem Cell Biol 82:482–489

Duncan EM, Muratore-Schroeder TL, Cook RG, Garcia BA, Shabanowitz J, Hunt DF, Allis CD (2008) Cathepsin L proteolytically processes histone H3 during mouse embryonic stem cell differentiation. Cell 135:284–294

Eickbush TH, Godfrey JE, Elia MC, Moudrianakis EN (1988) H2a-specific proteolysis as a unique probe in the analysis of the histone octamer. J Biol Chem 263:18972–18978

Eisenberg T et al (2009) Induction of autophagy by spermidine promotes longevity. Nat Cell Biol 11:1305–1314

Etzioni R, Legler JM, Feuer EJ, Merrill RM, Cronin KA, Hankey BF (1999) Cancer surveillance series: interpreting trends in prostate cancer–part III: Quantifying the link between population prostate-specific antigen testing and recent declines in prostate cancer mortality. J Natl Cancer Inst 91:1033–1039

Falk MM, Grigera PR, Bergmann IE, Zibert A, Multhaup G, Beck E (1990) Foot-and-mouth disease virus protease 3C induces specific proteolytic cleavage of host cell histone H3. J Virol 64:748–756

Faulkner RD, Bhatnagar YM (1987) A protease activity is associated with testicular chromatin of the mouse. Biol Reprod 36:471–480

Feng Q, Wang H, Ng HH, Erdjument-Bromage H, Tempst P, Struhl K, Zhang Y (2002) Methylation of H3-lysine 79 is mediated by a new family of HMTases without a SET domain. Curr Biol 12:1052–1058

Feser J, Truong D, Das C, Carson JJ, Kieft J, Harkness T, Tyler JK (2010) Elevated histone expression promotes life span extension. Mol Cell 39:724–735

Feser J, Tyler J (2011) Chromatin structure as a mediator of aging. FEBS Lett 585:2041–2048

Fletcher TM, Hansen JC (1996) The nucleosomal array: structure/function relationships. Crit Rev Eukaryot Gene Expr 6:149–188

Fugita OE, Chan DY, Roberts WW, Kavoussi LR, Jarrett TW (2004). Laparoscopic radical nephrectomy in obese patients: outcomes and technical considerations. Urology 63:247–252; discussion 252

Gary JD, Clarke S (1998) RNA and protein interactions modulated by protein arginine methylation. Prog Nucleic Acid Res Mol Biol 61:65–131

Gaubatz JW, Cutler RG (1990) Mouse satellite DNA is transcribed in senescent cardiac muscle. J Biol Chem 265:17753–17758

Gerber M, Shilatifard A (2003) Transcriptional elongation by RNA polymerase II and histone methylation. J Biol Chem 278:26303–26306

Gershey EL, Haslett GW, Vidali G, Allfrey VG (1969) Chemical studies of histone methylation. Evidence for the occurrence of 3-methylhistidine in avian erythrocyte histone fractions. J Biol Chem 244:4871–4877

Girdwood D, Bumpass D, Vaughan OA, Thain A, Anderson LA, Snowden AW, Garcia-Wilson E, Perkins ND, Hay RT (2003) P300 transcriptional repression is mediated by SUMO modification. Mol Cell 11:1043–1054

Glauche I, Thielecke L, Roeder I (2011) Cellular aging leads to functional heterogeneity of hematopoietic stem cells: a modeling perspective. Aging Cell 10:457–465

Gravina S, Vijg J (2010) Epigenetic factors in aging and longevity. Pflugers Arch 459:247–258

Greenwell PW, Kronmal SL, Porter SE, Gassenhuber J, Obermaier B, Petes TD (1995) TEL1, a gene involved in controlling telomere length in *S. cerevisiae*, is homologous to the human ataxia telangiectasia gene. Cell 82:823–829

Greer EL, Maures TJ, Ucar D, Hauswirth AG, Mancini E, Lim JP, Benayoun BA, Shi Y, Brunet A (2011) Transgenerational epigenetic inheritance of longevity in *Caenorhabditis elegans*. Nature 479:365–371

Grigera PR, Tisminetzky SG (1984) Histone H3 modification in BHK cells infected with foot-and-mouth disease virus. Virology 136:10–19

Grigoryev SA (2001) Higher-order folding of heterochromatin: protein bridges span the nucleosome arrays. Biochem Cell Biol 79:227–241

Grozinger CM, Chao ED, Blackwell HE, Moazed D, Schreiber SL (2001) Identification of a class of small molecule inhibitors of the sirtuin family of NAD-dependent deacetylases by phenotypic screening. J Biol Chem 276:38837–38843

Hakimi MA, Bochar DA, Chenoweth J, Lane WS, Mandel G, Shiekhattar R (2002) A core-BRAF35 complex containing histone deacetylase mediates repression of neuronal-specific genes. Proc Natl Acad Sci USA 99:7420–7425

Hampsey M, Reinberg D (2003) Tails of intrigue: phosphorylation of RNA polymerase II mediates histone methylation. Cell 113:429–432

Hanks SK, Rodriguez LV, Rao PN (1983) Relationship between histone phosphorylation and premature chromosome condensation. Exp Cell Res 148:293–302

Hansen JC (2002) Conformational dynamics of the chromatin fiber in solution: determinants, mechanisms, and functions. Annu Rev Biophys Biomol Struct 31:361–392

Hirano T (2000) Chromosome cohesion, condensation, and separation. Annu Rev Biochem 69:115–144

Hong L, Schroth GP, Matthews HR, Yau P, Bradbury EM (1993) Studies of the DNA binding properties of histone H4 amino terminus. Thermal denaturation studies reveal that acetylation markedly reduces the binding constant of the H4 "tail" to DNA. J Biol Chem 268:305–314

Hu Z, Chen K, Xia Z, Chavez M, Pal S, Seol JH, Chen CC, Li W, Tyler JK (2014) Nucleosome loss leads to global transcriptional up-regulation and genomic instability during yeast aging. Genes Dev 28:396–408

Imai S, Armstrong CM, Kaeberlein M, Guarente L (2000) Transcriptional silencing and longevity protein Sir2 is an NAD-dependent histone deacetylase. Nature 403:795–800

Jackson JP, Lindroth AM, Cao X, Jacobsen SE (2002) Control of CpNpG DNA methylation by the KRYPTONITE histone H3 methyltransferase. Nature 416:556–560

Jacobson MK, Jacobson EL (1999) Discovering new ADP-ribose polymer cycles: protecting the genome and more. Trends Biochem Sci 24:415–417

Jentsch S, Schlenker S (1995) Selective protein degradation: a journey's end within the proteasome. Cell 82:881–884

Jenuwein T, Allis CD (2001) Translating the histone code. Science 293:1074–1080

Jeyapalan JC, Ferreira M, Sedivy JM, Herbig U (2007) Accumulation of senescent cells in mitotic tissue of aging primates. Mech Ageing Dev 128:36–44

Jones PA, Baylin SB (2002) The fundamental role of epigenetic events in cancer. Nat Rev Genet 3:415–428

Jurkowski TP, Jeltsch A (2011) Burning off DNA methylation: new evidence for oxygen-dependent DNA demethylation. ChemBioChem 12:2543–2545

Kaeberlein M, McVey M, Guarente L (1999) The SIR2/3/4 complex and SIR2 alone promote longevity in Saccharomyces cerevisiae by two different mechanisms. Genes Dev 13:2570–2580

Kanfi Y, Naiman S, Amir G, Peshti V, Zinman G, Nahum L, Bar-Joseph Z, Cohen HY (2012) The sirtuin SIRT6 regulates lifespan in male mice. Nature 483:218–221

Kappus S, Apweiler R, White CJ, Whish WJ (1993) In vitro poly-(ADP-ribosyl)ation of chromatin proteins in the rat tapeworm, Hymenolepis diminuta. Comp Biochem Physiol B 104:711–716

Kennedy BK, Gotta M, Sinclair DA, Mills K, McNabb DS, Murthy M, Pak SM, Laroche T, Gasser SM, Guarente L (1997) Redistribution of silencing proteins from telomeres to the nucleolus is associated with extension of life span in S. cerevisiae. Cell 89:381–391

Kornberg RD (1974) Chromatin structure: a repeating unit of histones and DNA. Science 184:868–871

Kornberg RD, Thomas JO (1974) Chromatin structure; oligomers of the histones. Science 184:865–868

Kosar M, Bartkova J, Hubackova S, Hodny Z, Lukas J, Bartek J (2011) Senescence-associated heterochromatin foci are dispensable for cellular senescence, occur in a cell type- and insult-dependent manner and follow expression of p16(ink4a). Cell Cycle 10:457–468

Kouzarides T (2002) Histone methylation in transcriptional control. Curr Opin Genet Dev 12:198–209

Krishnan V, Chow MZ, Wang Z, Zhang L, Liu B, Liu X, Zhou Z (2011) Histone H4 lysine 16 hypoacetylation is associated with defective DNA repair and premature senescence in Zmpste24-deficient mice. Proc Natl Acad Sci USA 108:12325–12330

Krogan NJ, Dover J, Khorrami S, Greenblatt JF, Schneider J, Johnston M, Shilatifard A (2002) COMPASS, a histone H3 (Lysine 4) methyltransferase required for telomeric silencing of gene expression. J Biol Chem 277:10753–10755

Krogan NJ et al (2003) Methylation of histone H3 by Set2 in Saccharomyces cerevisiae is linked to transcriptional elongation by RNA polymerase II. Mol Cell Biol 23:4207–4218

Kuo MH, Brownell JE, Sobel RE, Ranalli TA, Cook RG, Edmondson DG, Roth SY, Allis CD (1996) Transcription-linked acetylation by Gcn5p of histones H3 and H4 at specific lysines. Nature 383:269–272

Lachner M, O'Sullivan RJ, Jenuwein T (2003) An epigenetic road map for histone lysine methylation. J Cell Sci 116:2117–2124

Lacoste N, Utley RT, Hunter JM, Poirier GG, Cote J (2002) Disruptor of telomeric silencing-1 is a chromatin-specific histone H3 methyltransferase. J Biol Chem 277:30421–30424

Larson K, Yan SJ, Tsurumi A, Liu J, Zhou J, Gaur K, Guo D, Eickbush TH, Li WX (2012) Heterochromatin formation promotes longevity and represses ribosomal RNA synthesis. PLoS Genet 8:e1002473

Le Rhun Y, Kirkland JB, Shah GM (1998) Cellular responses to DNA damage in the absence of Poly(ADP-ribose) polymerase. Biochem Biophys Res Commun 245:1–10

Lee DY, Hayes JJ, Pruss D, Wolffe AP (1993) A positive role for histone acetylation in transcription factor access to nucleosomal DNA. Cell 72:73–84

Lehnertz B, Ueda Y, Derijck AA, Braunschweig U, Perez-Burgos L, Kubicek S, Chen T, Li E, Jenuwein T, Peters AH (2003) Suv39 h-mediated histone H3 lysine 9 methylation directs DNA methylation to major satellite repeats at pericentric heterochromatin. Curr Biol 13:1192–1200

Lesur I, Campbell JL (2004) The transcriptome of prematurely aging yeast cells is similar to that of telomerase-deficient cells. Mol Biol Cell 15:1297–1312

Li G, Reinberg D (2011) Chromatin higher-order structures and gene regulation. Curr Opin Genet Dev 21:175–186

Li X et al (2014) Histone demethylase KDM5B is a key regulator of genome stability. Proc Natl Acad Sci USA 111:7096–7101

Lin R, Cook RG, Allis CD (1991) Proteolytic removal of core histone amino termini and dephosphorylation of histone H1 correlate with the formation of condensed chromatin and transcriptional silencing during Tetrahymena macronuclear development. Genes Dev 5:1601–1610

Liu L, Cheung TH, Charville GW, Hurgo BM, Leavitt T, Shih J, Brunet A, Rando TA (2013) Chromatin modifications as determinants of muscle stem cell quiescence and chronological aging. Cell Rep 4:189–204

Luger K, Dechassa ML, Tremethick DJ (2012) New insights into nucleosome and chromatin structure: an ordered state or a disordered affair? Nat Rev Mol Cell Biol 13:436–447

Luger K, Rechsteiner TJ, Flaus AJ, Waye MM, Richmond TJ (1997) Characterization of nucleosome core particles containing histone proteins made in bacteria. J Mol Biol 272:301–311

Macieira-Coelho A, Puvion-Dutilleul F (1989) Evaluation of the reorganization in the high-order structure of DNA occurring during cell senescence. Mutat Res 219:165–170

Mahendra G, Gupta S, Kanungo MS (1999) Effect of 17beta estradiol and progesterone on the conformation of the chromatin of the liver of female Japanese quail during aging. Arch Gerontol Geriatr 28:149–158

Mahendra G, Kanungo MS (2000) Age-related and steroid induced changes in the histones of the quail liver. Arch Gerontol Geriatr 30:109–114

Marchal C, Miotto B (2015) Emerging concept in DNA methylation: role of transcription factors in shaping DNA methylation patterns. J Cell Physiol 230:743–751

Marushige K, Marushige Y, Wong TK (1976) Complete displacement of somatic histones during transformation of spermatid chromatin: a model experiment. Biochemistry 15:2047–2053

Marushige Y, Marushige K (1983) Proteolysis of somatic type histones in transforming rat spermatid chromatin. Biochim Biophys Acta 761:48–57

Maures TJ, Greer EL, Hauswirth AG, Brunet A (2011) The H3K27 demethylase UTX-1 regulates C. elegans lifespan in a germline-independent, insulin-dependent manner. Aging Cell 10:980–990

McColl G, Killilea DW, Hubbard AE, Vantipalli MC, Melov S, Lithgow GJ (2008) Pharmacogenetic analysis of lithium-induced delayed aging in *Caenorhabditis elegans*. J Biol Chem 283:350–357

Melchior F (2000) SUMO–nonclassical ubiquitin. Annu Rev Cell Dev Biol 16:591–626

Meshorer E, Gruenbaum Y (2009) NURD keeps chromatin young. Nat Cell Biol 11:1176–1177

Metzger E, Wissmann M, Schule R (2006) Histone demethylation and androgen-dependent transcription. Curr Opin Genet Dev 16:513–517

Miller T, Krogan NJ, Dover J, Erdjument-Bromage H, Tempst P, Johnston M, Greenblatt JF, Shilatifard A (2001) COMPASS: a complex of proteins associated with a trithorax-related SET domain protein. Proc Natl Acad Sci USA 98:12902–12907

Mishra RN, Kanungo MS (1994) Alterations in histones of the liver and oviduct of Japanese quail during aging. Mol Biol Rep 20:15–18

Morrison AJ et al (2007) Mec1/Tel1 phosphorylation of the INO80 chromatin remodeling complex influences DNA damage checkpoint responses. Cell 130:499–511

Muller J, Hart CM, Francis NJ, Vargas ML, Sengupta A, Wild B, Miller EL, O'Connor MB, Kingston RE, Simon JA (2002) Histone methyltransferase activity of a Drosophila Polycomb group repressor complex. Cell 111:197–208

Murray K (1964) The Occurrence of Epsilon-N-Methyl Lysine in Histones. Biochemistry 3:10–15

Nakayama J, Rice JC, Strahl BD, Allis CD, Grewal SI (2001) Role of histone H3 lysine 9 methylation in epigenetic control of heterochromatin assembly. Science 292:110–113

Nathan D, Sterner DE, Berger SL (2003) Histone modifications: Now summoning sumoylation. Proc Natl Acad Sci USA 100:13118–13120

Ng HH, Feng Q, Wang H, Erdjument-Bromage H, Tempst P, Zhang Y, Struhl K (2002) Lysine methylation within the globular domain of histone H3 by Dot1 is important for telomeric silencing and Sir protein association. Genes Dev 16:1518–1527

Ni Z, Ebata A, Alipanahiramandi E, Lee SS (2012) Two SET domain containing genes link epigenetic changes and aging in Caenorhabditis elegans. Aging Cell 11:315–325

Nielsen PR, Nietlispach D, Mott HR, Callaghan J, Bannister A, Kouzarides T, Murzin AG, Murzina NV, Laue ED (2002) Structure of the HP1 chromodomain bound to histone H3 methylated at lysine 9. Nature 416:103–107

O'Carroll D et al (2000) Isolation and characterization of Suv39h2, a second histone H3 methyltransferase gene that displays testis-specific expression. Mol Cell Biol 20:9423–9433

Oberdoerffer P, Sinclair DA (2007) The role of nuclear architecture in genomic instability and ageing. Nat Rev Mol Cell Biol 8:692–702

Ogryzko VV, Schiltz RL, Russanova V, Howard BH, Nakatani Y (1996) The transcriptional coactivators p300 and CBP are histone acetyltransferases. Cell 87:953–959

Okawa Y, Takada K, Minami J, Aoki K, Shibayama H, Ohkawa K (2003) Purification of N-terminally truncated histone H2A-monoubiquitin conjugates from leukemic cell nuclei: probable proteolytic products of ubiquitinated H2A. Int J Biochem Cell Biol 35:1588–1600

Olins DE, Wright EB (1973) Glutaraldehyde fixation of isolated eucaryotic nuclei. Evidence for histone-histone proximity. J Cell Biol 59:304–317

Padilla PA, Garcia AM, Ladage ML, Toni LS (2014) *Caenorhabditis elegans*: an old genetic model can learn new epigenetic tricks. Integr Comp Biol 54:52–60

Panda P, Chaturvedi MM, Panda AK, Suar M, Purohit JS (2013) Purification and characterization of a novel histone H2A specific protease (H2Asp) from chicken liver nuclear extract. Gene 512:47–54

Panigrahi AK, Tomar RS, Chaturvedi MM (2003) A SWI/SNF-like factor from chicken liver that disrupts nucleosomes and transfers histone octamers in cis and trans. Arch Biochem Biophys 414:24–33

Patterson BD, Davies DD (1969) Specificity of the enzymatic methylation of pea histone. Biochem Biophys Res Commun 34:791–794

Pegoraro G, Kubben N, Wickert U, Gohler H, Hoffmann K, Misteli T (2009) Ageing-related chromatin defects through loss of the NURD complex. Nat Cell Biol 11:1261–1267

Peters AH et al (2003) Partitioning and plasticity of repressive histone methylation states in mammalian chromatin. Mol Cell 12:1577–1589

Peters AH et al (2001) Loss of the Suv39h histone methyltransferases impairs mammalian heterochromatin and genome stability. Cell 107:323–337

Pfaffeneder T, Hackner B, Truss M, Munzel M, Muller M, Deiml CA, Hagemeier C, Carell T (2011) The discovery of 5-formylcytosine in embryonic stem cell DNA. Angew Chem Int Ed Engl 50:7008–7012

Pina B, Suau P (1987) Changes in histones H2A and H3 variant composition in differentiating and mature rat brain cortical neurons. Dev Biol 123:51–58

Pray-Grant MG, Daniel JA, Schieltz D, Yates JR 3rd, Grant PA (2005) Chd1 chromodomain links histone H3 methylation with SAGA- and SLIK-dependent acetylation. Nature 433:434–438

Pu M, Ni Z, Wang M, Wang X, Wood JG, Helfand SL, Yu H, Lee SS (2015) Trimethylation of Lys36 on H3 restricts gene expression change during aging and impacts life span. Genes Dev 29:718–731

Purohit JS, Chaturvedi MM, Panda P (2012) Histone proteases: the tale of tail clippers. Int J Intgv Sci Innov Tech 1:51–60

Purohit JS, Tomar RS, Panigrahi AK, Pandey SM, Singh D, Chaturvedi MM (2013) Chicken liver glutamate dehydrogenase (GDH) demonstrates a histone H3 specific protease (H3ase) activity in vitro. Biochimie 95:1999–2009

Rea S et al (2000) Regulation of chromatin structure by site-specific histone H3 methyltransferases. Nature 406:593–599

Rice JC, Briggs SD, Ueberheide B, Barber CM, Shabanowitz J, Hunt DF, Shinkai Y, Allis CD (2003) Histone methyltransferases direct different degrees of methylation to define distinct chromatin domains. Mol Cell 12:1591–1598

Ringrose L, Paro R (2004) Epigenetic regulation of cellular memory by the Polycomb and Trithorax group proteins. Annu Rev Genet 38:413–443

Robinson PJ, Fairall L, Huynh VA, Rhodes D (2006) EM measurements define the dimensions of the "30-nm" chromatin fiber: evidence for a compact, interdigitated structure. Proc Natl Acad Sci U S A 103:6506–6511

Robzyk K, Recht J, Osley MA (2000) Rad6-dependent ubiquitination of histone H2B in yeast. Science 287:501–504

Roguev A, Schaft D, Shevchenko A, Pijnappel WW, Wilm M, Aasland R, Stewart AF (2001) The Saccharomyces cerevisiae Set1 complex includes an Ash2 homologue and methylates histone 3 lysine 4. EMBO J 20:7137–7148

Roth SY, Denu JM, Allis CD (2001) Histone acetyltransferases. Annu Rev Biochem 70:81–120

Sarraf SA, Stancheva I (2004) Methyl-CpG binding protein MBD1 couples histone H3 methylation at lysine 9 by SETDB1 to DNA replication and chromatin assembly. Mol Cell 15:595–605

Scaffidi P, Misteli T (2006) Lamin A-dependent nuclear defects in human aging. Science 312:1059–1063

Schneider J, Wood A, Lee JS, Schuster R, Dueker J, Maguire C, Swanson SK, Florens L, Washburn MP, Shilatifard A (2005) Molecular regulation of histone H3 trimethylation by COMPASS and the regulation of gene expression. Mol Cell 19:849–856

Schotta G, Lachner M, Sarma K, Ebert A, Sengupta R, Reuter G, Reinberg D, Jenuwein T (2004) A silencing pathway to induce H3-K9 and H4-K20 trimethylation at constitutive heterochromatin. Genes Dev 18:1251–1262

Sedivy JM, Banumathy G, Adams PD (2008) Aging by epigenetics–a consequence of chromatin damage? Exp Cell Res 314:1909–1917

Sen P et al (2015) H3K36 methylation promotes longevity by enhancing transcriptional fidelity. Genes Dev 29:1362–1376

Shah PP et al (2013) Lamin B1 depletion in senescent cells triggers large-scale changes in gene expression and the chromatin landscape. Genes Dev 27:1787–1799

Sharma R, Nakamura A, Takahashi R, Nakamoto H, Goto S (2006) Carbonyl modification in rat liver histones: decrease with age and increase by dietary restriction. Free Radic Biol Med 40:1179–1184

Shenker N, Flanagan JM (2011) Intragenic DNA methylation: implications of this epigenetic mechanism for cancer research. Br J Cancer 106:248–253

Shi Y, Lan F, Matson C, Mulligan P, Whetstine JR, Cole PA, Casero RA (2004) Histone demethylation mediated by the nuclear amine oxidase homolog LSD1. Cell 119:941–953

Shi Y, Sawada J, Sui G, el Affar B, Whetstine JR, Lan F, Ogawa H, Luke MP, Nakatani Y (2003) Coordinated histone modifications mediated by a CtBP co-repressor complex. Nature 422:735–738

Shiio Y, Eisenman RN (2003) Histone sumoylation is associated with transcriptional repression. Proc Natl Acad Sci U S A 100:13225–13230

Shogren-Knaak M, Ishii H, Sun JM, Pazin MJ, Davie JR, Peterson CL (2006) Histone H4-K16 acetylation controls chromatin structure and protein interactions. Science 311:844–847

Singer MS, Kahana A, Wolf AJ, Meisinger LL, Peterson SE, Goggin C, Mahowald M, Gottschling DE (1998) Identification of high-copy disruptors of telomeric silencing in Saccharomyces cerevisiae. Genetics 150:613–632

Smeal T, Claus J, Kennedy B, Cole F, Guarente L (1996) Loss of transcriptional silencing causes sterility in old mother cells of S. cerevisiae. Cell 84:633–642

Smith ST, Petruk S, Sedkov Y, Cho E, Tillib S, Canaani E, Mazo A (2004) Modulation of heat shock gene expression by the TAC1 chromatin-modifying complex. Nat Cell Biol 6:162–167

Strahl BD, Allis CD (2000) The language of covalent histone modifications. Nature 403:41–45

Strahl BD et al (2002) Set2 is a nucleosomal histone H3-selective methyltransferase that mediates transcriptional repression. Mol Cell Biol 22:1298–1306

Straight AF, Shou W, Dowd GJ, Turck CW, Deshaies RJ, Johnson AD, Moazed D (1999) Net1, a Sir2-associated nucleolar protein required for rDNA silencing and nucleolar integrity. Cell 97:245–256

Struhl K (1998) Histone acetylation and transcriptional regulatory mechanisms. Genes Dev 12:599–606

Suka N, Suka Y, Carmen AA, Wu J, Grunstein M (2001) Highly specific antibodies determine histone acetylation site usage in yeast heterochromatin and euchromatin. Mol Cell 8:473–479

Sun ZW, Allis CD (2002) Ubiquitination of histone H2B regulates H3 methylation and gene silencing in yeast. Nature 418:104–108

Tachibana M, Sugimoto K, Fukushima T, Shinkai Y (2001) Set domain-containing protein, G9a, is a novel lysine-preferring mammalian histone methyltransferase with hyperactivity and specific selectivity to lysines 9 and 27 of histone H3. J Biol Chem 276:25309–25317

Tamaru H, Selker EU (2001) A histone H3 methyltransferase controls DNA methylation in Neurospora crassa. Nature 414:277–283

Tanaka Y, Fujii T, Yamana H, Kato S, Morimatsu M, Shirouzu K (2004) Experimental gene therapy using p21Waf1 gene for esophageal squamous cell carcinoma by gene gun technology. Int J Mol Med 14:545–551

Tesar M, Marquardt O (1990) Foot-and-mouth disease virus protease 3C inhibits cellular transcription and mediates cleavage of histone H3. Virology 174:364–374
Timmermann S, Lehrmann H, Polesskaya A, Harel-Bellan A (2001) Histone acetylation and disease. Cell Mol Life Sci 58:728–736
Timmers HT, Tora L (2005) SAGA unveiled. Trends Biochem Sci 30:7–10
Tsukada Y, Fang J, Erdjument-Bromage H, Warren ME, Borchers CH, Tempst P, Zhang Y (2006) Histone demethylation by a family of JmjC domain-containing proteins. Nature 439:811–816
Tsurumi A, Li WX (2012) Global heterochromatin loss: a unifying theory of aging? Epigenetics 7:680–688
Turner BM (2000) Histone acetylation and an epigenetic code. BioEssays 22:836–845
Urban MK, Zweidler A (1983) Changes in nucleosomal core histone variants during chicken development and maturation. Dev Biol 95:421–428
Vakoc CR, Mandat SA, Olenchock BA, Blobel GA (2005) Histone H3 lysine 9 methylation and HP1gamma are associated with transcription elongation through mammalian chromatin. Mol Cell 19:381–391
van Holde K, Zlatanova J (1995) Chromatin higher order structure: chasing a mirage? J Biol Chem 270:8373–8376
van Leeuwen F, Gafken PR, Gottschling DE (2002) Dot1p modulates silencing in yeast by methylation of the nucleosome core. Cell 109:745–756
Vaquero A, Loyola A, Reinberg D (2003) The constantly changing face of chromatin. Sci Aging Knowl Environ 2003 RE4
Vignali M, Hassan AH, Neely KE, Workman JL (2000) ATP-dependent chromatin-remodeling complexes. Mol Cell Biol 20:1899–1910
Vire E et al (2006) The Polycomb group protein EZH2 directly controls DNA methylation. Nature 439:871–874
Vogelauer M, Wu J, Suka N, Grunstein M (2000) Global histone acetylation and deacetylation in yeast. Nature 408:495–498
Whetstine JR et al (2006) Reversal of histone lysine trimethylation by the JMJD2 family of histone demethylases. Cell 125:467–481
Wood JG, Helfand SL (2013) Chromatin structure and transposable elements in organismal aging. Front Genet 4:274
Woodage T, Basrai MA, Baxevanis AD, Hieter P, Collins FS (1997) Characterization of the CHD family of proteins. Proc Natl Acad Sci USA 94:11472–11477
Woodcock CL, Ghosh RP (2010) Chromatin Higher-order Structure and Dynamics. Cold Spring Harbor Perspectives in Biology 2
Woodcock CL, Grigoryev SA, Horowitz RA, Whitaker N (1993) A chromatin folding model that incorporates linker variability generates fibers resembling the native structures. Proc Natl Acad Sci USA 90:9021–9025
Wysocka J, Swigut T, Milne TA, Dou Y, Zhang X, Burlingame AL, Roeder RG, Brivanlou AH, Allis CD (2005) WDR5 associates with histone H3 methylated at K4 and is essential for H3 K4 methylation and vertebrate development. Cell 121:859–872
Xu W, Edmondson DG, Roth SY (1998) Mammalian GCN5 and P/CAF acetyltransferases have homologous amino-terminal domains important for recognition of nucleosomal substrates. Mol Cell Biol 18:5659–5669
Xu Y et al (2011) Genome-wide regulation of 5hmC, 5mC, and gene expression by Tet1 hydroxylase in mouse embryonic stem cells. Mol Cell 42:451–464
Xue Y, Wong J, Moreno GT, Young MK, Cote J, Wang W (1998) NURD, a novel complex with both ATP-dependent chromatin-remodeling and histone deacetylase activities. Mol Cell 2:851–861
Yamane K, Toumazou C, Tsukada Y, Erdjument-Bromage H, Tempst P, Wong J, Zhang Y (2006) JHDM2A, a JmjC-containing H3K9 demethylase, facilitates transcription activation by androgen receptor. Cell 125:483–495

Yang L, Xia L, Wu DY, Wang H, Chansky HA, Schubach WH, Hickstein DD, Zhang Y (2002) Molecular cloning of ESET, a novel histone H3-specific methyltransferase that interacts with ERG transcription factor. Oncogene 21:148–152

Yang SY, Yang XL, Yao LF, Wang HB, Sun CK (2013) Effect of CpG methylation on DNA binding protein: molecular dynamics simulations of the homeodomain PITX2 bound to the methylated DNA. J Mol Graph Model 29:920–927

Yang XJ, Ogryzko VV, Nishikawa J, Howard BH, Nakatani Y (1996) A p300/CBP-associated factor that competes with the adenoviral oncoprotein E1A. Nature 382:319–324

You A, Tong JK, Grozinger CM, Schreiber SL (2001) CoREST is an integral component of the CoREST—human histone deacetylase complex. Proc Natl Acad Sci U S A 98:1454–1458

Zeng L, Zhou MM (2002) Bromodomain: an acetyl-lysine binding domain. FEBS Lett 513:124–128

Zhang Y, Reinberg D (2001) Transcription regulation by histone methylation: interplay between different covalent modifications of the core histone tails. Genes Dev 15:2343–2360

Zou J, Lippert C, Heckerman D, Aryee M, Listgarten J (2014) Epigenome-wide association studies without the need for cell-type composition. Nat Methods 11:309–311

Part III
Neuromodulations in Aging

Changing Population of Neurons and Glia in the Human Cochlear Nucleus During Aging

Saroj Sharma, Tony G. Jacob, Tapas C. Nag, Daya N. Bhardwaj and T.S. Roy

Abstract The human cochlear nucleus (CN) is composed of dorsal and the ventral subdivisions. Neurons of this nucleus are unique because they respond to specific impulses arising from the cochlea. This nucleus and the neurons of the cochlea are continuously exposed to external stimuli and can degenerate with aging. This may be one of the causes of presbycusis—an age related hearing loss. One of the therapeutic interventions for this condition is a brainstem implant. An understanding of the morphology of the CN with aging changes will help making better implants for these patients. In the present study we have investigated the qualitative and quantitative changes in the neuronal and glial population of various sub-divisions of human CN at different ages. Forty one (1st to 9th decade) CN were processed for morphometry and stereology. Although the volume of the nucleus does not change with aging, neuronal number showed significant changes with aging. The data from the current morphometric report may facilitate in understanding the physiology of prebycusis, provide insight for cochlear nuclear implantation and eventually help clinicians to rehabilitate geriatric patients with hearing loss.

Keywords Presbycusis · Stereology · Hearing · Cochlear implant · Auditory pathway

S. Sharma
Dr. B.S. Ambedkar Medical College and Hospital, Rohini, Delhi 110085, India
e-mail: drsaroj2001@gmail.com

T.G. Jacob · T.C. Nag
Department of Anatomy, All India Institute of Medical Sciences,
New Delhi 110029, India
e-mail: tapas_nag@yahoo.com

D.N. Bhardwaj
Department of Forensic Medicine, All India Institute of Medical Sciences,
New Delhi 110029, India
e-mail: dn_11@rediffmail.com

T.S. Roy (✉)
Department of Anatomy, All India Institute of Medical Sciences, Room no. 1018,
Ansari Nagar, New Delhi 110029, India
e-mail: tsroyaiims@gmail.com; tsroy@aiims.edu

© Springer Science+Business Media Singapore 2017

P.C. Rath et al. (eds.), *Topics in Biomedical Gerontology*,
DOI 10.1007/978-981-10-2155-8_12

1 Introduction

The cochlear nucleus (CN) is a well organized structure located in the brainstem at the junction of the pons and medulla. It contains groups of neurons with specific morphological and physiological properties, related to the sense of hearing. Similar to the cochlea and spiral ganglion a tonotopic organization has been described in the CN. The CN is made of two major parts-ventral and dorsal (VCN and DCN) (Moore and Osen 1979; Idrizbegovic et al. 2006). These parts have exclusive groups of neurons that have distinct physiological and biochemical properties and respond specifically to auditory stimulation (Webster and Trune 1982). The CN receives afferent from all cochlear nerve fibers (Ray et al. 2005); hence, it is highly susceptible to physiological and pathological changes occurring in the internal ear.

Like all other regions of the brain, the CN also shows physiological changes with aging (Frisina and Frisina 1997). Presbycusis is hearing loss that occurs with increasing age. It leads to problems in communication and is a chronic medical condition seen in our geriatric population (Weinstein and Ventry 1982). Preventive and curative biomedical interventions can be conceptualized and actualized only by increasing our understanding of the neural and molecular basis of this sensory deficit. Hence, in the present study we have investigated the morphological changes that occur with aging in the human CN. This will provide baseline data that may us help in understanding the pathophysiology of prebycusis and eventually help clinicians to rehabilitate patients with this condition.

2 Materials and Methods

2.1 Brain Stem Collection and Tissue Processing

Brain specimens used for this study were collected with proper permission from the institutional ethical committee for the use of cadaveric brain for research (Institutional Ethics Committee, All India Institute of Medical Sciences, New Delhi, India). Since the tissue was obtained post-mortem, the ethics committee waived off the need for taking consent from the relatives of the deceased individual. Thereafter, forty-one brainstems from persons, who were aged between 2 days and 90 years, were used for the study. The brains were divided into four groups according to their biological age: birth to 20 years (group 1, n = 9), 21–40 years (group 2, n = 10), 41–60 years (group 3, n = 10) and 61–90 years (group 4, n = 12). Tissue fixation was done using a solution of 4 % paraformaldehyde in 0.1 M phosphate buffer (pH 7.4) and for further use it was kept in fresh fixative at 4 °C that diminish post-mortem changes.

The segment of the brainstem with CN was cut, cryopreserved in 30 % sucrose and sectioned serially to obtain 40 μm thick transverse sections. The sections were stained with 1 % cresyl violet acetate and mounted with DPX.

2.2 Stereological Estimates

Cavalieri probe (Fig. 1) was used to measure the entire volume of the CN. For estimating the total number of neurons and glia the optical fractionator program of the Stereo Investigator software was used (MicroBrightfield Inc. VT, USA) (Howard and Reed 1998; Glaser et al. 2007).

2.3 Statistical Analysis

For statistical comparison and analysis a mean was computed from raw data. The data were expressed as mean ± SD (standard deviation). All analysis was done using SPSS 20 version 20.0 (IBM Corp. 2011) software. For analysis of total neuronal and glial numbers and nuclear volume Kruskal-Wallis test was used, which followed post-hoc Dunnett's test for multiple comparisons. An overall 'p' value of less than 0.05 was accepted to be statistically significant.

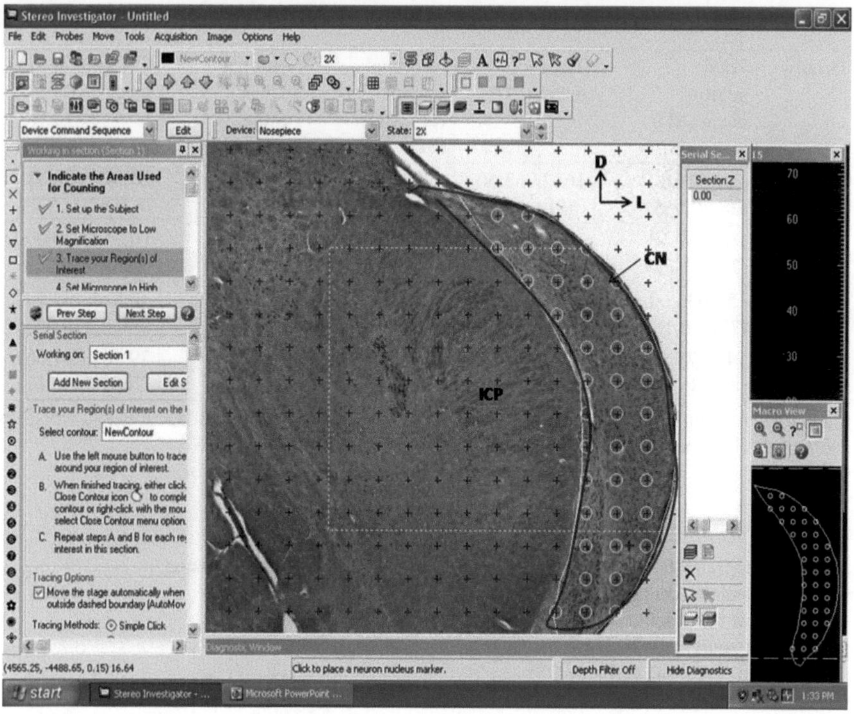

Fig. 1 Screenshot of the Stereo Investigator workflow with the Cavalieri tool's grid in place (*crosses*) and the cochlear nucleus (CN) of the left side marked out in the *contour line*. *D* Dorsal, *ICP* Inferior cerebellar peduncle; *L* Lateral. The spacing of the grid (*crosses*) = 250 μm

3 Results

After the removal of cerebellum and choroid plexus from the brainstem, the CN was identified as a continuous structure at the junction of the pons and medulla, near the lateral recess of the fourth ventricle. The CN partly encircled the inferior cerebellar peduncle (ICP), making it look like a hook.

Under the microscope, each neuron that was counted was also identified according to the description given by Moore and Osen (1979).

3.1 *Neuron Number*

The cochlear nucleus neuron population was observed to be highest in Group 2. This was significant with respect to Group 1 ($p = 0.03$). No significant difference between groups 2, 3 and 4 was observed ($p > 0.05$) (Table 1 and Fig. 2).

Table 1 Stereological estimates of the CN

Groups	Total neurons mean (SD)	Total glia mean (SD)	Neuronal volume (μ^3) Mean (SD)	Neuronal nuclear volume (μ^3) mean (SD)	CN volume (μ^3) mean (SD)
Group 1 (n = 9)	102,349.80 (28,202.55)*	1632,963.60 (551,654.20)	3910.75 (964.73)	577.72 (161.14)	24,573,608,889 (6718,035,963)
Group 2 (n = 10)	164,421.60 (78,761.88)	1606,848.33 (792,139.81)	4088.01 (645.52)	663.22 (162.46)	30,043,030,000 (7984,857,873)
Group 3 (n = 10)	150,036.60 (77,446.18)	1814,679.74 (533,035.74)	4470.33 (1180.53)	689.91 (283.77)	30,826,490,000 (8810,473,565)
Group 4 (n = 12)	140,277.80 (42,385.46)	2042,367.18 (662,439.70)	4539.72 (809.40)	690.52 (171.82)	29,162,841,667 (7687,749,361)

Stereological estimation of cochlear nucleus (CN) parameters by Optical fractionator, Nucleator and Cavalieri probe. (*–$p < 0.05$)

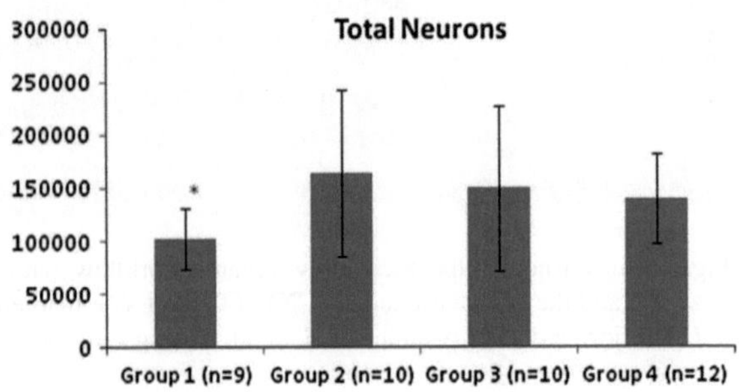

Fig. 2 Histogram showing mean total estimated number of neurons in the Cochlear nucleus in Group 1 (0–20 year), Group 2 (21–40 year), Group 3 (41–60 year) and Group 4 (>61 years). Error bars represent standard deviation. (*–$p < 0.05$)

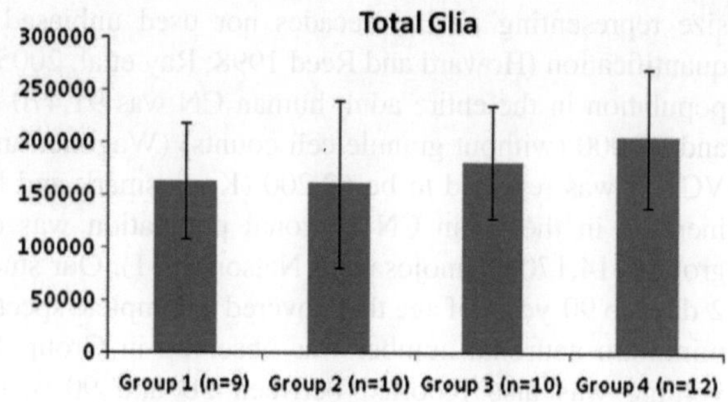

Fig. 3 Histogram showing mean total estimated number of glia in the Cochlear nucleus in Group 1 (0–20 year), Group 2 (21–40 year), Group 3 (41–60 year) and Group 4 (>61 years). Error bars represent standard deviation

3.2 *Glial Number*

Though there was an apparent increase in glial population with age, no significant difference was observed between different groups ($p > 0.05$) (Table 1 and Fig. 3).

3.3 *Neuronal and CN Volume*

Neuronal, neuronal nucleus volume and CN volume did not show any significant change over the decades ($p > 0.05$) (Table 1).

4 Discussion

To the best of our knowledge, this is the first stereological study of the age related changes in the human CN, ranging from birth to 90 years of age. We observed that among all age groups studied, Group 1 had the least number of neurons. A trend of increasing glial numbers was seen across the groups. The neuronal volume increased in groups 3 and 4; however the neuronal nuclear volume was relatively constant across groups 2, 3 and 4. The CN volume did not show much change across groups 2, 3 and 4.

There are contradictory reports regarding the age related morphological changes in the CN in animal species and humans. A decrease in the volume, neuronal number and neuron size of the CN and preference for particular type of neurons have been reported in animal models of age related hearing loss (Willott and Bross 1996). The qualitative and quantitative studies on human CN showed variable results with unremarkable changes and/or cell loss (Hinojosa and Nelson 2011). A higher range of number of glial cells and neurons, in the volume of neurons and the corresponding nucleus and in the entire CN was noted in the present study in comparison to the earlier reports (Wagoner and Kulesza 2009). This discrepancy may be because the previous studies on human CN neither included a large sample size representing all the decades nor used unbiased stereology as a method of quantification (Howard and Reed 1998; Ray et al. 2005). The reported mean neuron population in the entire adult human CN was 91,470 (Hinojosa and Nelson 2011) and 97,000 (without granule cell counts) (Wagoner and Kulesza 2009) whereas in VCN it was reported to be 63,200 (Konigsmark and Murphy 1972). A significant increase in the mean CN neuronal population was observed in the presbycusis group (114,170) (Hinojosa and Nelson 2011). Our study included brainstems from 2 days to 90 years of age that covered a complete spectrum of different ages and the minimum neuronal number was observed in Group 1. A decrease in human CN volume was also reported between 50 and 90 years of age (Konigsmark and Murphy 1972). In our study, the mean CN volume was significantly increased in Groups 2, 3 and 4 with respect to Group 1. Hence, there was a relatively constant volume across the adult age groups, which is also reflected in relatively unchanging populations of neurons and glia (Table 1).

There may be initial proliferation of the neuronal processes (dendrites and axons) followed by pruning and/or plastic changes (Sharma et al. 2014a). None of the previous studies have reported changes in glial cell population counts with progressive age.

There are few studies showing increase in glial cell population in the rat and human CN with aging (Jalenques et al. 1995). We have previously observed that in later decades there is a two folds increase in total glial cells as shown by significant increase in GFAP immunoreactivity (Sharma et al. 2014a). This increase in glial number maybe due to an increase in the number of astrocytes due to modifications in the CN (Jalenques et al. 1995) such as degeneration of synapses, pruning of dendrites or of entire neurons as has been seen in the aging CN (Adams and Jones 1982).

Knowledge of changes in the specific groups of neurons with age will help in engineering a more physiological brainstem implant. We had previously described eight neuronal clusters in the CN that show changes within themselves across ages. It raises the possibility of some compensatory mechanism and/or plastic response within the CN neurons without much alteration in various morphometric parameters with age (Sharma et al. 2014b). Therefore, we need more studies with a larger number of cases and further investigations on physiological, neurochemical and behavioural parameters to understand fully how these diverse and specialized cells undergo morphological changes and contribute to the complex functioning of the CN within creasing age. The results of the present study may add new insight in understanding the pathophysiology of hearing loss with age.

5 Conclusion

The data support the concept that hearing loss with aging in not only due to physiological changes in the CN, but there are age related changes in the neuronal and glial cells also.

Acknowledgements The authors would like to acknowledge the help provided by Prof. Alok Thakar of the Department of Otorhinolaryngology and the Department of Forensic Medicine and Toxicology, AIIMS, New Delhi. This work was partially funded by Indian Council of Medical Research, New Delhi (5/8/10-2(oto)/2009-NCD-1).

References

Adams I, Jones DG (1982) Synaptic remodelling and astrocytic hypertrophy in rat cerebral cortex from early to late adulthood. Neurobiol Aging 3:179–186

Frisina DR, Frisina RD (1997) Speech recognition in noise and presbycusis: relations to possible neural mechanisms. Hear Res 106:95–104

Glaser JR, Greene G, Hendricks S (2007) Stereololgy for biological research. M.B.F. Press, Willistin, VT

Hinojosa R, Nelson EG (2011) Cochlear nucleus neuron analysis in individuals with presbycusis. Laryngoscope 121:2641–2648

Howard CV, Reed MG (1998) Unbiased stereology: three dimensional measurement in microscopy. Springer, New York

Idrizbegovic El, Salman H, Niu X, Canlon B (2006) Presbyacusis and calcium-binding protein immunoreactivity in the cochlear nucleus of BALB/c mice. Hear Res 216–217:198–206

Jalenques I, Albuisson E, Despres G, Romand R (1995) Distribution of glial fibrillary acidic protein (GFAP) in the cochlear nucleus of adult and aged rats. Brain Res 686:223–232

Konigsmark BW, Murphy EA (1972) Volume of the cochlear nucleus in man: its relationship to neuronal population and age. J Neuropathol Exp Neurol 31:304–316

Moore JK, Osen KK (1979) The cochlear nuclei in man. Am J Anat 154:393–418

Ray B, Roy TS, Wadhwa S, Roy KK (2005) Development of the human fetal cochlear nerve: a morphometric study. Hear Res 202:74–86

Sharma S, Nag TC, Thakar A, Bhardwaj DN, Roy TS (2014a) The aging human cochlear nucleus: changes in the glial fibrillary acidic proteins, intracellular calcium regulatory proteins, GABAergic neurotransmitter and cholinergic receptors. J Chem Neuroanat 56:1–12

Sharma S, Nag TC, Bhardwaj DN, Vanamail P, Roy TS (2014b) Changing population of neurons and glia in the human cochlear nucleus with progressive age–A stereological study. J Anat Soc India 63(2):142–150

Wagoner JL, Kulesza RJ (2009) Topographical and cellular distribution of perineuronal nets in the human cochlear nucleus. Hear Res 254:42–53

Webster DB, Trune DR (1982) Cochlear nuclear complex of mice. Am J Anat 163(2):103–130

Weinstein BE, Ventry IM (1982) Hearing impairment and social isolation in the elderly. J Speech Hear Res 25:593–599

Willott JF, Bross LS (1996) Morphological changes in the anteroventral cochlear nucleus that accompany sensorineural hearing loss in DBA/2 J and C57BL/6 J mice. Brain Res Dev Brain Res 9:218–226

Changes in Social Behavior Associated with Alzheimer's Disease-Related Aβ Pathology

Neha Mishra, Rameshwar Singh and Deepak Sharma

Abstract Aβ pathology was produced in the rat brain by injecting aggregated $A\beta_{40}$ or $A\beta_{42}$ into the amygdala or hippocampus. Social behavior alterations in the rat with this pathology were studied by three-chamber social behavior test method. The data obtained show that sociability is impaired in rats in which the brain is afflicted with Aβ pathology, a condition similar to that in Alzheimer's disease brain. It is to be noted that human-specific behaviors such as empathy, sympathy and pro-social behaviors exist in rodents, and in humans afflicted with Alzheimer's disease social behaviors decline severely. The present study, thus, suggests that Aβ pathology/neurotoxicity may be involved in the genesis of social behavior changes.

Keywords Amyloid beta · Social behavior · Three-chamber social behavior test

1 Introduction

Alzheimer's disease (AD) is the most common dementia and it is essentially a disease of the elderly, for which age is an important risk factor. Apart from several cognitive deficits, AD brain is also afflicted with alterations in sociability. Social behaviors such as empathy, sympathy, social cue perception, and experience sharing have been shown to decline in AD; and behaviors such as social withdrawal and even social phobias have also been shown to occur in many cases. In a study conducted by Bózzola et al. (1992), increasing apathy was found to be the most common personality change in AD patients (61.3 %), followed by abnegation of hobbies (55.0 %), and rigidity (41.3 %). In another study Mega et al. (1996), showed that 88 % of AD patients had measurable non-cognitive behavioral

N. Mishra · R. Singh
School of Life Sciences, Jawaharlal Nehru University, New Delhi, India

D. Sharma (✉)
School of Life Sciences, Jawaharlal Nehru University, Room 321,
New Delhi 110067, Delhi, India
e-mail: deepak57in@yahoo.co.in

© Springer Science+Business Media Singapore 2017

P.C. Rath et al. (eds.), *Topics in Biomedical Gerontology*,
DOI 10.1007/978-981-10-2155-8_13

changes. Apathy was the most common behavior, showed by 72 % of patients. Agitation was also found to be increased in AD patients (60 %), followed by anxiety (48 %), irritability (42 %), motor behavioral anomalies (both 38 %), disinhibition (36 %), delusions (22 %), and hallucinations (10 %). As can be derived from the litrature, apathy and other associated social behavioral anomalies are also increasingly associated with AD. Apart from eliciting apathy, AD patiens are socially awkward, and withdrawn, and also shown to be mistrusful of their surroundings.

How are these changes in sociability (social behavior) produced? What pathology underlies their etiology? There are no easy answers to these questions. It is known that oxidative stress may be causally related to cognitive behaviors (Borbély et al. 2014; Ginsberg et al. 2013; Hritcu et al. 2014), and impairment in social behaviors may be mediated by alterations in endocannabinoid system (Ben-Ami Bartal et al. 2014; Kerr et al. 2013).

AD is characterized by the presence (in the brain) of extracellular Aβ plaques and intracellular tau neurofibrillary tangles (Selkoe 2001). The generation, deposition and accumulation of Aβ fragments generated via the amyloidogenic pathway in neuritic plaques and neurotoxicity of Aβ are considered central to the pathogenesis of AD (Hardy and Allsop 1991; Hardy and Higgins 1992). $A\beta_{40}$ (Irvine et al. 2008) and $A\beta_{42}$ (Jarrett et al. 1993) are the two amyloid forms that are of importance in the pathogenesis of AD. $A\beta_{42}$ is, however, considered to be more toxic than $A\beta_{40}$. The AD brain is under sever oxidative stress and Aβ protein deposition is causally related to the genesis of oxidative stress. It is therefore possible that Aβ pathology (oxidative stress and other consequences) is also linked to or involved in the impairment of sociability in AD.

The hippocampus and amygdala are among the brain regions that are severely affected in AD. Hippocampus is intricately involved in memory and cognition and is one of the primary centers for accumulation of amyloid plaques. Amygdala also shows severe pathology in AD and is reported to be one of the sites where the density of senile plaques is very high, in the brain of AD patients (Shoghi-Jadid et al. 2002). It together with the hippocampus is a part of the limbic system, and is involved in emotional conditioning. Hippocampus (Du 2001; Jack et al. 1997; West et al. 1994), and amygdala (Cuénod 1993) also shows physiological modifications in AD.

The present experimental study was therefore aimed at determining the social behavior anomalies that may be produced in an experimental animal model in which Aβ pathology was generated by injecting $A\beta_{40}$ or $A\beta_{42}$ into the amygdala or hippocampus. Exogenous Aβ injected into the rat brain provides an animal model that is commonly used in Alzheimer's disease-related research studies (Brouillette et al. 2012; Cleary et al. 2005). It is also important to note here that human-specific behaviors such as empathy, sympathy and pro-social behaviors also exist in rodents (Mogil 2012). Therefore results from animal models are appropriate for social behavior studies.

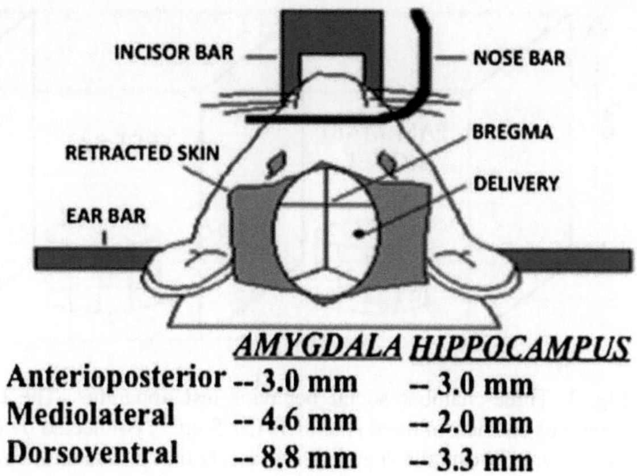

Fig. 1 Stereotaxic co-ordinates. Stereotaxic co-ordinates for administration of amyloid in hippocampus and amygdala of the rat brain for generation of Alzheimer's disease-like rat model

2 Method

2.1 Animals and Production of Ab Pathology

Male Wistar rats of 6 months of age were used. Amyloid-peptides ($A\beta_{40}$ or $A\beta_{42}$) were injected stereotaxically into amygdala and hippocampus according to methods described previously (Mishra et al. 2016) at co-ordinates outlined in Fig. 1. Animals were given a recovery period of 2 weeks after the surgery, before the commencement of behavioral experiments. All the experimental protocols were approved by the Institutional Animal Ethical Committee (IAEC) Jawaharlal Nehru University, New Delhi.

2.2 Statistics

Data were expressed as mean ± standard deviation (S.D.). Statistical comparison was performed by two-way ANOVA followed by Holm-Sidak pairwise analysis. Calculated probabilities of <0.05 were considered to be of significance and <0.001 highly significant, respectively. Pearson's correlation analysis was performed for the analysis of correlation between all the parameters of three-chamber social behavior test. Co-efficient (r) values >0.5 were considered indicators of strong correlation, $0.4 < r < 0.5$ of medium correlation, and <4 of week correlation. Positive values indicated the factors tend to increase together and negative values indicated inverse relationship between the factors. All tests were performed using Sigma Plot software version 11.0.

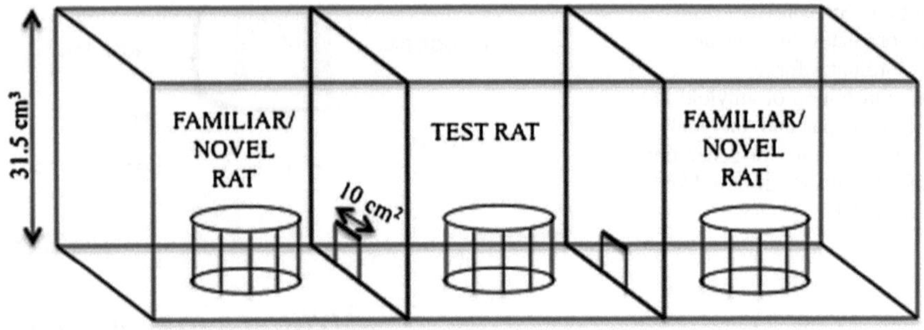

Fig. 2 Three-chamber social behavior test apparatus. The three-chamber social behavior test apparatus consists of three chambers (31.5 cm^3), connected by two doors (10 cm^2), between which the test animal (middle chamber) can freely move. The familiar/novel animals are placed in the bar cages in the left or right chamber. All recordings of the test animals are made by an over-head camera

2.3 Measurement of Social Behavior

The three-chamber social behavior test was conducted to assess the social and individual recognition/avoidance of the subjects. The test is based on the principle established by J. N. Crawley of sociability in rodents (Crawley 2004) and performed according to the protocol of Mishra et al. (2016). The apparatus was made of three identical chambers ($31.5 \times 31.5 \times 31.5$ cm^3) made of transparent plexiglas, connected by a 10×10 cm^2 doors (Fig. 2). Test subject was housed with a same sex, age matched conspecific animal 4 days prior to the beginning of experiment and throughout the duration of the experiment. In the left and right chambers, familiar rat (animal the subject is housed with) and unfamiliar/novel rat (animal the subject has had no encounter with previous to the experiment day) are kept in bar-wired cages. The bars in the left and right cages are such that the animals can interact without fighting or harming each other. Test subject was placed in the middle chamber and allowed to explore the apparatus.

Parameters monitored were duration of time spent by the test animal in each chamber, the chamber of first entry (primary entry) that is the chamber the test animal first enters, and close contacts made by the test animal with the familiar or unfamiliar/novel animal. The chamber of primary entry was considered as a parameter because Nadler et al. (2004), had shown that the first minute of the trial was more important and usually the chamber of primary entry is the chamber where the rat spends more time later in the trial. Close contacts were defined as the tactile contacts made by the test animal with the wire-caged conspecifics, either with the nose (sniffs) or fore paws or any other body part actively touching the wired cage. Close contacts reflect actual physical acts of exploration of, and curiosity towards, the conspecific versus nonsocial exploration of other areas of the chamber.

If the time spent in the middle chamber is greater than the time spent in the left or right chambers (combined), the animal is said to be non-sociable as opposed to sociable if it prefers to stay in the left or right chamber. A greater affinity towards the familiar animal is marked as social familiarity and towards the unfamiliar/novel (used interchangeably in the text) is marked as social novelty. Prior to the commencement of the experiment, the subject is placed in the middle chamber for 10 min (habituation period), during which the central area becomes a familiar "home base". The familiar and novel rats are also previously habituated to the wired cage, so that they are generally inactive and sit quietly in the wire cage during the test sessions. The position of the familiar and unfamiliar animals is changed periodically to eliminate position effects. The test was conducted for 2 consecutive days with three trials of 5 min each per day per animal.

3 Results

3.1 *Three Chamber Social Behavior Test*

The results pertaining to each parameter assessed (measured) by the three chamber social behavior test are discussed individually, that is, brain-region wise and amyloid-peptide wise:

(A) Sociability (Fig. 3a): Both $A\beta_{40}$- and $A\beta_{42}$-intrahippocampally or intraamygdalarly-injected animals showed increased duration of time spent by them (test animals) in the middle chamber as compared to their respective controls. This indicates that sociability was decreased in the amyloid-administered animals. The interaction between brain region of amyloid administration and the amyloid peptide (Brain region X Treatment) was found to be significant [$F_{(3, 40)} = 4.6$; $p < 0.05$]. This indicated that the amyloid-peptides had a differential action on the brain region, and there is different susceptibility of different brain regions to the same amyloid-peptide.

(B) Chamber of primary entry (Fig. 3b): Both $A\beta_{40}$- and $A\beta_{42}$ intrahippocampally or intraamygdalarly-injected animals made decreased number of primary entries into the chamber containing the familiar animal as compared to their respective controls. The interaction between brain region of amyloid administration and the amyloid peptide (Brain region X Treatment) was not significant [$F_{(3, 40)} = 1.6$; $p < 0.21$]. The effect on brain region was found to be significant [$F_{(1, 40)} = 4.4$; $p < 0.05$] and treatment was also found to be significant [$F_{(3, 40)} = 9.8$; $p < 0.001$].

(C) Time spent by the test animal in the chamber of the familiar animal (Fig. 3c) versus that in the chamber of the novel (unfamiliar) animal (Fig. 3d): Both $A\beta_{40}$- and $A\beta_{42}$-intrahippocampaly or intraamygdalarly-injected animals showed decrease in the time spent by the test animals in the chamber containing the familiar animal as compared to the respective controls. The amyloid-injected animals also showed increases in the time spent in the chamber containing the unfamiliar animal as compared to their respective

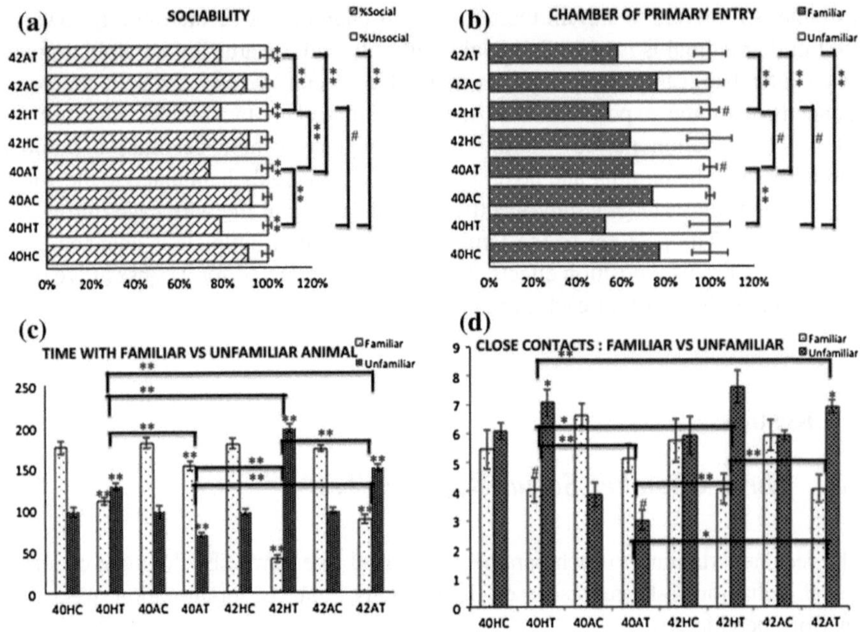

Fig. 3 Three-chamber social behavior test. Effect of Aβ toxicity on three-chamber social behavior. **a** Sociability is decreased in amyloid-injected animals **b** chamber of primary entry **c** time spent in the chamber of familiar animals is decreased and time spent in the chamber of unfamiliar animal is increased in the amyloid-injected animals **d** close contacts made by the amyloid injected animals with the unfamiliar animal is increased and with the familiar animals is decreased (*40HT* $A\beta_{40}$ injected-intrahippocampally; *40HC* Control for 40HT; *40AT* $A\beta_{40}$ injected-intraamygdalarly; *40AC* Control for 40AT; *42HT* $A\beta_{42}$injected-intrahippocampally; *42HC* Control for 42HT; *42AT* $A\beta_{42}$ injected-intraamygdalarly; *42AC* Control for 42AT). Each data point represents the mean ± SD of n = 6 rats. The symbols on the error bars represent significance of amyloid test values compared to their respective controls. The symbols on top horizontal bars represent the significance level between the various treatment groups. All statistical analysis by Two-way ANOVA. $^{**}p < 0.001$, $^{*}p < 0.05$, #Not significant

controls. This is indicative of decline in sociability towards the familiar animals (social familiarity), and increase in sociability towards the unfamiliar animals (social novelty) in the amyloid-injected animals. The interaction between brain region of amyloid administration and the amyloid peptide (Brain region X Treatment) was found to be significant for time spent with familiar rat [$F_{(3,40)} = 65.6$; $p < 0.001$]; and was also found to be significant for the time spent with unfamiliar rat [$F_{(3,40)} = 122.1$; $p < 0.001$. This again indicated that there was differential susceptibility of brain regions to amyloid-peptides.

(D) Close contacts with the familiar animal (Fig. 4e) versus novel (unfamiliar) animal (Fig. 4f): Both $A\beta_{40}$- and $A\beta_{42}$ intrahippocampally and intraamygdalarly-injected animals showed decreases in the number of close contacts made by the test animals with the familiar animal as compared to the respective controls. The amyloid-injected animals also showed increases in the number of close contacts made by the test animals with the unfamiliar animal as compared to their respective controls. This is also indicative of a decline in sociability towards the familiar animal and increase in sociability towards the unfamiliar animals. The interaction between brain region of amyloid administration and the amyloid peptide (Brain region X Treatment) was found to be significant for the close contacts made by the test animals with the familiar rat [$F_{(3,\ 40)} = 5.0$; $p < 0.001$] and close contacts made by the test animals with the unfamiliar rat [$F_{(3,\ 40)} = 24.0$; $p < 0.001$]. This indicates differential susceptibility of brain regions to different peptides.

The overall results indicate quantitatively greater decline in sociability in the $A\beta_{42}$ peptide-injected animals.

3.2 *Intra-hippocampal Injection*

Sociability (Fig. 3a): Both $A\beta_{40}$- and $A\beta_{42}$-intra-hippocampally-injected rats exhibited decreased sociability as compared to their respective controls. $A\beta_{40}$-injected rats showed 1.2-fold and Aβ42-injected ones showed a 1.2-fold decrease in the mean percentage sociability as compared to their controls.

Chamber of primary entry (Fig. 3b): Both $A\beta_{40}$- and $A\beta_{42}$-intra-hippocampally-injected rats exhibited decreases in the mean percentage of

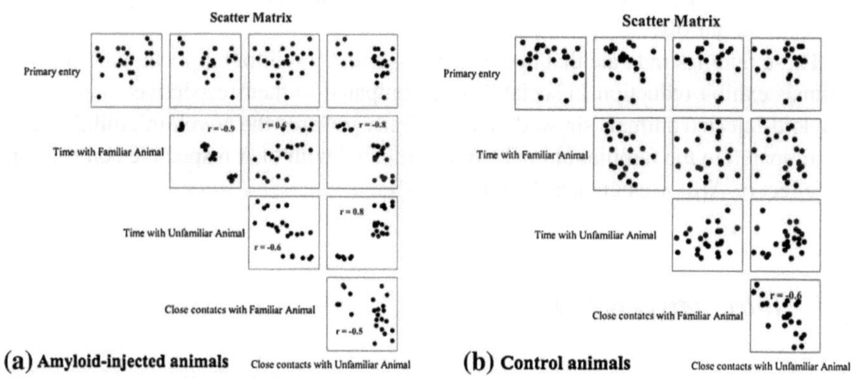

Fig. 4 Pearson's correlation plot between the parametres of three-chamber social behavior for amyloid-injected test animals. Pearson's correlation scatter matrix chart depicting relationship between three-chamber social behavior test parameters in amyloid-injected animals (**a**) and control animals (**b**). These results indicate that the amyloid-injected animals have a preference towards the unfamiliar animals as opposed to the familiar ones. Also, the chamber of primary entry of the animal does not provide any indication towards the social inclination (social novelty or familiarity) or the animal. These results indicate that the control animals show a negative correlation between close contacts for the familiar and unfamiliar animals

primary entry into the chamber of the familiar animal as compared to the controls. $A\beta_{40}$-injected animals showed 1.5-fold and Aβ42-injected ones showed a 1.2-fold decrease in the mean percentage primary entry into the chamber containing the familiar rat as compared to their controls.

Time spent by the test animals in the chamber of the familiar animal (Fig. 3c) versus that spent by the test animals in the chamber of the novel (unfamiliar) animal (Fig. 3d): Both $A\beta_{40}$- and $A\beta_{42}$-intra-hippocampally-injected animals spent a greater duration of time in the chamber of the novel animals as compared to the chamber containing the familiar animals; whereas their respective controls spent a greater duration of time in the chamber containing the familiar animals as compared to the time the control animals spent in the chamber containing the novel animals. The mean time spent in the chamber of the familiar animal by $A\beta_{40}$-injected rats showed a 1.6-fold and $A\beta_{42}$-injected ones showed a 4.4-fold decrease as compared to their respective controls, subsequently, $A\beta_{40}$-injected rats showed a 1.3-fold and $A\beta_{42}$-injected ones showed a 2.0-fold increase in the time spent in the chamber of the novel animal as compared to their respective controls.

Close contacts made by the test animals with the familiar animal (Fig. 3e) versus that made by the test animals with the novel (unfamiliar) animal (Fig. 3f): Both $A\beta_{40}$- and $A\beta_{42}$-intra-hippocampally-injected rats exhibited increases in the number of mean close contacts made by them with the novel animals whereas, the mean close contacts made by the control animals were greater towards the familiar animals. The close contacts made by the test animals with the familiar animal in $A\beta_{1\text{-}40}$-injected rats showed a 1.1-fold and $A\beta_{42}$-injected ones showed a 1.4-fold decrease as compared to their respective controls. The close contacts made by the test animals with the novel (unfamiliar) animal in $A\beta_{40}$-injected rats showed a 1.2-fold and $A\beta_{42}$-injected ones showed a 1.3-fold increase as compared to their respective controls.

These findings indicate that both the $A\beta_{40}$- and $A\beta_{42}$-intra-hippocampally-injected animals exhibit reduction in sociability as compared to their respective controls. The amyloid-injected animals showed greater affinity towards the novel/unfamiliar animal as compared to the familiar animal when compared with their respective controls, and the effect of $A\beta_{42}$ was greater than that of $A\beta_{40}$.

3.3 Intra-amygdalar Injection

Sociability (Fig. 3a): Both $A\beta_{40}$- and $A\beta_{42}$-intra-amygdalarly-injected rats exhibited decreased sociability as compared to their respective controls. $A\beta_{40}$-injected rats showed a 1.2-fold and $A\beta_{42}$-injected ones showed a 1.1-fold decrease in mean percentage sociability as compared to their controls.

Chamber of primary entry (Fig. 3b): Both $A\beta_{40}$- and $A\beta_{42}$-intra-amygdalarly-injected rats exhibited decreases in the mean percentage of primary entry into the chamber of the familiar animal as compared to the controls. $A\beta_{40}$-injected animals showed a 1.1-fold and $A\beta_{42}$-injected ones showed a 1.3-fold decrease in mean percentage primary entry into the chamber containing the familiar animal as compared to their controls.

Time spent by the test animals in the chamber of the familiar animal (Fig. 3c) versus that spent by the test animals in the chamber of the novel (unfamiliar) animal (Fig. 3d): Both Aβ_{40}- and Aβ_{42}-intra-amygdalarly-injected animals spent a greater duration of time in the chamber containing the novel (unfamiliar) animals as compared to the time spent by them in the chamber containing the familiar animals; whereas their respective controls spent a greater duration of time in the chamber containing the familiar animals as compared to the chamber containing the novel animals. The mean time spent by the test animals in the chamber containing the familiar animal for Aβ_{40}-injected rats showed a 1.2-fold and Aβ_{42}-injected ones showed a 2.0-fold decrease as compared to their respective controls, subsequently, Aβ_{40}-injected rats showed a 1.4-fold and Aβ_{42}-injected ones showed a 1.5-fold increase in the time spent by them in the chamber containing the novel animal as compared to their respective controls.

Close contacts made by the test animals with the familiar animal (Fig. 3e) versus that made by the test animals towards the novel (unfamiliar) animal (Fig. 3f): Both Aβ_{40}-and Aβ_{42}-intra-amygdalarly-injected rats exhibited increases in the number of mean close contacts made by them towards the novel (unfamiliar) animals; whereas, the mean close contacts made by the control animals was found to be greater towards the familiar animals. The close contacts made by the test animals with the familiar animal in Aβ_{40}-injected rats showed a 1.3-fold and Aβ_{42}-injected ones showed a 1.5-fold decrease as compared to their respective controls. The close contacts made by the test animals with the novel animal in Aβ_{40}-injected rats showed a 1.3-fold and Aβ_{42}-injected ones showed a 1.2-fold increase as compared to their respective controls.

These findings indicate that both the Aβ_{40}- and Aβ_{42}-intra-amygdalarly-injected animals exhibited reduction in sociability as compared to their respective controls. The amyloid-injected animals showed greater affinity towards the novel (unfamiliar) animal as compared to the familiar animal when compared with their respective controls, and the effect of Aβ_{42} was greater than that of Aβ_{40}.

3.4 Correlation Analysis

As can be seen from the data obtained by the three-chamber social behavior test experiment, the amyloid injected test animals showed decreased sociability as compared to their respective controls; also there is differential preference in the amyloid injected animals towards the novel animals, that is, the amyloid injected animals show a preference for social novelty; whereas the control animals have a differential preference for the familiar animals, i.e. they show social familiarity. For further analysis of this observation, Pearson's correlation analysis was done to measure the degree of sociability by calculating the interaction between intra-social behavior parameters:

Table 1 Pearson's correlation for three-chamber social behavior parameters

Amyloid-injected animals	**Time with familiar**	**Time with unfamiliar**	**Close contacts with familiar**	**Close contacts with unfamiliar**
Primary entry	r = 0.295 p = 0.162	r = −0.341 P = 0.103	R = 0.105 P = 0.627	**r = −0.411** **p = 0.0462**
Time with familiar		**r= −0.983** **p = 1.07e^{-17}**	**r = 0.634** **p = 8.77e^{-4}**	**r = −0.812** **p = 1.48e^{-6}**
Time with unfamiliar			**r = −0.654** **p = 5.27e^{-4}**	**r = 0.851** **p = 1.41e^{-7}**
Close contacts with familiar				**r = −0.531** **p = 7.62e^{-3}**
Control animals	**Time with familiar**	**Time with unfamiliar**	**Close contacts with familiar**	**Close contacts with unfamiliar**
Primary entry	r = −0.233 p = 0.274	r = −0.22 p = 0.302	r = −0.278 p = 0.743	r = −0.0707 p = 0.743
Time with familiar		r = −0.38 p = 0.066	r = 0.0962 p = 0.655	r = −0.301 p = −0.152
Time with unfamiliar			r = 0.217 p = 0.309	r = 0.122 p = 0.571
Close contacts with familiar				**r = −0.638** **p = 7.92e^{-4}**

Pearson's correlation coefficients (r) and p values for three-chamber social behavior parameters in amyloid-injected animals and their controls. The values in bold are significant interactions; the values in grey colored boxes indicate a strong correlation between the variables. $r > 0.5$ indicates strong correlation, $0.4 < r < 0.5$ indicates medium correlation, $r < 0.4$ indicates weak correlation. $p < 0.05$ indicates significant correlation interaction

1. Chamber of primary entry and time with familiar animal.
2. Chamber of primary entry and time with unfamiliar animal.
3. Chamber of primary entry and close contacts with familiar animal.
4. Chamber of primary entry and close contacts with unfamiliar animal.
5. Time with familiar animal and time with unfamiliar animal.
6. Time with familiar animal and close contacts with familiar animal.
7. Time with familiar animal and close contacts with unfamiliar animal.
8. Time with unfamiliar animal and close contacts with familiar animal.
9. Time with unfamiliar animal and close contacts with unfamiliar animal.
10. Close contacts with familiar animal and close contacts with unfamiliar animal.

It was observed in the three-chamber social behavior test that the amyloid-injected animals showed greater inclination towards the unfamiliar animals (spent more time with unfamiliar as compared to the familiar animal, larger number of close contacts towards the unfamiliar rat) as opposed to the control rats which had a greater affinity for the familiar conspecifics (spent more time with familiar as compared to the unfamiliar animal, larger number of close contacts towards the familiar rat). The correlations obtained are given in Table 1. From these it is apparent that time spent with the unfamiliar animal is strongly correlated to close contacts made with the unfamiliar animal ($r = 0.851$; $p = 1.41e^{-7}$) and inversely correlated with the time spent with the familiar animal ($r = -0.983$; $p = 1.07e^{-17}$) as well as close contacts made with the familiar animal ($r = -0.654$; $p = 5.27e^{-4}$) for the amyloid injected animals (Fig. 4a). The control animals on the other hand showed inverse relationship between the close contacts with familiar rat and the close contacts with unfamiliar rat ($r = -0.638$, $p = 7.92e^{-4}$) (Fig. 4b).

4 Discussion

The present results obtained from the three-chamber social behavior test demonstrate that the amyloid-injected rats (both intrahippocampally- and intraamygdalar-injected ones) and their respective controls have differential preferences for familiar animal or novel animal (Mishra et al. 2016). Familiarity is known to be crucial to empathetic responses in rodents such as mice (Mogil 2012). Many studies show that rodents become friendly with and enjoy the company of the other rodents they are caged with, they will even share their food peacefully with their cage mates, and have been demonstrated to learn tricks to free their cage mates when trapped and share their food with them (Ben-Ami Bartal et al. 2014). It was observed that the amyloid injected animals showed greater inclination towards the unfamiliar animals (spent more time with unfamiliar as compared to the familiar animal, larger number of close contacts towards the unfamiliar rat) as opposed to the control rats which had a greater affinity for the familiar conspecifics (spent more time with familiar as compared to the unfamiliar animal, larger number of close contacts towards the familiar rat). Also, the control animals showed greater curiosity than the amyloid-injected subjects, as was indicated by the greater number of total close contacts made by them. In Aβ_{40}- and Aβ_{42}-injected animals all the sociability parameters were significantly different as these animals preferred unfamiliar animals to associate with. To further strengthen the social preference by the amyloid-injected and control animals, Pearson's correlation coefficient was calculated for intra-social behavior parameters (chamber of primary entry, time with familiar, time with unfamiliar, close contacts with familiar, and close contacts with unfamiliar rat) (Fig. 4a, b). The correlations obtained are given in Table 1. It can be observed that time spent with the unfamiliar animal is strongly correlated to close contacts made with the unfamiliar animal and inversely correlated with the time spent with the familiar animal as well as close contacts made with the familiar animal (Fig. 4a) for the amyloid-injected animals. The control animals on the other hand showed inverse relationship between close contacts with familiar rat and close contacts with unfamiliar rat (Fig. 4b).

This social behavior test has not been previously applied to Aβ peptide-administered experimental animal models. There are reports of problems with sociability in Alzheimer's disease patients (Coen et al. 1997; Gauthier et al. 1996; Rayner et al. 2006), the present study would thus suggest that Aβ-related neurotoxicity may be responsible also for sociability impairment. It is also of interest to point out here that impaired social behavior is exhibited in rats prenatally exposed to valproic acid, and the impairment is mediated by alterations in endocannabinoid system (Ben-Ami Bartal et al. 2014; Kerr et al. 2013).

References

Ben-Ami Bartal I, Rodgers DA, Bernardez Sarria MS, Decety J, Mason P (2014) Pro-social behavior in rats is modulated by social experience. Elife 3:e01385

Borbély E, Horváth J, Furdan S, Bozsó Z, Penke B, Fülöp L (2014) Simultaneous changes of spatial memory and spine density after intrahippocampal administration of fibrillar Aβ1–42 to the rat brain. BioMed Res Int 2014:1–9

Bózzola FG, Gorelick PB, Freels S (1992) Personality changes in Alzheimer's disease. Arch. Neurol. 49:297–300

Brouillette J, Caillierez R, Zommer N, Alves-Pires C, Benilova I, Blum D, De Strooper B, Bue′e L (2012) Memory deficits induced by soluble low-molecular-weight amyloid-b1–42 oligomers are revealed in vivo by using a novel animal model. J Neurosci 32:7852–7861

Cleary JP, Walsh DM, Hofmeister JJ, Shankar GM, Kuskowski MA, Selkoe DJ, Ashe KH (2005) Natural oligomers of the amyloid-beta protein specifically disrupt cognitive function. Nat Neurosci 8:79–84

Coen RF, Swanwick GR, O'Boyle CA, Coakley D (1997) Behavior disturbance and other predictors of care burden in Alzheimer's disease. Int J Geriatr Psychiatry 12:331–336

Crawley JN (2004) Designing mouse behavioral tasks relevant to autistic-like behaviors. Ment Retard Dev Disabil Res Rev 10:248–258

Cuénod CA (1993) Amygdala atrophy in alzheimer's disease. Arch Neurol 50:941–945

Du AT (2001) Magnetic resonance imaging of the entorhinal cortex and hippocampus in mild cognitive impairment and Alzheimer's disease. J Neurol Neurosurg Psychiartry 71:441–447

Gauthier S, Baumgarten M, Becker R (1996) Dementia behavior disturbance scale. Int Psychogeriatr 8:325–327

Ginsberg SD, Zussy C, Brureau A, Keller E, Marchal S, Blayo C, Delair B, Ixart G, Maurice T, Givalois L (2013) Alzheimer's disease related markers, cellular toxicity and behavioral deficits induced six weeks after oligomeric amyloid-β peptide injection in rats. PLoS ONE 8:e53117

Hardy J, Allsop D (1991) Amyloid deposition as the central event in the aetiology of Alzheimer's disease. Trend Pharmacol Sci 12:383–388

Hardy JA, Higgins GA (1992) Alzheimer's disease: the amyloid cascade hypothesis. Science 256:184–185

Hritcu L, Noumedem JA, Cioanca O, Hancianu M, Kuete V, Mihasan M (2014) Methanolic extract of Piper nigrum fruits improves memory impairment by decreasing brain oxidative stress in amyloid beta(1–42) rat model of Alzheimer's disease. Cel Mol Neurobiol 34:437–449

Irvine G, El-Agnaf O, Shankar GM, Walsh DM (2008) Protein aggregation in the brain: the molecular basis for Alzheimer's and Parkinson's diseases. Mol Med 14:451–464

Jack CR, Petersen RC, Xu YC, Waring SC, O'Brien PC, Tangalos EG, Smith GE, Ivnik RJ, Kokmen E (1997) Medial temporal atrophy on MRI in normal aging and very mild Alzheimer's disease. Neurology 49:786–794

Jarrett JT, Berger EP, Lansbury PT (1993) The carboxy terminus of the beta amyloid protein is critical for the seeding of amyloid formation: implications for the pathogenesis of Alzheimer's disease. Biochemistry 32:4693–4697

Kerr DM, Downey L, Conboy M, Finn DP, Roche M (2013) Alterations in the endocannabinoid system in the rat valproic acid model of autism. Behav Brain Res 249:124–132

Mega MS, Cummings JL, Fiorello T, Gornbein J (1996) The spectrum of behavioral changes in Alzheimer's disease. Neurology 46:130–135

Mishra N, Singh R, Sharma D (2016) Differential responses of amygdala and hippocampus consequent to Aβ40 and Aβ42 induced toxicity in the rat brain: a comparative study. Int J Adv Res 4:959–994

Mogil JS (2012) The surprising empathic abilities of rodents. Trends Cogn Sci 16:143–144

Nadler JJ, Moy SS, Dold G, Simmons N, Perez A, Young NB, Barbaro RP, Piven J, Magnuson TR, Crawley JN (2004) Automated apparatus for quantitation of social approach behaviors in mice. Gene Brain Behav 3:303–314

Rayner AV, O'Brien JG, Schoenbachler B (2006) Behavior disorders of dementia: recognition and treatment. Am Fam Physician 73:647–652

Selkoe DJ (2001) Alzheimer's disease: genes, proteins, and therapy. Physiol Rev 81:741–766

Shoghi-Jadid K, Small GW, Agdeppa ED, Kepe V, Ercoli LM, Siddarth P, Read S, Satyamurthy N, Petric A, Huang SC, Barrio JR (2002) Localization of neurofibrillary tangles and beta-amyloid plaques in the brains of living patients with alzheimer disease. Am J Geriatr Psych 10:24–35

West MJ, Coleman PD, Flood DG, Troncoso JC (1994) Differences in the pattern of hippocampal neuronal loss in normal ageing and Alzheimer's disease. Lancet 344:769–772

Recovery of Age-Related Memory Loss: Hopes and Challenges

Mahendra Kumar Thakur, Arpita Konar, Dhiraj Kumar, Meghraj Singh Baghel and Padmanabh Singh

Abstract Advancing age is associated with drastic decline in memory and is a predisposing factor for neurodegenerative and neuropsychiatric disorders. Such decline results from aging of the brain involving loss of morphological integrity, alterations at the level of genes, enzymes and hormones, metabolism, oxidative stress, protein processing and synaptic function. Multiple biological scales ranging from genes to neural network and behavior and the individual variability that span age associated memory loss have added complexity to the recovery strategies. However, recent advancement in neuroscience research has not only removed the myth of unrecoverable memory loss during aging but also proposed a multitude of recovery approaches. These approaches include herbal interventions, dietary restrictions, antioxidant supplementation, environmental enrichment, lifestyle modulation and molecular targeting. Our laboratory is particularly interested in unraveling the molecular mechanism of age related memory loss and delineate therapeutic targets. Studies on animal models and humans reveal drastic changes in the expression and function of a wide array of molecules including chromatin modifying enzymes, immediate early genes, neurotrophins, presynaptic and post-synaptic proteins and neurite growth markers in vulnerable brain regions of cerebral cortex and hippocampus during aging. Such molecular changes are well translated into behavioral paradigms of memory impairment. In this chapter, we review age associated changes in brain, mechanisms of memory loss and recovery strategies. Essentially, we highlight the molecular correlates of brain aging and their potential as therapeutic targets for age associated memory loss.

Keywords Aging · Cognitive decline · Therapeutic interventions

M.K. Thakur (✉) · A. Konar · D. Kumar · M.S. Baghel · P. Singh
Biochemistry and Molecular Biology Laboratory, Department of Zoology,
Brain Research Centre, Banaras Hindu University, Varanasi 221005, India
e-mail: mkt_bhu@yahoo.com

A. Konar
CSIR-Institute of Genomics and Integrative Biology, New Delhi 110025, India

© Springer Science+Business Media Singapore 2017
P.C. Rath et al. (eds.), *Topics in Biomedical Gerontology*,
DOI 10.1007/978-981-10-2155-8_14

1 Introduction

Every individual craves for aging successfully and staying mentally sharp. Such attainment depends largely on maintaining healthy brain with advancing age. However, the number of aged people with reduced brain functions, particularly cognitive abilities and associated neurological disorders, is on rise and has overshadowed the hope of healthy brain aging. According to the demographic survey, about 25 million individuals aged 60–70 years worldwide suffer from neurodegenerative disorders particularly dementia and are expected to increase up to 80 million by 2040. The deteriorating geriatric mental health has imposed serious challenges to scientists and clinicians throughout the world. Further, the multifaceted mechanisms of brain aging ranging from genes and neural circuitry to behavior have added tremendous complexity to this global issue.

However, every cloud has a silver lining as certain inspiring examples of healthy brain of old individuals have raised immense hope among the neuroscientists. A landmark study on brain aging was published from the University of Groningen, Netherlands, stating the case of a 113 year old Dutch woman with improved cognitive abilities and incredibly healthy brain till her death at 115 years (den Dunnen et al. 2008). In addition, we have examples of many eminent scientists including Stephen Hawkins, Paul Greengard and Eric Kandel who are above 80 years and still actively engaged in scientific pursuits. These observations have made researchers to believe that there is no limit to the healthy life of brain and it is possible to maintain cognitively intact brain till old age.

Recent advancements in neuroscience research have successfully confronted the challenge of retarding age-associated memory loss. It is well established now that anatomical and functional changes associated with the brain aging underlie cognitive decline, particularly deficits in memory. Further, mechanistic investigations with the aid of animal models coupled with genomics and neuro-imaging techniques have deciphered that gene expression alteration and driving epigenetic modifications are pivotal for age associated memory loss. Having determined the underlying mechanisms, multiple approaches have come up to delay or reverse brain aging associated deteriorations and thus recover memory impairment. These interventions include intake of herbal extracts, calorie restriction and dietary supplementation with antioxidants, life style changes such as exercise and brain training activities and molecular targeting. Our laboratory is particularly interested to elucidate the molecular correlates including gene expression and epigenetic changes of age associated memory loss and delineate therapeutic targets. Here, we review the present knowledge on brain aging including anatomical and functional alterations, the causative factors of memory decline and recovery strategies adopted till date. In particular, we highlight our findings of molecular basis of memory decline with aging and potential recovery targets.

2 Age Associated Changes During Aging

Aging is a natural process which slows down several functions including memory. It is characterized by colossal loss of cell homeostasis due to Ca^{2+} deregulation, generation of reactive oxygen species (ROS) and reduction of antioxidants which result into mitochondrial dysfunctions ultimately causing cell death. Moreover, during aging the brain undergoes morphological and anatomical, biochemical, physiological and synaptic plasticity changes in region specific manner. Such changes are mainly responsible for age associated memory impairment (AAMI) and neurological disorders like Alzheimer's disease (AD) and Parkinson's disease (PD).

2.1 *Morphological and Anatomical Changes*

The brain goes through several morphological and anatomical changes that lead to cognitive deficiency and vulnerability to neurological disorders (Hermann et al. 2014). This influences the survival of neuronal cells and thus dynamics of synaptic connections, which are important for brain functions (Burke and Barnes 2006). Moreover, such changes predominantly affect the cerebral cortex and hippocampus, which are crucial brain regions for memory formation (Burger 2010). Region specific alterations in dendritic spine density, arborization, myelin dystrophy and synaptic transmission are influenced by increasing age. Dendritic spines are crucial for the formation, maintenance and strengthening of neural network, transmission of electrical inputs, long term potentiation and synaptic plasticity (deToledo-Morrell et al. 1988; Peinado et al. 1997). Earlier reports have shown reduction in dendritic length and spine count in frontal cortex and temporal cortex of aged animals (Sorra and Harris 2000; Jacobs et al. 2001). MRI reports have also revealed decrease in grey matter and increase in white matter volume in frontal, parietal and temporal cortices of brain with advancing age (Berti et al. 2011; Schmidt et al. 2011).

2.2 *Biochemical Changes*

Aging shows several biochemical changes in different regions of brain. Such changes include modulation of hormones, neurotransmitters and their receptors (Perry et al. 1981). Cognitive deficit and motor dysfunction are correlated with alterations in level of hormones, synthesis and reuptake of neurotransmitters and their receptors in region specific manner (Nagata et al. 1998; Smith et al. 2005). The level of 5-hydroxytryptamine, a stress neurotransmitter, is reported to increase in prefrontal cortex, hypothalamus and hippocampus during aging (van Luijtelaar et al. 1992). Glutamate is an excitatory neurotransmitter and is also involved in memory consolidation. Its function is mediated through N-methyl-D-aspartate (NMDA) and α-amino-3-hydroxy-5-methyl-4-isoxazolepropionic acid (AMPA) receptors. The dopamine synthesis decreases in striatum and extrastriatal region of aged human postmortem brain (Ota et al. 2006). Previous reports have shown decline in glutamate and its receptor in the hippocampus of aged brain, causing impairment of memory through decrease in synaptic plasticity (Wenk and Barnes 2000; Yamamoto 2001; Chang et al. 2009). Further, the loss of cholinergic neurons and acetylcholine level are key regulators of memory in aging and age associated neurological disorders. Thus, decline in cognitive function can be recovered by supplementation of agonist of cholinergic system (Drachman et al. 1979).

Another important factor for memory decline is hormone which undergoes alterations during aging process (Konar et al. 2015b). The reduced level of estrogen, progesterone, thyroxine and melatonin, and increased level of stress hormones cortisol and cortisone are known to speed up the aging process and cause cognitive deficiency (Schumacher et al. 2003; Veiga et al. 2004; Lupien et al. 2009). The low level of melatonin is also reported in neurological disorder like AD as compared to age matched normal subjects (Sandyk 1997).

2.3 Physiological Changes

The physiological changes occur asynchronously in different regions of the brain and are known to be modulated by various environmental factors including life style (Mora et al. 2007). Particularly, the intracellular Ca^{2+} level affects homeostasis of neuronal cells and disrupts neurotransmitter release and activity of several kinases leading to cognitive deficits during aging (Foster 1999, 2007). The deregulation of Ca^{2+} efflux from endoplasmic reticulum (ER) promotes cytochrome c release from mitochondria leading to apoptosis (Pinton et al. 2001). Mitochondria generate intracellular ROS during oxidation process (Turrens 2003). Further, oxidative stress causes oxidation of several molecules and modulates various cell signaling pathways during aging. Moreover, mitochondrial dysfunction, reduction of antioxidant enzymes and generation of ROS are markers of neurodegeneration and cognitive dysfunction with advancing age (Chakrabarti et al. 2007). Therefore, cognitive impairment during aging can be ameliorated by supplementation of antioxidants, environmental enrichment and exercise during aging (Radak et al. 2013).

A combination of above mentioned anatomical, biochemical and physiological changes in brain during aging are associated with decline in cognitive functions and increase in the incidence of neurological disorders. Hence, the strategies to reverse such changes can help to develop the therapeutic avenues to improve the cognitive functions during aging and delay or prevent the occurrence of age associated neurological disorders.

3 Cognitive Changes During Normal Aging

Age associated anatomical, biochemical and physiological changes in brain lead to impairment of psychological and cognitive functions, which include attention, language, reasoning, problem solving and memory (Deary et al. 2009). Memory mainly consists of three stages: acquisition- acquires the information, consolidation- stabilizes or stores the information, and retrieval- recalls the stored information (Abel and Lattal 2001). The loss of memory during aging is characterized by the inability to store the new information or loss of previously encoded events. This results from a combination of multiple factors including alterations in neurotransmitter, hormone, oxidative stress, mitochondrial dysfunction, DNA damage, protein aggregation and cell death (Yankner et al. 2008).

Accumulating age associated studies have shown decline in working and long term memory during aging in rodents as well as humans. Working memory is the ability to process the information for executing a particular task and is known to be regulated primarily by prefrontal cortex. Different behavioral paradigms in rodents, primates and humans showed decline in working memory with aging (Deary et al. 2009; Peleg et al. 2010; Penner et al. 2011). To elucidate the mechanism of such decline in working memory, many theories have been proposed. These include decrease in attention, slow information processing and failure in command for irrelevant information (Glisky 2007). Earlier reports have shown that the short term memory remains normal but its conversion into long term memory is predominantly affected during aging. Similarly, the hippocampal dependent consolidation of short term memory into long term memory, associative, fear, recognition and spatial memory are diminished with advancing age (Peleg et al. 2010; Penner et al. 2011; Singh and Thakur 2014). Also, different forms of long term memory including episodic, semantic, autobiographical, procedural, implicit and prospective are reduced during aging (Glisky 2007).

Age associated decline in memory has been observed even in the absence of pathological conditions. However, the detailed molecular mechanism underlying such changes is poorly understood. Studies using animal models have shown the importance of gene expression in formation and maintenance of memory. Microarray analysis of transcriptome has revealed several genes implicated in memory loss. It further shows increase in stress and inflammation related gene expression, but decrease in growth/trophic factors, energy metabolism, protein turnover and synaptic plasticity gene expression in old (Blalock et al. 2003; Lu et al. 2004; Rowe et al. 2007).

Among many genes, those associated with synaptic plasticity gained special attention because they are involved in remodeling of synaptic structure and memory formation. They include immediate early response, neurogenesis, growth/neurotrophic factors, transcription factors, vesicular transporters, microtubule structure, kinases and receptors associated genes (Thakur et al. 2012). The expression of majority of these genes is down regulated in the cerebral cortex and hippocampus during aging. Recently, we have reported that the expression of synaptic plasticity genes like neuropsin, neurexin, neuroligin and nogoA is altered during aging (Kumar and Thakur 2015; Kumari and Thakur 2014; Konar and Thakur 2015). Interestingly, the expression of neuroligin3 is increased in the hippocampus of old mice, suggesting that it might play a compensatory role in improving the memory loss during aging (Kumar and Thakur 2015). On the other hand, the expression of cyclic AMP responsive element binding protein (CREB), brain derived neurotrophic factor (BDNF), nerve growth factor (NGF) and glia derived neurothrophic factor (GDNF) was reduced during aging (Paramanik and Thakur 2013, Budni et al. 2015). This might also be responsible for decrease in hippocampal volume and neurogenesis during aging (Tapia-Arancibia et al. 2008; Erickson et al. 2010). Neuronal immediate early gene which codes for activity regulated cytoskeletal associated protein (Arc) shows immediate response after neuronal stimulation and regulates the remodeling of synaptic structure during memory consolidation (Kubik et al. 2007; Okuno 2011). Arc interacts with endophilin and dynamin to regulate actin dynamics as well as AMPA receptor trafficking at synapse (Tzingounis and Nicoll 2006; Bramham et al. 2008). Arc mRNA level is down regulated in CA1 and DG regions of hippocampus in old (Penner et al. 2011). Early growth response protein1 (EGR1) is a transcription factor, which regulates the expression of late response genes including neuropeptide Y, synapsin I and II, nicotinic acetylocholine receptor (a7 subunit), and p75 nerve growth factor (p75 NGF) receptor (Knapskaa and Kaczmarek 2004) and is reduced in old mice (Desjardins et al. 1997).

The transcriptional regulation of synaptic plasticity genes during aging is still elusive. Studies during the last decade have explored the importance of epigenetics in the regulation of expression of synaptic plasticity genes. DNA methylation and histone acetylation regulate gene expression at transcriptional level and play a crucial role in learning and memory (Guan et al. 2009; Feng et al. 2010). DNA methylation is regulated by DNA methyltranasferase (DNMT), whereas histone acetylation is regulated by histone acetylatransferase (HAT) and histone deacetylase (HDAC). We have reported that the recognition memory in old is associated with increased HDAC2 and decreased DNMT1 expression in hippocampus (Singh and Thakur 2014). Furthermore, increase in DNA methylation and decrease in histone acetylation at the promoter region of synaptic plasticity genes result in reduced expression and memory consolidation during aging (Peleg et al. 2010; Penner et al. 2011). Thus, changes in chromatin modification and underlying gene expression during aging could be a potential target to recover aging and neurodegeneration associated decline in memory.

4 Age Associated Memory Loss: Recovery Strategies

Age associated memory loss and its recovery during aging is a major challenge today, particularly in the light of increasing elderly population. Mostly, the cognitive impairment speeds up around the age of 50 (Verhaeghen and Salthouse 1997;

Salthouse 2003) and subjective complaints about cognitive capacity increases with age (Martin and Zimprich 2003; Newson and Kemps 2006). Several approaches such as memory training, physical exercise, social and nutritional interventions are the traditional strategies to improve memory during aging. Memory training classes in elderly improve the performance of cognitive tasks including perceptual discrimination, visual search, recognition, recall and spatial perception (Kramer et al. 2004). It typically deals with mnemonic strategies, concentration, attention, relaxation, personal insight, self monitoring, motivation, feedback and problem solving approaches that improve memory (McDougall 2000). Regular exercise programme among elderly can slow down or prevent functional decline associated with aging and improve health. Participation in endurance, balance and resistance training programmes provides health benefits, which include improved muscle mass, arterial compliance, energy metabolism, cardiovascular fitness, muscle strength, overall functional capacity (Lemura et al. 2000) and the maintenance or even enhancement of cognitive function (Colcombe and Kramer 2003). Similarly, engagement in social activity affects cognition in aging. Elderly with higher social activity are at lower risk of cognitive decline as compared to those with fewer social ties. Nutrition is another very critical factor in successful cognitive aging. Studies have identified nutritional intake and metabolic indicators which are correlated with cognitive performance in aging. Supplements such as antioxidants, vitamins, fats and minerals are known to improve cognitive functions when provided during common chronic diseases such as hypertension, hyperlipidemia and diabetes. Antioxidants, such as vitamins C and E, are potential compounds for limiting vascular inflammation and are supplemented to reduce cognitive decline for aging adults (Morris 2006).

Now several modern approaches like herbal intervention, life style, hormonal replacement therapy, epigenetic modifiers and stem cell based regenerative medicine are popularly used for preventing cognitive decline during aging (Thakur and Rattan 2012; Konar et al. 2015b). Our laboratory has investigated the nootropic potential of Ashwagandha and Brahmi extract as natural factors for improving memory in drug induced amnesic mouse model (Konar et al. 2011; Gautam et al. 2013, 2015; Konar et al. 2015a). These extracts exhibited remarkable potential to increase the expression of plasticity genes (Arc, neuropsin, BDNF), promote neurite growth and reverse memory loss in animal models mimicking age related cognitive deficits (Thakur and Konar 2014). Epigenetic modifiers are becoming more promising as a key therapeutic avenue in aging and neurodegenerative diseases. In aged rodents, administration of suberoylanilidine hydroxamic acid (SAHA) and HDAC inhibitors is reported to recover memory through chromatin modifications (Reolon et al. 2011; Intlekofer et al. 2013). Moreover, selective inhibition of HDAC2 by antisense oligonucleotide showed recovery of memory through alteration in histone acetylation at the promoter region of synaptic plasticity genes (Singh et al. 2015).

Thus, the expression of genes can also be targeted as recovery strategies for age associated memory loss. Several protein kinases including protein kinase A and PI3K/AKT/mTOR pathway have been identified as crucial molecular targets in reversal of age associated deficits of memory consolidation. CREB dependent transcriptional regulation of memory linked genes is considered to be a principal therapeutic target for the recovery of central nervous system disorder during aging (Thakur et al. 2012). The increased expression of synaptophysin in hippocampus of old mice is reported to play a compensatory role in age related functional deficits (Benice et al. 2006). Thus, these genes can be targeted to develop new therapeutic avenues to improve the age associated memory loss.

5 Conclusion

Age related deficits in cognitive abilities, particularly learning and memory, deteriorate the quality of life as well as increase the vulnerability for neurological disorders. Accumulative changes in brain with advancing age coupled with accelerating factors like stress switch normal aging to neurodegenerative pathologies such as AD and PD. It is well established that brain aging is multi-factorial involving several anatomical and biochemical deteriorations and altered molecular cascades eventually leading to memory loss. Accordingly, integration of several approaches like herbal interventions, dietary constraints and life style improvement programs have been adopted to prevent or reverse brain aging accompanied changes and recover memory loss. Amongst several trajectories that influence age related memory loss, it is clear that molecular changes including gene expression and upstream epigenetic modifications is pivotal. Majority of gene expression studies including our laboratory findings reveal alterations in stress, inflammation,

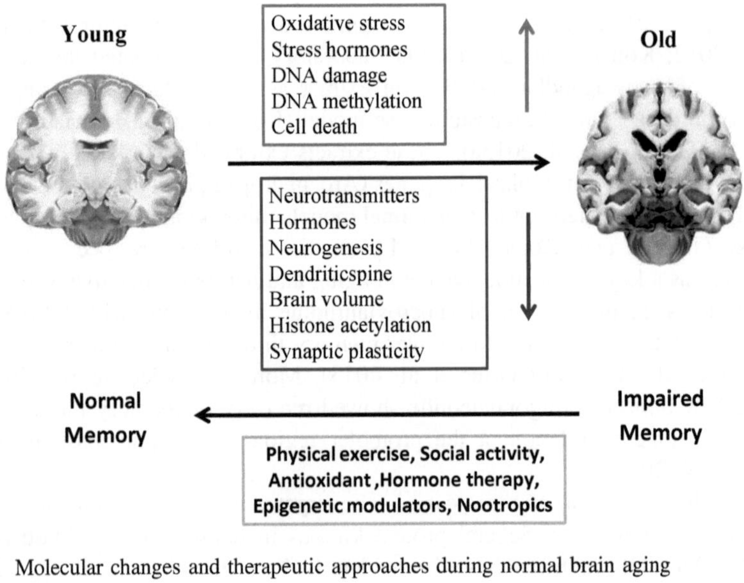

Fig. 1 Molecular changes and therapeutic approaches during normal brain aging

immune response, mitochondrial functions, growth factors, neuronal survival, synaptic plasticity and calcium homeostasis genes. Further these gene expression changes are the eventual consequence of chromatin remodeling mechanisms and accumulation of aberrant epigenetic marks including histone and DNA modifications. Therefore, targeting gene expression changes and developing epigenetic modulators can prove to be the most potent therapeutic measure of age associated memory loss (Fig. 1).

Acknowledgments The authors thank Department of Science and Technology (DST), Government of India for INSPIRE faculty award to AK. PS and DK are recipient of Senior Research Fellowship from the Council of Scientific and Industrial Research (CSIR) Government of India, and MSB is a recipient of Junior Research Fellowship from Department of Biotechnology (DBT) Government of India. The work cited in this article has been supported by grants from the DBT, DST, CSIR, ICMR, Government of India, to MKT.

References

Abel T, Lattal KM (2001) Molecular mechanisms of memory acquisition, consolidation and retrieval. Curr Opin Neurobiol 11:180–187

Benice TS, Rizk A, Kohama S, Pfankuch T, Raber J (2006) Sex differences in age-related cognitive decline in C57BL/6 J mice associated with increased brain microtubule-associated protein 2 and synaptophysin immunoreactivity. Neuroscience 137:413–423

Berti V, Mosconi L, Glodzik L, Li Y, Murray J et al (2011) Structural brain changes in normal individuals with a maternal history of Alzheimer's. Neurobiol Aging 32:2325

Blalock EM, Chen K, Sharrow K, Herman JP, Porter NM, Foster TC, Landfield PW (2003) Gene microarrays in hippocampal aging: statistical profiling identifies novel processes correlated with cognitive impairment. J Neurosci 23:3807–3819

Bramham CR, Worley PF, Moore MJ, Guzowski JF (2008) The immediate early gene Arc/Arg3.1: regulation, mechanisms, and function. J Neurosci 28:11760–11767

Budni J, Bellettini-Santos T, Mina F, Garcez ML, Zugno AI (2015) The involvement of BDNF. NGF and GDNF in aging and Alzheimer's disease. Aging Dis. doi:10.14336/AD.2015.0825

Burger C (2010) Region-specific genetic alterations in the aging hippocampus: implications for cognitive aging. Front Aging Neurosci 2:1–8

Burke SN, Barnes CA (2006) Neural plasticity in the ageing brain. Nat Rev Neurosci 7:30–40

Chakrabarti S, Munshi S, Banerjee K, Thakurta IG, Sinha M, Bagh MB (2007) Mitochondrial dysfunction during brain aging: Role of oxidative stress and modulation by antioxidant supplementation. Aging Dis 2:242–256

Chang L, Jiang CS, Ernst T (2009) Effects of age and sex on brain glutamate and other metabolites. Mag Res Ima 27:142–145

Colcombe S, Kramer AF (2003) Fitness effects on the cognitive function of older adults; a meta-analytic study. Psychol Sci 14:125–130

Deary IJ, Corley J, Aj Gow, Harris SE, Houlihan LM et al (2009) Age-associated cognitive decline. Brit Med Bull 92:135–152

den Dunnen WF, Brouwer WH, Bijlard E, Kamphuis J, van Linschoten K, Eggens-Meijer E, Holstege G (2008) No disease in the brain of a 115-year-old woman. Neurobiol Aging 29:1127–1132

Desjardins S, Mayo W, Vallee M, Hancock D, Le Moal M, Simon H, Abrous DN (1997) Effect of aging on the basal expression of c-Fos, c-Jun, and Egr-1 proteins in the hippocampus. Neurobiol Aging 18:37–44 deToledo-Morrell L, Geinisman Y, Morrell F (1988) Age-dependent alterations in haippocampal synaptic plasticity: Relation to memory disorders. Neurobiol Aging 9:581–590

Drachman DA, Noffsinger D, Sahakian BJ, Kurdziel S, Fleming P (1979) Aging, memory and cholinergic system: A study of dichotic listening. Ann Neurol 6:157–158

Erickson KI, Prakash RS, Voss MW, Chaddock L, Heo S et al (2010) Brain-derived neurotrophic factor is associated with age-related decline in hippocampal volume. J Neurosci 30:5368–5375

Feng J, Zhou Y, Campbell SL, Le T, Li E et al (2010) Dnmt1 and Dnmt3a maintain DNA methylation and regulate synaptic function in adult forebrain neurons. Nat Neurosci 13:423–430

Foster TC (1999) Involvement of hippocampal synaptic plasticity in age related memory decline. Brain Res Rev 30:236–249

Foster TC (2007) Calcium homeostasis and modulation of synaptic plasticity in the aged brain. Aging Cell 6:319–325

Gautam A, Kaul S, Thakur MK (2015) Alcoholic extract of ashwagandha leaves protects against amnesia by regulation of Arc function. Neurobiol, Mol. doi:10.1007/s12035-015-9117-2

Gautam A, Wadhwa R, Thakur MK (2013) Involvement of hippocampal Arc in amnesia and its recovery by alcoholic extract of Ashwagandha leaves. Neurobiol Learn Mem 106:177–184

Glisky EL (2007) Changes in cognitive function in human aging; in Brain Aging: models, methods, and mechanisms. In: Riddle DR (ed). CRC Press

Guan JS, Haggarty SJ, Giacometti E, Dannenberg JH, Joseph N et al (2009) HDAC2 negatively regulates memory formation and synaptic plasticity. Nature 459:55–60

Hermann PM, Watson SN and Wildering WC (2014) Phospholipase A2- nexus of aging, oxidative stress, neuronal excitability, and functional decline of the aging nervous system? Insights from a snail model system of neuronal aging and age-associated memory impairment. Front Genet. doi:10.3389/fgene.2014.00419

Intlekofer KA, Berchtold NC, Malvaez M, Carlos AJ, McQuown SC et al (2013) Exercise and sodium butyrate transform a subthreshold learning event into long-term memory via a brain-derived neurotrophic factor- dependent mechanism. Neuropsychopharmacology 38:2027–2034

Jacobs B, Schall M, Prather M, Kapler E, Driscoll L et al (2001) Regional dendritic and spine variations in human cerebral cortex: a quantitative Golgi study. Cereb Cortex 11:558–571

Knapskaa E and Kaczmarek L (2004) A gene for neuronal plasticity in the mammalian brain: Zif268/Egr-1/NGFI-A/Krox-24/TIS8/ZENK? Prog Neurobiol 74:183–211

Konar A, Thakur MK (2015) Neuropsin expression correlates with dendritic marker MAP2c level in different brain regions of aging mice. Mol Neurobiol 51:1130–1138

Konar A, Gautam A, Thakur MK (2015a) Bacopa monniera (CDRI-08) upregulates the expression of neuronal and glial plasticity markers in the brain of scopolamine-induced amnesic mice. Evi Based Compl Alt Med id:837012

Konar A, Shah N, Singh R, Saxena N, Kaul SC, Wadhwa R, Thakur MK (2011) Protective role of Ashwagandha leaf extract and its component withanone on scopolamine induced changes in the brain and brain-derived cells. PLoS ONE. doi:10.1371/journal.pone.0027265

Konar A, Singh P, Thakur MK (2015b) Age-associated cognitive decline: Insights into molecular switches and recovery avenues. Aging Dis. doi:10.14336/AD.2015.1004

Kramer AF, Bherer L, Colcombe SJ, Dong W, Greenough WT (2004) Environmental influences on cognitive and brain plasticity during aging. J Gerontol A Biol Sci Med Sci 59:940–957

Kubik S, Miyashita T, Guzowski JF (2007) Using immediate-early genes to map hippocampal subregional functions. Learn Mem 14:758–770

Kumar D, Thakur MK (2015) Age-related expression of Neurexin1 and Neuroligin3 is correlated with presynaptic density in the cerebral cortex and hippocampus of male mice. Age. doi:10.1007/s11357-015-9752-6

Kumari A, Thakur MK (2014) Age-dependent decline of nogo-a protein in the mouse cerebrum. Cell Mol Neurobiol 34:1131–1141

Lemura LM, von Duvillard SP, Mookerjee S (2000) The effects of physical training of functional capacity in adults. Ages 46 to 90: a meta-analysis. J Sports Med Phys Fitness 40:1–10

Lu T, Pan Y, Kao S, Li C, Kohane I, Chan J, Yankner BA (2004) Gene regulation and DNA damage in the aging human brain. Nature 429:883–891

Lupien SJ, McEwen BS, Gunnar MR, Heim C (2009) Effects of stress throughout the lifespan on the brain, behaviour and cognition. Nat Rev Neurosci 10:434–445

Martin M, Zimprich D (2003) Are changes in cognitive functioning in older adults related to changes in subjective complaints? Exp Aging Res 29:335–352

McDougall GJ (2000) Memory improvement in assisted living elders. Issues Ment Health Nurs 21:217–233

Mora F, Segovia G, del Arco A (2007) Aging, plasticity and environmental enrichment: Structural changes and neurotransmitter dynamics in several areas of the brain. Brain Res Rev 55:78–88

Morris MC (2006) Nutrition and cognitive aging. Modifiers of cognitive aging; friedman conference St. Louis: Washington University Center for Aging 1–30

Nagata Y, Uehara T, Kitamura Y, Nomura Y, Horiike K (1998) D-serine and D-[3H]serine binding in the brain regions of senescence-accelerated mouse. Mech Aging Dev 104:115–124

Newson RS, Kemps EB (2006) The nature of subjective cognitive complaints of older adults. Int J Aging Hum Dev 63:139–151

Okuno H (2011) Regulation and function of immediate-early genes in the brain: Beyond neuronal activity markers. Neurosci Res 69:175–186

Ota M, Yasuno F, Ito H, Seki C, Nozaki S, Asada T, Suhara T (2006) Age-related decline of dopamine synthesis in the living human brain measured by positron emission tomography with L-[beta-11C]DOPA. Life Sci 79:730–736

Paramanik V, Thakur MK (2013) Role of CREB signaling in aging brain. Arch Ital Biol 151:33–42

Peinado MA, Quesada A, Pedrosa JA, Martinez M, Esteban FJ, Del Moral ML, Peinado JM (1997) Light Microscopic quantification of morphological changes during aging in neurons and glia of the rat parietal cortex. Anat Rec 247:420–425

Peleg S, Sananbenesi F, Zovoilis A, Burkhardt S, Bahari-Javan S et al (2010) Altered histone acetylation is associated with age-dependent memory impairment in mice. Science 753:328–333

Penner MR, Roth TL, Chawla MK, Hoanga LT, Roth ED et al (2011) Age-related changes in Arc transcription and DNA methylation within the hippocampus. Neurobiol Aging 32:2198–2210

Perry EK, Blessed G, Tomlinson BE, Perry RH, Crow TJ et al (1981) Neurochemical activities in human temporal lobe related to aging and Alzheimer-type changes. Neurobiol Aging 2:251–256

Pinton P, Ferrari D, Rapizzi E, Di Virgilio F, Pozzan T, Rizzuto R (2001) The Ca^{2+} concentration of the endoplasmic reticulum is a key determinant of ceramide-induced apoptosis: significance for the molecular mechanism of Bcl-2 action. EMBO J 20:2690–2701

Radak Z, Marton O, Nagy E, Koltai E, Goto S (2013) The complex role of physical exercise and reactive oxygen species on brain. J Sport Health Sci 2:87–93

Reolon GK, Maurmann N, Werenicz A, Garcia VA, Schröder N, Wood MA et al (2011) Posttraining systemic administration of the histone deacetylase inhibitor sodium butyrate ameliorates aging-related memory decline in rats. Behav Brain Res 221:329–332

Rowe WB, Blalock EM, Chen KC, Kadish I, Wang D, Barrett JF, Thibault O, Porter NM, Rose GM, Landfield PW (2007) Hippocampal expression analyses reveal selective association of immediate-early, neuroenergetic, and myelinogenic pathways with cognitive impairment in aged rats. J Neurosci 27:3098–3110

Salthouse TA (2003) Memory aging from 18 to 80. Alzheimer Dis Assoc Disord 17:162–167

Sandyk R (1997) The accelerated aging hypothesis of Parkinson's disease is not supported by the pattern of circadian melatonin secretion. Int J Neurosci 90:271–275

Schmidt R, Schmidt H, Haybaeck J, Loitfelder M, Weis S, Cavalieri M, Seiler S, Enzinger C, Ropele S et al (2011) Heterogeneity in age-related white matter changes. Acta Neuropathol 122:171–185

Schumacher M, Weill-Engerer S, Liere P, Robert F, Franklin RJM, Garcia-Segura LM, Lambert JJ, Mayoe W, Melcangi RC et al (2003) Steroid hormones and neurosteroids in normal and pathological aging of the nervous system. Prog Neurobiol 71:3–29

Singh P, Thakur MK (2014) Reduced recognition memory is correlated with decrease in DNA methyltransferase1 and increase in histone deacetylase2 protein expression in old male mice. Biogerontology 15:339–346

Singh P, Konar A, Kumar A, Srivas S, Thakur MK (2015) Hippocampal chromatin-modifying enzymes are pivotal for scopolamine-induced synaptic plasticity gene expression changes and memory impairment. J Neurochem. doi:10.1111/jnc.13171

Smith RG, Betancourt L, Sun Y (2005) Molecular endocrinology and physiology of the aging central nervous system. Endocr Rev 26:203–250

Sorra KE, Harris KM (2000) Overview on the structure, composition, function, development, and plasticity of hippocampal dendritic spines. Hippocampus 10:501–551

Tapia-Arancibia L, Aliaga E, Silhol M, Arancibia S (2008) New insights into brain BDNF function in normal aging and Alzheimer disease. Brain Res Rev 59:201–220

Thakur MK, Rattan SIS (2012) Brain aging and therapeutic interventions. Springer Publication

Thakur MK, Konar A (2014) Brain Aging and associated diseases; in Text book on geriatric medicine. In: Sanchetee P (ed). Paras Medical Publisher, pp 32–36

Thakur MK, Konar A, Gautam A (2012) Brain aging: A critical reappraisal; in Brain Aging and Therapeutic Interventions. In: Thakur MK, Rattan SIS (eds). Springer Science, pp 1–18

Turrens JF (2003) Mitochondrial formation of reactive oxygen species. J Physiol 15:335–344

Tzingounis A, Nicoll RA (2006) Arc/Arg3.1: Linking gene expression to synaptic plasticity and memory. Neuron 52:403–407

van Luijtelaar MG, Tonnaer JA, Steinbusch HW (1992) Aging of the serotonergic system in the rat forebrain: An immunocytochemical and neurochemical study. Neurobiol Aging 13:201–215

Veiga S, Melcangib RC, Doncarlosc LL, Garcia-Seguraa LM, Azcoitiad I (2004) Sex hormones and brain aging. Exp Gerontol 39:1623–1631

Verhaeghen P, Salthouse TA (1997) Meta-analyses of age-cognition relations in adulthood: estimates of linear and nonlinear age effects and structural models. Psycho Bull 122:231–249

Wenk GL, Barnes CA (2000) Regional changes in the hippocampal density of AMPA and NMDA receptors across the lifespan of the rat. Brain Res 885:1–5

Yamamoto M (2001) Depression in Parkinson's disease: its prevalence, diagnosis, and neurochemical background. J Neurol 248:5–11

Yankner BA, Lu T, Loerch P (2008) The aging brain. Annu Rev Pathol Mech Dis 3:41–66

Expression and Regulation of Pax6 in Brain of Aging Mice

Shashank Kumar Maurya, Ratnakar Tripathi, Suman Mishra and Rajnikant Mishra

Abstract Aging, an inevitable complex phenomenon among organisms, exhibit progressive loss of functional anatomy but understanding on molecular mechanisms of aging remains elusive. This report presents analysis of complex network of genes and proteins, their functional cascades and hierarchy, and regulation of regulators like Pax6 (*Pa*ired bo*x* 6) because symptoms of age-associated changes match with patients having mutation in *Pax6*. The Pax6 has been observed in the olfactory bulb, amygdala, thalamus, and the cerebellum. It is pro-neurogenic in adult neural progenitors and specifies neuronal subtypes in developing brain and the adult. During aging in brain, the reduction in Pax6-positive cells indicates loss of neurons and affects on adult neuronal stem cells. The alteration in co-localization of Pax6, p53 and SPARC may lead to the loss of plasticity, and p53 mediated cell death pathway during aging. Since the TGF-β, interacts with Pax6 and also gets influenced by Catalase, the Pax6 appears influencing immunological surveillance of brain. The alteration in levels of S100β and genes of oxidative stress management by Pax6 also indicates involvement of Pax6-TGF-β-Catalase axis in aging. The microRNA based regulation is also reflected because the miR335 suppresses the Pax6 expression and acts as anti-oncogenic target in glioma. The implication of Pax6 could also be explored towards possibilities of modifying the rate of aging and reversal of aging clock because the Pax6 interacts with p53 and gets altered during aging.

1 Introduction

Organisms observe unavoidable progressive loss of functions of cells, tissues and organs during process of aging. They try to adapt diverse molecular modulations during aging but remain sensitive to risks of diseases and death (Fig. 1). The

S.K. Maurya · R. Tripathi · S. Mishra · R. Mishra (✉)
Biochemistry and Molecular Biology Laboratory, Department of Zoology,
Banaras Hindu University, Varanasi 221005, India
e-mail: rmishraa@yahoo.com

© Springer Science+Business Media Singapore 2017

P.C. Rath et al. (eds.), *Topics in Biomedical Gerontology*,
DOI 10.1007/978-981-10-2155-8_15

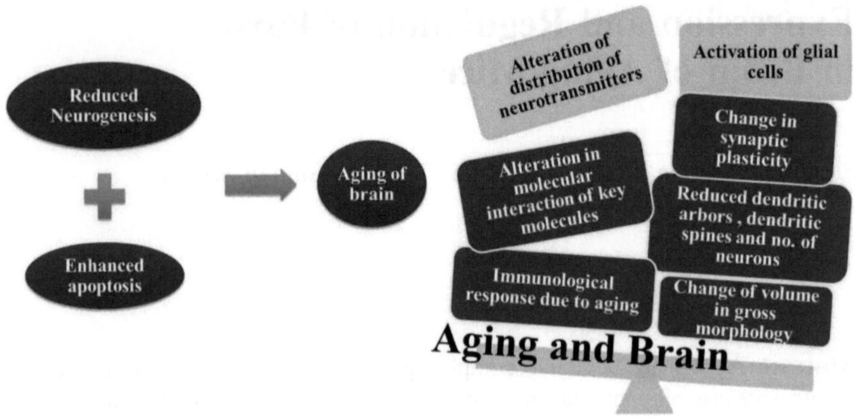

Fig. 1 In aging brain, reduced neurogenesis, increased apoptosis, reduces synaptic plasticity, neuronal loss along with change in gross morphology

molecular alterations could be either cause of loss of age-associated functional anatomy of organs including brain, or affect of expression and interactions of transcriptional regulator. A school of thoughts believes that genes during aging sequentially get switch on and off to influence the nervous, endocrine and immune systems (Kanungo 1975; Weinert and Timiras 2003). Others believe that environmental insults to living organisms induce progressive mitochondrial DNA damage, and accumulation of oxygen radicals (Harman 2003; Brian and Poala 2003). Physiological changes with advancing age have been grouped as loss of the functional reserve of the body's systems, changes in cellular homeostatic mechanisms, decrease in organ mass or their functional anatomy (Dodds 2006). Therefore, to understand molecular mechanisms of aging, analysis of complex network of genes and proteins, their functional cascades and hierarchy, and regulation of regulators like Pax6 (*Pa*ired bo*x* 6) appears important (Fig. 2).

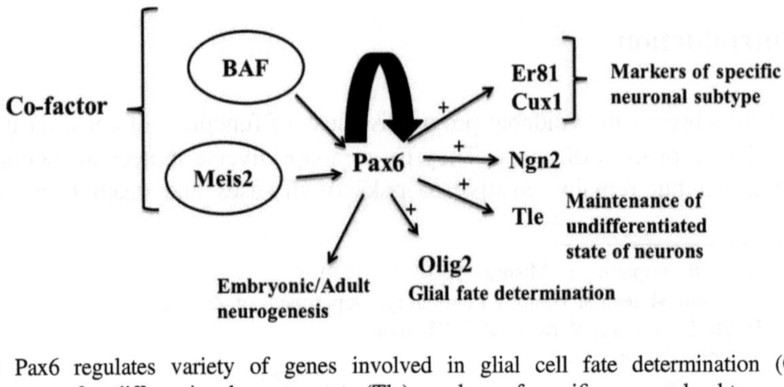

Fig. 2 Pax6 regulates variety of genes involved in glial cell fate determination (Olig2), maintenance of undifferentiated neuron state (Tle), markers of specific neuronal subtypes

Symptoms of age-associated changes like mental retardation, loss of memory, loss of synaptic plasticity, mood fluctuation, anxiety and depression also match with phenotypes observed in patients having mutation in *Pax6* (Glaser et al. 1994; Yasuda et al. 2002). The *Pax6* critically regulates axon guidance (Mastick et al. 1997), migration and differentiation of neurons (Engelkamp et al. 1999; Gotz et al. 1998), and regionalization of cerebral cortex (Bishop et al. 2002). The *Pax6* mutant mice show reduced number of inter-neurons in the olfactory bulb (Parrish-Aungst et al. 2007; Haba et al. 2009), defects in patterning of the forebrain, cortical neurons, and thickness of cortical plate, reduction in the proliferation of neural progenitor cells (Maekawa et al. 2005) or dopaminergic periglomerular cells (Dellovade et al. 1998; Kohwi et al. 2005). However, the over-expression of Pax6 increases number of dopaminergic periglomerular cells (Hack et al. 2005). An optimal dose of Pax6 seems essential for cell the maintenance of various olfactory inter-neurons (Nomura et al. 2007) and adult neurogenesis.

2 Pax6 and Adult Neurogenesis

The expression of Pax6 has been observed in dentate gyrus (DG), sub-granular zone (SGZ) and sub-ventricular zone (SVZ) (Stoykova and Gruss 1994; Kawano et al. 1999; Yamasaki et al. 2001; Tole et al. 2005; Tripathi and Mishra 2012). It cooperates with Dlx2 (Hack et al. 2005; Brill et al. 2008) to specify dopaminergic identity of periglomerular neurons (PGNs) where Pax6 is also found (Bastien-Dionne et al. 2010). It serves as molecular marker for newly generated cells in the DG, hilar mature neurons, and astrocytes of the adult hippocampus (Nacher et al. 2005). It acts as pro-neurogenic in adult neural progenitors and specifies neuronal subtype (Figs. 2 and 3) in the developing and adult brain (Heins et al. 2002; Hack et al. 2005; Stoykova et al. 2000; Kohwi et al. 2005; Kroll and O'Leary 2005; Nikoletopoulou et al. 2007; Tuoc et al. 2009). The physical association of Pax6 with p53 and SPARC in brain of mice suggests Pax6-dependent proliferation and differentiation of neurons, and glia (Tripathi and Mishra 2010).

During aging, neurogenesis and Pax6 (Fig. 4a, b) both get affected (Sisodiya et al. 2001; Gage 2002; Zechel et al. 2010; Bohlen 2010; Ming and Song 2011; Tripathi and Mishra 2012). The less number of Pax6-positive and more p53-positive cells also indicate neuronal loss. It is suggested that modulation of Pax6 may affect interaction with p53 and SPARC to cause defects in neural plasticity, and p53 mediated cell death pathway during aging (Tripathi and Mishra 2010). The observation indicates inverse regulation of Pax6 and p53 in brain of mice. Some report also suggest that increased dose of p53 delays age-associated functions of central nervous system (Carrasco et al. 2015) and the interplay between SOD2 and p53 regulates level of oxidative and nitrative stress (Barone et al. 2015).

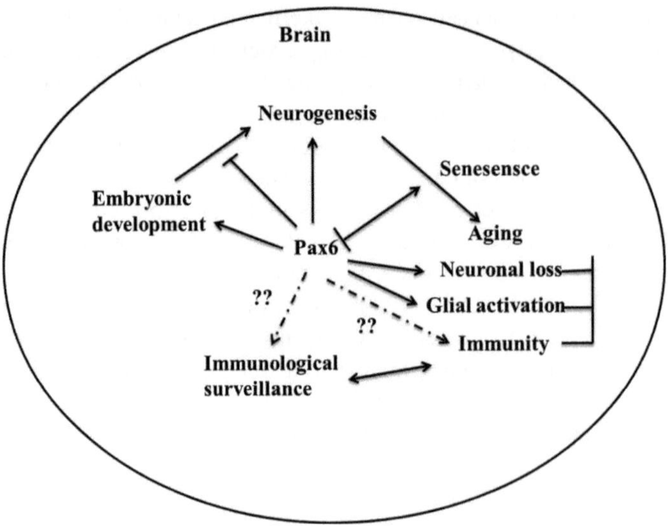

Fig. 3 Pax6 is considered as a central molecule responsible for adult neurogenesis and maintaining homeostasis in adult brain by regulating cell proliferation of neurons and glia. With senescence, as Pax6 expression decreases in brain leading towards neuronal loss and glial activation disturbing homeostasis balance of brain results in various neurological disease conditions. Brain immunity is also known to get affected with increase in age and play major role in disturbed homeostasis in aging brain

3 Pax6 and Immunological Surveillance

Brain was considered as immuno-privileged organ but recent studies suggest presence of lymphatic vessels in brain meningeal compartment (Louveau et al. 2015). It provides basis of not only brain specific immune system but also humoral immunity in brain. The glial cells were previously known as support system of neurons because they provide support during unfavourable (Nimmerjahn et al. 2005) cellular process. The microglia, a specific type of glial cell, has been explained as primary immune cells of the brain with unique multi-receptors. When neurons get injured, microglia becomes activated. During aging, significant increase in glial activation, presence of primed microglia, altered cytokine profile and complement factors, and inflammatory response has been observed in aging brain as comparison to younger brains (Lu et al. 2004; Streit et al. 2008). Some studies also suggest regulation of trophic factors by glia. The levels of BDNF production by hippocampal neurons are associated with CNS specific T-cell activity. The level of TGF-β becomes higher in brain of old mice (Yousef et al. 2015) and increases synaptic plasticity (Caraci et al. 2015).

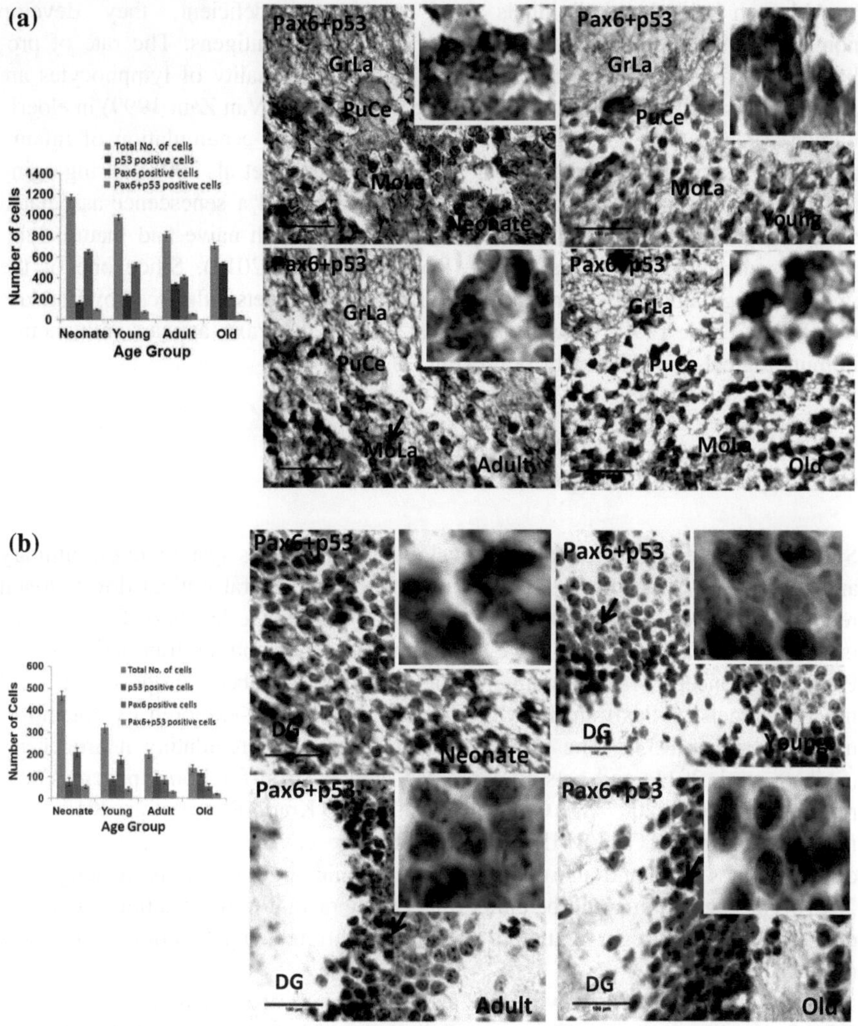

Fig. 4 a It shows the photomicrograph of immuno-histochemistry for expression and co-localization of Pax6 and p53 in cerebellum region in brain of aging mice, neonate, young, adult and old. The histogram shows the number of Pax6 and p53-positive co-localized cells from neonate to old age aged mice ($P < 0.001$). Brain from all age group mice were removed, fixed and processed for paraffin embedding, and sectioning. Sections were rehydrated and processed for immunohistochemical staining using standard protocol for double antigen labeling with anti-Pax6+anti p53 antibodies (Image were adapted from PhD thesis of Tripathi R. 2012). **b** It shows the photomicrograph of immuno-histochemistry for expression and co-localization of Pax6 and p53 in Dentate Gyrus (DG) region in brain of aging mice, neonate, young, adult and old. The histogram shows the number of Pax6 and p53-positive co-localized cells from neonate to old age aged mice ($P < 0.001$). Brain from all age group mice were removed, fixed and processed for paraffin embedding, and sectioning. Sections were rehydrated and processed for immunohistochemical staining using standard protocol for double antigen labeling with anti-Pax6+anti p53 antibodies. Pax6-positive cells were observed less than those of p53-positive cells (Images were taken from PhD thesis of Dr. Ratnakar Tripathi)

Although elderly individuals are not immune-deficient, they develop non-responsiveness to new or previously encountered antigens. The rate of proliferation and migration of naive B- and T-cells, and quality of lymphocytes are affected during aging (Harrison et al. 1989; De Haan and Van Zant 1999) in elderly humans (Pang et al. 2011; Rodriguez et al. 2013). The accumulation of inflammatory mediators (Inflammaging) in tissues (Franceschi et al. 2007) during aging has been proposed (Coppe et al. 2010) due to acquired a senescence-associated secretory phenotype (SASP). Mainly the abilities of both naive and mature lymphocytes during aging get affected. (Rodriguez et al. 2013). Since one of the important regulator TGF-β, interacts with Pax6 and also gets influenced by Catalase (De Bleser et al. 1999; Chamberlain et al. 2009), the Pax6 appears influencing, immunological surveillance of brain (Fig. 3).

4 Regulation of Regulator, Pax6

Since Pax6 knock-out mice fails to develop central nervous system, eyes, pituitary and alpha-cells of pancreas, Pax6 seems to regulate several critical down-stream regulators (Glaser et al. 1994; Tuoc et al. 2009). Since the levels of Pax6 and its isoforms (Mishra et al. 2002) have also been critical for optimal transactivation by Pax6, the concept of auto-regulation of Pax6 has also been proposed. Several miRNA targets of Pax6 are known but remain mostly un-explored except miR9, miR135b and miR335. The miR9 has been involved in regulating neurogenesis (Shibata et al. 2011) and gets co-expressed with Pax6 in forebrain, pancreas, and retina (Arora et al. 2007; De Chevigny et al. 2012; Kredo-Russo et al. 2012). The miR135b inhibits TGFβ/BMP signalling (Bhinge et al. 2014) and the miR335 downregulates Pax6 and also acts as anti-oncogenic target in glioma (Cheng et al. 2014). The regulation could be either directly interacting to the promoter sequence elements of down-stream regulators or indirectly through co-activators or repressors (Fig. 2).

The interaction of Pax6 with p53 and TGF-β indicates regulation of Pax6 by TGF-β signalling pathway. The crosstalk between p53, TGF-β and SPARC (Figs. 5 and 7) also supports SPARC-dependent regulation of Pax6 (Tripathi and Mishra 2010; Shubham and Mishra 2012). Reports also indicate that impairment of TGF-β signalling often leads to neuroinflammation, neuronal dysfunction and pathogenesis of neurodegenerative diseases (Tesseur and Wyss-Coray 2006; Tichauer et al. 2014). Thus, regulation of TGF-β signalling in aged mice could be a therapeutic approach (Fig. 5). A report based on knock-down studies of Pax6, also indicates down regulation of S100β, GFAP, BDNF, NGN2, p73α, p73δ (Mishra et al. 2015). The regulation of S100β by Pax6 (Fig. 6) seems critical for aging and age-related

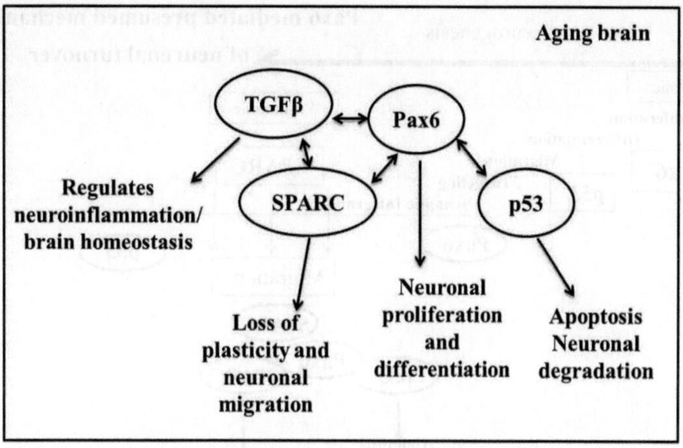

Fig. 5 In aging brain, regulation of Pax6 expression is achieved by physical interaction with p53, SPARC, and TGFβ known to play essential role from neurogenesis to neuronal degeneration

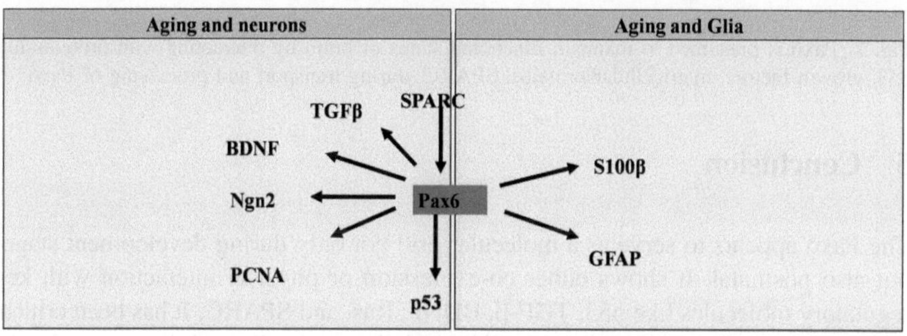

Fig. 6 Pax6 interaction with neuronal (BDNF, Ngn2), glial (S100β, GFAP) markers provide novel aspect of Pax6 involvement in aging and age modulated diseases

diseases and alternate and indirect regulation of oxidative stress through Pax6 seems mediated by Pax6-TGF-β-Catalase axis.

Some studies suggest that the process of aging gets affected by environmental and genetic factors but aging clock may be reversed (Thomas et al. 2012). Among several genes and protein, master regulators like Pax6 and p53 get co-localized and show physical interaction. Therefore, we presume that Pax6 could either be involved directly, or indirectly in brain aging and maintenance of functional anatomy of brain by interacting with proteins like growth factors, p53, TGF-β, TGIF, matricellular proteins, SPARC (Fig. 7). The functional analysis of Pax6 associated genes would be helpful to understand their association with neurological disorders during aging brain.

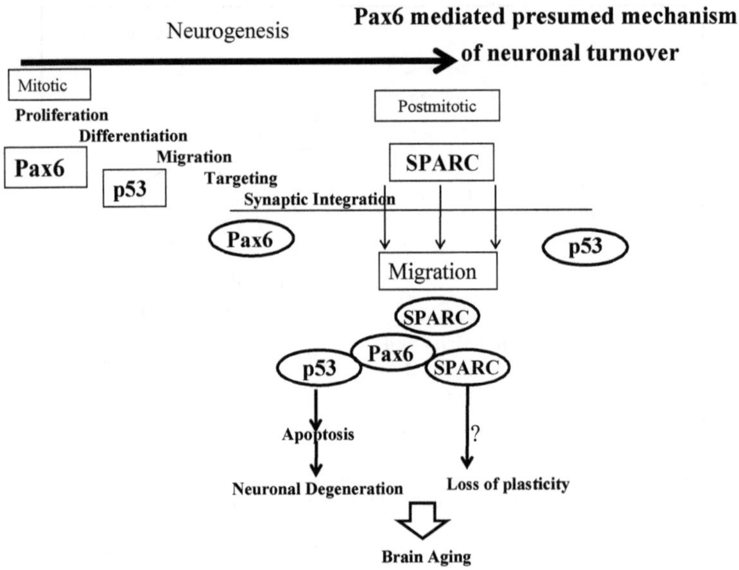

Fig. 7 Pax6 is presumed to maintain functional status of brain by interacting with proteins like p53, growth factors, matricellular proteins, SPARC, during transport and processing of Pax6

5 Conclusion

The Pax6 appears to serve as a molecular grid not only during development stages but also postnatal. It shows either co-expression or physical interaction with key regulatory molecules like p53, TGF-β, BDNF, Ras, and SPARC. It has been critical for regulation of cell cycle, growth, differentiation and cell death in neuronal cells and organs including brain and eyes. The siRNA based results indicate influence of Pax6 on expression of p53, S100β, GFAP, BDNF, NGN2, PCNA, p73α, and p73δ. The multifunctional regulator, Pax6, gets down regulated in brain during aging that may be either cause or effect of Pax6-dependent adult neurogenesis and neurodegeneration in brain. Reports also support Pax6 as indirect molecular switch for genes and/or proteins of oxidative stress management and immunological surveillance in brain. Since, Pax6 interacts to sequence elements and proteins of several regulators and Pax6 appears to be auto-regulated, management of Pax6 gene or protein may serve as potential target for maintaining functional status of brain and vital organs including possibilities of reversal or aging-clock.

Acknowledgments Authors gratefully acknowledge financial support from the ICMR (54/2/CFP/GER/2011-NCD-II).

References

Arora A, McKay GJ, Simpson DA (2007) Prediction and verification of miRNA expression in human and rat retinas. Invest Ophthalmol Vis Sci 48:3962–3967

Barone E, Cenini G, DiDomenico F, Noel T, Wang C, perluigi M, StClair DK, Butterfield DA (2015) Basal brain oxidative and nitrosative stress levels are finely regulated by interplay between superoxide dismutase2 and p 53. Neurosci Res 93:1728–1739

Bastien-Dionne PO, David LS, Parent A, Saghatelyan A (2010) Role of sensory activity on chemospecific populations of interneurons in the adult olfactory bulb. J Comp Neurol. 518:1847–1861

Bhinge A, Poschmann J, Namboori SC, Tian X, JiaHuiLoh S, Traczyk A, Prabhakar S, Stanton LW (2014) MiR-135b is a direct PAX6 target and specifies human neuroectoderm by inhibiting TGF-β/BMP signalling. EMBO J 33:1271–1283

Bishop KM, Rubenstein JL, O'Leary DD (2002) Distinct actions of Emx1, Emx2 and Pax6 in regulating the specification of areas in the developing neocortex. J Neurosci 22:7627–7638

Bohlen V (2010) Dendritic spine abnormalities in mental retardation. Cell Tissue Res 342:317–323

Brian TW, Poala ST (2003) Theories of aging. J Appl Physiol 95:1706–1716

Brill MS, Snapyan M, Wohlfrom H, Ninkovic J, Jawerka M, Mastick GS, Ashery-Padan R, Saghatelyan A, Berninger B, Gotz M (2008) A dlx2- and pax6-dependent transcriptional code for periglomerular neuron specification in the adult olfactory bulb. J Neurosci 28:6439–6452

Caraci F, Gulisano W, Guida CA, Inpellizzeri AA, Drago F, Puzzo D, Palmeri A (2015) A key role for TGF-β1 in hippocampal synaptic plasticity and memory. Sci Rep 5:1–10

Carrasco GE, Arrizabalago O, Serrano M, Lovell BR, Matheu A (2015) Increased gene dosage of Ink4/Arf and p53 delays age-associated central nervous system functional decline. Aging Cell 14:710–714

Chamberlain CG, Mansfield KJ, Cerra A (2009) Glutathione and catalase suppress TGFβ-induced cataract-related changes in cultured rat lenses and lens epithelial explants. Mol Vis 15:895–905

Cheng Q, Cao H, Chen Z, Ma Z, Wan X, Peng R, Jiang B (2014) PAX6, a novel target of miR-335, inhibits cell proliferation and invasion in glioma cells. Mol Med Rep 10:399–404

Coppe JP, Desprez PY, Krtolica A, Campisi J (2010) The senescence-associated secretory phenotype: the dark side of tumor suppression. Annu Rev Pathol 5:99–118

De Bleser PJ, Xu G, Rombouts K, Rogiers V, Geerts A (1999) Glutathione levels discriminate between oxidative stress and transforming growth factor-beta signaling in activated rat hepatic stellate cells. J Biol Chem 274:33881–33887

De Chevigny A, Coré N, Follert P, Gaudin M, Barbry P, Béclin C, Cremer H (2012) miR-7a regulation of Pax6 controls spatial origin of forebrain dopaminergic neurons. Nat Neurosci 15:1120–1126

De Haan G, Van Zant G (1999) Dynamic changes in mouse hematopoietic stem cell numbers during aging. Blood 93:3294–3301

Dellovade TL, Pfaff DW, Schwanzel-Fukuda M (1998) Olfactory bulb development is altered in small-eye (Sey) mice. J Comp Neuro 402:402–418

Dodds C (2006) Physiology of ageing. Anaesthea Intense Care Med 7:456–458

Engelkamp D, Rashbass P, Seawrigth A, Van Heyningen V (1999) Role of Pax6 in development of the cerebellar system. Development 126:3585–3596

Franceschi C, Capri M, Monti D, Giunta S, Olivieri F, Sevini F, Panourgia MP, Incidia L, Celani L, Scurti M, Cevenini E, Castellani GC, Salvioli S (2007) Inflammaging and anti-inflammaging: a systemic perspective on aging and longevity emerged from studies in humans. Mech Ageing Develop 128:92–105

Gage FH (2002) Neurogenesis in the adult brain. J Neurosci 22:612–613

Glaser T, Jepeal L, Edwards JG, Young SR, Favor J, Maas RL (1994) Pax6 gene dosage effect in a family with congenital cataracts, aniridia, anophthalmia and central-nervous-system defects. Nat Genet 7:463–471

Gotz M, Stoykova A, Gruss P (1998) Pax6 controls radial glia differentiation in the cerebral cortex. Neuron 5:1031–1044

Haba H, Nomura T, Suto F, Osumi N (2009) Subtype-specific reduction of olfactory bulb interneurons in Pax6 heterozygous mutant mice. Neurosci Res 65:115–121

Hack MA, Saghatelyan A, de Chevigny A, Pfeifer A, Ashery-Padan R, Lledo PM, Gotz M (2005) Neuronal fate determinants of adult olfactory bulb neurogenesis. Nat Neurosci 8:865–872

Harman D (2003) The free radical theory of aging. Antioxid Redox Signal 5:557–561

Harrison D, Astle C, Stone M (1989) Numbers and functions of transplantable primitive immunohematopoietic stem cells. Effects of age. J Immunol 142:3833–3840

Heins N, Malatesta P, Cecconi F, Nakafuku M, Tucker KL, Hack MA, Chapouton P, Barde YA, Gotz M (2002) Glial cells generate neurons: the role of the transcription factor Pax6. Nat Neurosci 4:308–315

Kanungo MS (1975) A model for aging. J Theor Biol 53:253–261

Kawano H, Fukuda T, Kubo K, Horie M, Uyemura K, Takeuchi K, Osumi N, Eto K, Kawamura K (1999) Pax-6 is required for thalamocortical pathway formation in fetal rats. J Comp Neuro 408:147–160

Kohwi M, Osumi N, Rubenstein JLR, Alvarez-Buylla A (2005) Pax6 is required for making specific subpopulations of granule and periglomerular neurons in the olfactory bulb. J Neurosci 25:6997–7003

Kredo-Russo S, Mandelbaum AD, Ness A et al (2012) Pancreas-enriched miRNA refines endocrine cell differentiation. Development 139:3021–3031

Kroll TT, O'Leary DD (2005) Ventralized dorsal telencephalic progenitors in Pax6 mutant mice generate GABA interneurons of a lateral ganglionic eminence fate. Proc Natl Acad Sci 102:7374–7379

Louveau A, Smirnov I, Keyes TJ, Eccles JD, Rouhani SJ, Peske D, Derecki NC, Castle D, Mandell JW, Lee KS, Harris TH, Kipnis J (2015) Structural and functional features of central nervous system lymphatic vessels. Nautre. doi:10.1038/nature14432

Lu T, Pan Y, Kao SY, Li C, Kohane I, Chan J, Yankner BA (2004) Gene regulation and DNA damage in the ageing human brain. Nature 429:883–889

Maekawa M, Takashima N, Arai Y, Nomura T, Inokuchi K, Yuasa S, Osumi N (2005) Pax6 is required for production and maintenance of progenitor cells in postnatal hippocampal neurogenesis. Genes Cells 10:1001–1014

Mastick GS, Davis NM, Andrew GL, Easter SS (1997) Pax6 functions in boundary formation and axon guidance in the embryonic mouse forebrain. Development 124:1985–1997

Ming GL, Song H (2011) Adult neurogenesis in the mammalian brain: significant answers and significant questions. Neuron 70:687–702

Mishra R, Gorlov IP, Chao LY, Singh S, Saunders GF (2002) PAX6, paired domain influences sequence recognition by the homeodomain. J Biol Chem 277:49488–49494

Mishra S, Maurya SK, Srivastava K, Shukla S, Mishra R (2015) Pax6 influence expression patterns of genes involved in neurodegeneration. Annals of Neuroscience 22:26–231

Nacher J, Varea E, Blasco-Ibanez JM, Castillo-Gomez E, Crespo C, Martinez-Guijarro FJ, Mcewen BS (2005) Expression of the transcription factor Pax6 in the adult rat dentate gyrus. J Neurosci Res 81:753–761

Nikoletopoulou V, Plachta N, Allen ND, Pinto L, Gotz M, Barde YA (2007) Neurotrophin receptor-mediated death of misspecified neurons generated from embryonic stem cells lacking Pax6. Cell Stem Cell 1:529–540

Nimmerjahn A, Kirchhoff F, Helmchen F (2005) Resting microglial cells are highly dynamic surveillants of brain parenchyma in vivo. Science 308:1314–1318

Nomura T, Haba H, Osumi N (2007) Role of a transcription factor Pax6 in the developing vertebrate olfactory system. Dev Growth Differ 49:683–686

Pang WW, Price AE, Sahoo D, Beerman I, Maloney JW, Rossi JD, Schrier LS, Weissman LI (2011) Human bone marrow hematopoietic stem cells are increased in frequency and myeloid-biased with age. Proc Natl Acad Sci 108:20012–20017

Parrish-Aungst S, Shipley MT, Erdelyi F, Szabo G, Puche AC (2007) Quantitative analysis of neuronal diversity in the mouse olfactory bulb. J Comp Neurol 501:825–836

Rodriguez E, Maoz BB, Dorshkind K (2013) Causes, consequences, and reversal of immune system aging. J Clin Invest 123:958–965

Shibata M, Nakao H, Kiyonari H, Abe T, Aizawa S (2011) MicroRNA-9 regulates neurogenesis in mouse telencephalon by targeting multiple transcription factors. J Neurosci 31:3407–3422

Shubham K, Mishra R (2012) Pax6 interacts with SPARC and TGF-β in murine eyes. Mol Vis 18:951–956

Sisodiya SM, Free SL, Williamson KA, Mitchell TN, Willis C, Stevens JM, Kendall BE, Shorvon SD, Hanson IM, Moore AT, Van Heyningen V (2001) PAX6 haploinsufficiency causes cerebral malformation and olfactory dysfunction in humans. Nat Genet 28:214–216

Stoykova A, Gruss P (1994) Roles of Pax genes in developing and adult brain as suggested by expression patterns. J Neurosci 14:1395–1412

Stoykova A, Treichel D, Hallonet M, Gruss P (2000) Pax6 modulates the dorsoventral patterning of the mammalian telencephalon. J Neurosci 20:8042–8050

Streit WJ, Miller KR, Lopes KO, Njie E (2008) Microglial degeneration in the aging brain–bad news for neurons? Front Biosci 13:3423–3438

Tesseur I, Wyss-Coray T (2006) A role for TGF-beta signaling in neurodegeneration: evidence from genetically engineered models. Curr Alzheimer Res 3:505–513

Thomas A, Rando L, Chang HY (2012) Aging, rejuvenation, and epigenetic reprogramming: resetting the aging clock. Cell 148:46–57

Tichauer JE, Flores B, Soler B, Eugenín-von Bernhardi L, Ramírez G, Von Bernhardi R (2014) Age-dependent changes on TGFbeta1 Smad3 pathway modify the pattern of microglial cell activation. Brain Behav Immun 37:187–196

Tole S, Remedios R, Saha B, Stoykova A (2005) Selective requirement of pax6, but not Emx2, in the specification and development of several nuclei of the amygdaloid complex. J Neurosci 25:2753–2760

Tripathi R, Mishra R (2010) Interaction of Pax6 with SPARC and p53 in brain of mice indicates Smad3 dependent auto-regulation. J Mol Neurosci 41:397–403

Tripathi R (2012) Studies on expression and interaction of Pax6 with p 53 and SPARC in brain of aging mice, Ph.D. thesis, Banaras Hindu University, Varanasi

Tripathi R, Mishra R (2012) Aging-associated modulation in the expression of Pax6 in mouse brain. Cell Mol Neurobiol 32:209–218

Tuoc TC, Radyushkin K, Tonchev AB, Piñon MC, Ashery-Padan R, Molnar Z, Davidoff MS, Stoykova A (2009) Selective cortical layering abnormalities and behavioral deficits in cortex-specific Pax6 knock-out mice. J Neurosci 26:8335–8349

Weinert BT, Timiras PS (2003) Theories of aging. J Appl Physiol 95:1706–1716

Yamasaki T, Kawaji K, Ono K, Bito H, Hirano T, Osumi N, Kengaku M (2001) Pax6 regulates granule cell polarization during parallel fiber formation in the developing cerebellum. Development 128:3133–3144

Yasuda T, Kajimoto Y, Fujitani Y, Watada H, Yamamoto S, Watarai T, Umayahara Y, Matsuhisa M, Gorogawa S, Kuwayama Y, Tano Y, Yamasaki Y, Hori M (2002) PAX6 mutation as a genetic factor common to aniridia and glucose intolerance. Diabetes 51:224–230

Yousef H, Condoy MJ, Morgenthalar A, Schlesinger C, Buqaj L, Paliwal P, Greer C, Conboy IM, Schaffer D (2015) Systemic attenuation of TGF-beta pathway by single drug simultaneously rejuvenates hippocampal neurogenesis and myogenesis in the same old mammal. Oncotarget 6:11959–11978

Zechel S, Werner S, Unsicker K, Halbach O (2010) Expression and functions of fibroblast growth factor 2 (FGF-2) in hippocampal formation. Neuroscientist 4:357–373

Electrophysiological Ageing of the Brain: Ageing-Related Impairments in Neural and Cognitive Functions

Rameshwar Singh

Abstract Electrical activity is the functional basis of the nervous system. Even the most complicated forms of information processing such as sensory detection, higher mental function, consciousness, has its basis in electrical signals produced by changes in voltage across the plasma membrane of particular collections of neurons. Ageing of the brain/nervous system involves changes in its electrophysiological activity and age-related disorganization of electrophysiological activity may cause derangement of neural and cognitive functions. Ageing effects on electrical signals will therefore determine the physiological, cognitive and behavioral outcomes of normal ageing as computational processing of information is adversely affected. There is thus a potential link between the electrophysiological and behavioral consequences of ageing. Ageing seems to impact electrophysiological parameters rather selectively. While the resting membrane potential is unaffected, the synaptosomal resting membrane potential is reduced during ageing. The amplitude of action potential is not affected by ageing, the duration and after-hyperpolarization may increase. Normal ageing involves decreased synaptic excitation and increased synaptic inhibitory processes leading to cognitive impairments. The information encoded in hippocampal place field firing may be altered with impaired spatial learning in the aged. The intensity and frequency of stimulation necessary to produce long-term potentiation is increased with advanced age and long-term potentiation decays more rapidly in the aged indicating forgetting or deficiency of memory formation. The electroencephalogram undergoes significant frequency specific and topographically variable age-related changes. Ageing-related alterations in basal electrical neuronal firing may occur in brain regions. Ageing-related disorganization of neuroelectric activity related to age-associated cognitive neuronal outcomes can be restored by appropriately designed pharmacological treatments.

R. Singh (✉)
School of Life Sciences, Jawaharlal Nehru University, New Delhi 110067, India
e-mail: deepak57in@yahoo.co.in

© Springer Science+Business Media Singapore 2017

P.C. Rath et al. (eds.), *Topics in Biomedical Gerontology*,
DOI 10.1007/978-981-10-2155-8_16

1 Introduction

Electrical activity is the fundamental basis of brain functions. The brain is an information-processing system in which information is conducted and integrated as electrical signals. Even the most complicated forms of information processing such as that in consciousness, higher mental functions, sensory detection has its basis in electrical signals produced by changes in voltage across the plasma membrane of particular collections of neurons (Fain 1999). Ageing effects on brain neural electrical activity will therefore determine the physiological, cognitive and behavioral outcomes of normal ageing. It is, therefore, important to understand the electrophysiological consequences of ageing, that is, how electrical activity of the brain is altered over the course of normal ageing.

Changes occurring with normal ageing in neuronal electrophysiological parameters: the action potential, synaptic potentials, multiple-unit and unit activities, spontaneous field potentials (EEG) constitute electrophysiological ageing, which signify ageing-related impairment and disorganization of electrical signals leading to disordered neurological function (Singh 1994; Singh and Sharma 2005; Singh et al. 2006; Baquer et al. 2009). For example, deficiency of, or absence of synchronous action potential bursts in the aged nervous tissue will result in altered presynaptic neurotransmitter release which will impair synaptic plasticity and information processing (Lisman 1997; Apartis et al. 2000).

2 Ageing Effects on Action and Synaptic Potentials

The resting membrane potential of nerve cells is not generally affected by age (Frolkis et al. 1984; Potier et al. 1993; Chang et al. 2005; Rapp and Bachevalier 2013) after it has acquired its normal value during early postnatal days (Zhang 2004). The synaptic resting membrane potential however, decreases with ageing (Tanaka and Ando 1990). The latter will impact neurotransmitter release from the presynaptic terminal in the aged nervous tissue. Ageing may also alter dendritic input resistance leading to inhibitory signaling, for example in the hippocampus cells (Luebke and Rosena 2003; Luebke et al. 2004; Chang et al. 2005) contributing to age-associated cognitive impairment.

The action potential consists of a series of sequential transient changes in the membrane potential, that is: threshold, a rising spike phase, a falling phase and an after-hyperpolarization. These changes occur over a small period of time that constitutes the duration of action potential. Some components of the action potential are affected by the ageing process (Singh and Sharma 2005). While the action potential amplitude generally shows no change with age, the spike duration may be enhanced with age. The after-hyperpolarization following a single spike may not show an age-related change, but the hyperpolarization following a burst of spikes may exhibit an age-related change (Potier et al. 1992; Disterhoft and Oh 2006).

Such changes are associated with alteration in repetitive action potential firing rates. Increases in repetitive action potential firing rate may result in dysfunction (Potier et al. 1992). The rhythmic bursting activity may be impaired by ageing while the nonrhythmic activity is spared (Apartis et al. 2000). Some neurons are endowed with the ability to change their pattern of firing in the brain during different states of arousal and this functional plasticity may decrease with age (Apartis et al. 2000). Alteration in the pattern of firing action potential in the suprachiasmatic nucleus neurons in old age result in disturbance in endogenous biological rhythms (Nygard et al. 2005). The velocity of the nerve impulse (action potential) tends to decrease in some restricted set of fibres (Aston-Jones et al. 1985).

At synaptic junctions, interaction between neurotransmitter molecules and their receptors precedes the generation of postsynaptic potentials. During ageing, the generation of postsynaptic potentials is impacted by a number of events/factors: presynaptic neurotransmitter release, postsynaptic receptor density, synapse elimination without pre and postsynaptic death. There are changes in postsynaptic potentials in the aged nervous tissue and normal ageing may result in decreased synaptic excitation (i.e., excitatory postsynaptic potential EPSP) but increased synaptic inhibition (inhibitory postsynaptic potential IPSP) and this may contribute to age—associated cognitive impairment, for example as in the prefrontal cortex of monkeys (Luebke et al. 2004). The prefrontal cortex is the site of planning and decision making processes in human. Further, this area is responsible for our conscience (Winterer and Goldman 2003).

Aged synapse may exhibit an attenuation of EPSP amplitude together with an attenuation of both the amplitude and frequency of miniature EPSPs (Coggan et al. 2004). However, both EPSP and IPSP may exhibit age-related alteration. An age-related decrease in the amplitude and duration of GABA-related IPSPs was found in CA1 neurons of rats although glutamate-related EPSPs were not significantly altered (Potier et al. 1993).

Long-term potentiation (LTP) is an electrophysiological measure of population spike. It represents a form of synaptic plasticity and is considered as a candidate device for learning and memory. Ageing is associated with an impaired ability to maintain LTP which is a Ca^{2+} dependant process and is altered with age (Foster and Kumar 2002). Ca^{2+} dysregulation with age and consequent changes in Ca^{2+} signalling, together with alteration in balance of protein kinase and protein phosphatase activities mediate this impairment (Hsu et al. 2002; Biessels et al. 2002). LTP has been mostly studied in the hippocampus and particularly in its CA1 cells. Hippocampus dependant memory is age-sensitive (Foster and Kumar 2002; Rapp and Bachevalier 2013). Aged CA3 cells were found to fail to rapidly encode new spatial information compared with young CA3 cells (Wilson et al. 2005). Peak amplitude of LTP is not altered in aged hippocampus, the intensity and frequency of stimulation necessary to produce LTP is, however, increased with advanced age. Further, the strength of stimulus required to generate LTP in older subjects is more (Rapp and Bachevalier 2013). LTP decays more rapidly in aged and this may be related to forgetting and deficiency in memory formation. Furthermore, hippocampal encoding becomes more rigid with age and there is an impaired spatial learning in the aged, and the information encoded by the place field firing is altered. Status of memory mediated by the hippocampal formation and related temporal lobe structures varies substantially among older individuals as some aged individuals exhibit considerable impairment while others not (Rapp and Bachevalier 2013).

3 Ageing Effects on Excitability

The excitability of nerve cells refer to their ability or capacity to undergo excitation, that is respond to stimulation by producing electrical signals such as action potentials, receptor potentials and synaptic potentials. The basic excitability of the neuronal cell membrane seems to be achieved through modification of its ion channels (Na^+, K^+, Ca^{2+}, Cl^-). Because channel current contributes to signal generation, age-related alteration in the genetic expression of channels (number of channels) and their subunit composition (Murchison and Griffith 1995) affect potential signalling.

Excitability of neurons may decrease in aged neurons. Hippocampal CA1 pyramidal neurons from aged animals have enhanced voltage-gated Ca^{2+} entry and post-burst after-hyperpolarization. As a result there is decrease in their intrinsic excitability (Hemond and Jaffe 2005; Cingolani et al. 2002; Disterhoft and Oh 2006) which leads to age-associated impairment in cognitive functioning and modulation of firing properties of neurons. The axonal excitability (ability to generate nerve impulse by means of voltage-gated channels) can change due to alteration in nodal and individual ion channels, nodal width, electrical isolation between the internodal and nodal compartments, altered myelination, and membrane potential (Nodera et al. 2003). Species and regional differences, however, are often found. Age-related increases, in the excitability of CA1 and CA3 hippocampal (Barnes et al. 1987) and granule cells in the rat hippocampus were also reported (Papatheodoropolos and Kostopoulos 1996). The excitability of the cortical motor neurons in animals and humans have been reported to decrease with ageing (Yordanova et al. 2004). Blockage of the Na^+ channel, but not the input resistance or membrane potential was found to be responsible for the decrease in excitability of hippocampal neurons (Leung and Yim 1991).

4 Ageing Changes in Evoked Potential and Cognition

Sensory stimulus-evoked responses consisting of short train of waves (representing changes in the EEG) recorded from the scalp and corresponding to dendritic activity in the cerebral cortex called evoked potential or event-related potential provide a picture of information flow through synaptic tracts of the brain. This involves sensory processing, axonal conduction, synaptic transmission and cognitive processing (Crowley and Colrain 2004), and alterations in the latency and amplitude of the waveform of evoked responses provide clinically relevant information about neuropathic events in health and disease (Singh et al. 2004; Karthikkeyan et al. 2015). The components of evoked potentials reveal both early and late electrical brain processes. With increasing age, there is likely to be a decline in the performance of tasks and there are multiple and varying effects of age on evoked responses. Event-related potentials are also induced endogenously by processes such as decision making (Crowley and Colrain 2004). There are age-related changes in the components of evoked potential. Advancing age may be associated with elevated amplitude of early positive waves and significant delay of the major late positive waves in the old subjects (Pelosi and Blumbardt 1999). The components of evoked potential can be used as process-specific time markers in young and aged subjects (Pfutz et al. 2002; Boutros et al. 2000). Cognitive ability can be correlated with P300 waveform of the human auditory evoked potential which is mainly generated in the temporoparietal cortex (Winterer and Goldman 2003). The P300 amplitude and latency seems to be influenced by ageing (Boutros et al. 2000).

5 Ageing Effects on the Electroencephalogram

The cortical electroencephalogram (EEG) is derived mainly from summation of the synaptic activity in the pyramidal cells of the cerebral cortex and the EEG rhythms are driven by thalamic activity. Spectral content of the EEG does reflect changes occurring in brain subcortical structures, and several of its electrophysiological measures are genetically heritable (Winterer and Goldman 2003). The low frequency EEG may be indicative of a cognitive decline, and increases in the alpha and beta activities may reflect vigilance-promoting influences. Altered frequencies and increased synchronization have been found in the human EEG with ageing (Duffy et al. 1993, 1996). Age-related reduction of power in the human scalp EEG and frequency specific changes in the sleep EEG topography have also been reported (Landolt and Borbely 2001). EEG studies in animals have also indicated that spectral composition of the baseline EEG is likely to change with age.

Another point of interest is sleep and the process of ageing. Sleep is accompanied by changes in brain electrical activity. Age-impairment in sleep is often seen in the elderly and sleep has antiageing influence because it reduces oxidative stress (Singh et al. 2008). Therefore, lack of sleep in the elderly is likely to augment the ageing process.

Disordered neuronal (electrical) activity also is a characteristic feature of brain disorder epilepsy. Epilepsy is a brain disorder in which a group of neurons or groups of neurons become abnormal i.e. epileptic neurons which are in a heightened state of excitability and exhibit recurrent episodes of disordered hypersynchronous electrical firing that disrupt normal brain functioning (Rakhade and Loeb

2008; Sharma et al. 2007). The process of development of epilepsy (epileptogenesis) is complex and there is an increased incidence and prevalence of epilepsy in the elderly human population (Gupta and Sanchetee 2014). The aged central nervous system may become more susceptible to epileptogenesis (Jyoti et al. 2009) and therefore knowledge about the susceptibility of the brain to seizure disorder in ageing is of critical importance in geriatric medicine and public health (Shin et al. 2011). Experimentally it was observed that aged animals were more susceptible to seizure induction than young ones (Jyoti et al. 2009; Shin et al. 2011) and this may be due to age-related factors such as oxidative stress, oxidative deterioration of neurotransmitter transporters, androgen hormone levels etc. Dehydroepiandrosterone (DHEA) is a natural physiological androgen hormonal substance. The level of this hormone declines with age, and experimentally it was found to have antiageing and antiepileptic action (Mishra et al. 2010). DHEA upregulates glutamate transporters and counters epileptiform electrical activity (Mishra et al. 2013). The Na^+, K^+-ATPase inhibition is also implicated in epileptogenesis. Normal aging is also associated with depression of Na^+, K^+-ATPase activity. Decreased Na^+, K^+-ATPase activity elevates neuronal sensitivity to glutamate and NMDA receptors become more sensitive with ageing (Kaur et al. 2003a). These factors may elevate aged nervous tissues propensity for epileptogenesis. DHEA also has modulatory influences on excitatory neurotransmission, Na^+, K^+-ATPase activity and oxidative stress. Therefore, DHEA can be of therapeutic importance in the elderly with epilepsy.

6 Ageing and Neuronal Activity

Multiple-unit activity (MUA, action potentials simultaneously derived from many neurons) represents an electrophysiological marker of cellular electrical firing activity of the concerned neuronal population. MUA alterations may reflect the biochemical, physiological and behavioural alteration of neurons (Mizumori et al. 1996; Sharma et al. 1993). The decline in the spontaneous MUA with age in brain areas (Fig. 1) can be considered as a measure of neuronal impairment and thus as an important parameter of electrophysiological ageing (Singh and Sharma 2005; Baquer et al. 2009). Age-related declines in spontaneous MUA have been found to occur in several brain areas: locus coeruleus of rats (Olpe and Steinmann 1982), forebrain neurons of rats (Jones and Olpe 1984), the rat cerebral cortex (Roy and Singh 1988), the rat hippocampus, thalamus, striatum (Kaur et al. 1998, 2001, 2003b; Sharma et al. 1993). Age-related changes in membrane lipid peroxidation (Pellmar 1986) and sodium pump activity may contribute to changes in membrane electrophysiology (Sharma et al. 1993; Singh and Sharma 2005; Singh et al. 2012; Baquer et al. 2009)..

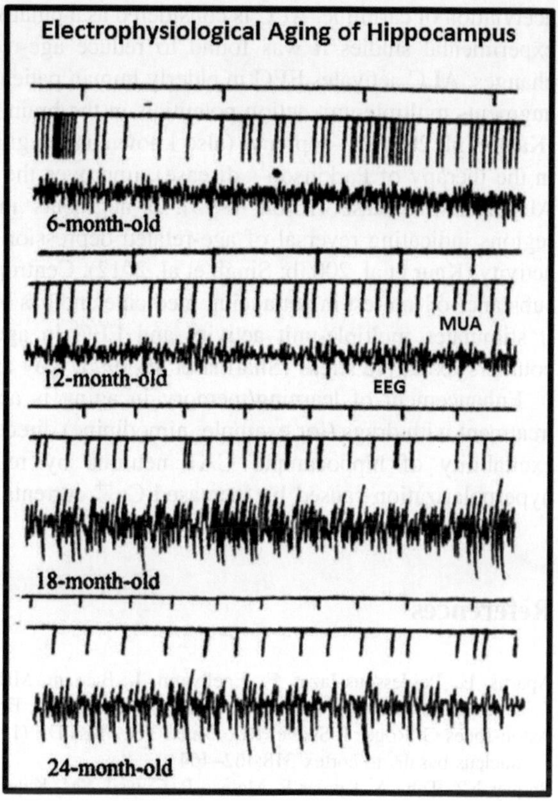

Fig. 1 The figure shows how basal firing rate (MUA) of neurons normally declines with age in rat hippocampus CA3 subfield. EEG slowing is also evident

7 Antiageing Pharmacological Strategies

Ageing-related deterioration/disorganization of neuroelectric activity and associated cognitive outcome can be restored by appropriately designed pharmacological treatment/interventions (Singh and Sharma 2005; Rapp and Bachevalier 2013). For example, memory-related decline in neuronal activity in the prefrontal cortex of aged monkeys can be substantially restored by pharmacological treatments that benefits behavioral performance (Wang et al. 2011). L-deprenyl was found to augment excitability/neuronal electrical activity possibly related to longer-term memory in CA1 pyramidal neurons (Singh et al. 2012). There have been some studies where vitamin E supplementation was shown to increase the amplitude of P_3 waveform of evoked potential in demented patients (Vaney et al. 2002) leading to improved cognitive pool of P_3 generator neurons.

There are many chemical substances/drugs which exert antiageing pharmacological action that involves modulation of electrophysiological phenomena. Acetyl-L-carnitine (ALC) is a normal physiological component of the inner mitochondrial membrane and is endogenously formed in the brain from the reversible acetylation of carnitine. ALC is considered as a putative antiageing substance and in experimental studies it was found to reduce age-related cognitive and neuronal changes. ALC activates EEG in elderly human patients (Herrmann et al. 1990) and augments multiple unit action potentials in the brain of aged experimental animals (Kaur et al. 2001). L-deprenyl (also known as selegiline, a drug used as an adjunct in the therapy of Parkinson's disease), improves the performance of patients with Alzheimer's disease (Knoll 1993). It augments multiple-unit activity in brain regions indicating reversal of age-related depression of brain electrophysiological activity (Kaur et al. 2003b; Singh et al. 2012). Centrophenoxine is another chemical substance of interest in antiageing medicine and has been studied by many workers. It stimulates multiple-unit activity and EEG in aged experimental animals and counters oxidative stress (Sharma et al. 1993; Roy and Singh 1988).

Enhancement of learning/memory in aging is achievable by pharmacological treatment with drugs (for example, nimodipine) directed at augmenting the intrinsic excitability of hippocampal CA1 neurons by reducing their post-burst after hyperpolarization caused by increased Ca^{++} currents (Disterhoft and Oh 2006).

References

Apartis E, Poidessous-Jazat F, Epelbaum J, Bassant MH (2000) Age-related changes in rhythmically bursting activity in medial septum of rats. Brain Res 876:37–47

Aston-Jones G, Roger J, Shaver RD, Dinen TG, Moss DE (1985) Age-related impulse flow from nucleus basalis to cortex 318:462–464

Baquer NZ, Taha A, Kumar P, Mclean P, Cowsik SM, Kale RK, Singh R, Sharma D (2009) A metabolic and functional overview of brain ageing linked to neurological disorders. Biogerontology 10:377–413

Barnes CA, Rao G, Mc Naughton BL (1987) Increased electrotonic coupling in aged rat hippocampus: a possible mechanism for cellular excitability change. J Comp Neurol 259:249–258

Biessels GT, Terlaak MP, Hamers FP, Gipsen WH (2002) Neuronal Ca^{2+} disregulation in diabetes mellitus. Eur J Pharmacol 447:201–209

Boutros NN, Reid MC, Petrakis I, Campbell D, Torello M, Krystal J (2000) Similarities in the disturbances in cortical information processing in alcoholism and aging: a pilot evoked potential study. Int Psychogeriatr 12:513–525

Chang YM, Rosene DL, Killiany RJ, Mangiamele LA, Luebke JI (2005) Increased action potential firing rates of layer 2/3 pyramidal cells in the prefrontal cortex are significantly related to cognitive performance in aged monkeys. Cereb Cortex 15:409–418

Cingolani LA, Gymnopoulos M, Boccaccio A, Stocker M, Pedarzani P (2002) Developmental regulation of small-conductance Ca^{2+} activated channel expression and function in rat Purkinje neurons. J Neurosci 22:4456–4467

Coggan JS, Grutzendler J, Bishop DL, Cook MR, Gen W, Heym J, Lichtman JW (2004) Age-associated synapse elimination in mouse parasympathetic ganglia. Neurobiology. 60:214–226

Crowley KE, Colrain IM (2004) A review of the evidence for P2 being an independent component process: age, sleep and modality. Clin Neurophysiiol 115:732–744

Disterhoft JF, Oh MM (2006) Pharmacological and molecular enhancement of learning in aging and Alzheimer's disease. J Physiol Paris 180–192

Duffy FH, Mc Anulty GB, Albert MS (1993) The pattern of age-related differences in electrophysiological activity of healthy males and females. Neurobiol Aging 14:73–84
Duffy FH, Mc Anulty GB, Albert MS (1996) Effect of age upon interhemispheric EEG coherence in normal adults. Neurobiol Aging 17:587–599
Fain GL (1999) Molecular and cellular physiology of Neurons: Prentice hall of India new Delhi 2005 pp 63-94
Foster TC, Kumar A (2002) Calcium dysfunction in the aging brain. Neuroscientist 8:297–301
Frolkis VV, Train SA, Martynenko DA, Bogatskaya LN, Bezrukov VV (1984) Aging of the neurons. In: Frolkis VV, Johnson JE (eds) Physiology of cell aging. Raven Press, New York, pp 149–185
Gupta S, Sanchetee P (2014) Epilepsy in elderly. In: Sanchetee P (ed) Text Book of Geriatrics Medicine. Paras Medical Publisher, Hyderabad India, pp 485–489
Hemond P, Jaffe DB (2005) Caloric restriction prevents aging associated changes in spike-mediated Ca^{2+} accumulation and the slow after hyperpolarization in hippocampal CA1 pyramidal neurons. Neuroscience 135:413–420
Herrmann WM, Dietrich B, Hiersemenzel R (1990) Pharmaco-electroencephalographic and clinical effects of the cholinergic substances- acetyl-L-carnitine-in patients with organic brain syndrome. Int J Clin Pharmacol Res 10:81–84
Hsu KS, Huang CC, Liang YC, Wu HM, Chen YL, Lo SW, Ho WC (2002) Alterations in the balance of protein kinase and phosphatase activities and age-related impairments of synaptic transmission and long-term potentiation. Hippocampus 12(787):802
Jones RSG, Olpe HR (1984) Multiple changes in the sensitivity of cingulated cortex neurons to putative neurotransmitters in ageing rats: substance P, acetylcholine and noradrenaline. Neurosci Lett 50:31–36
Jyoti A, Sethi P, Sharma D (2009) Aging accelerates the progression and manifestation of seizures in post-traumatic model of epilepsy. Neurosci Lett 453:86–91
Karthikkeyan K, Padma K, Rao B (2015) Evolution of visual evoked potential (VEP) in patients with chronic obstructive pulmonary disease. Indian J Pysiol Pharmcol 59:182–188
Kaur J, Sharma D, Singh R (1998) Regional effect of ageing on Na^+, K^+ ATPase in rat brain and correlation with multiple unit action potentials and lipid peroxidation. Indian J Biochem Biophys 35:364–371
Kaur J, Sharma D, Singh R (2001) Acetyl-L-carnitine enhances Na^+, K^+ ATPase, glutathion-S-transferase and multiple unit activity and reduces lipid peroxidation and lipofuscin concentration in aged at brain region. Neurosci Lett 301:1–4
Kaur J, Singh S, Sharma D, Singh R (2003a) Aluminium induced enhancement of ageing- related biochemical and electrophysiological parameters in rat brain regions. Indian J Biochem Biophys 40:330–339
Kaur J, Singh S, Sharma D, Singh R (2003b) Neurostimulatory and antioxidative effect of L-deprenyl in aged brain regions. Biogerontology 4:105–111
Knoll J (1993) The pharmacological basis of the beneficial effects of (-) deprenyl (selegiline) in Parkinson's and Alzheimer's disease. J Neural Transuppl 40:69–91
Landolt H-P, Borbely AA (2001) Age-dependent changes in sleep EEG topography. Clin Neurophysiol 112:369–377
Leung LS, Yim CC (1991) Intrinsic membrane potential oscillation in hippocampal neurons in vitro. Brain Res 553:261–274
Lisman JE (1997) Burst as a unit of neuronal information: making unreliable synapse reliable. Trends Neurosci 20:38–43
Luebke JI, Rosene DL (2003) Ageing alters dendritic morphology, input resistance, and inhibitory signalling in dentate granule cells in the rhesus monkey. J Camp Neurol 460:573–584
Luebke JI, Chang YM, Moore TL, Rosene DL (2004) Normal ageing results in decreased synaptic excitation and increased synaptic inhibition of layer 2/3 pyramidal cells in the monkey prefrontal cortex. Neuroscience 125:277–288
Mizumori SJY, Lovie AM, Kalyani A (1996) Redistribution of spatial representation in the hippocampus of aged rats performing a spatial memory task. Behav Neurosci 110:1006–1016

Murchison D, Griffith WH (1995) Low-voltage activated calcium current increase in basal forebrain neurons from aged rats. J Neurophysiol 74:876–887

Mishra M, Singh R, Mukherjee S, Sharma D (2013) Dehydroepiandrosterone, s antiepileptic action in $FeCl_3$ induced epileptogenesis involves upregulation of glutamate transporters. Epilepsy Res 106:83–91

Mishra M, Singh R, Sharma D (2010) Antiepileptic action of exogenous dehydroepiandrosterone in iron—induced epilepsy in rat brain. Epilepsy Behav 19:264–271

Nodera H, Bostock H, Kuwabara S, Sakamoto T, Asanuma K, Jia-Ying S, Ogawara K, Hattori N, Hirayama M, Sobue G, Kaji R (2003) Nerve excitability properties in Charcot- Marie-Tooth diseases type 1A. Brain 127:203–211

Nygard M, Hill RH, Wikstrom MA, Kristensson K (2005) Age-related changes in electrophysiological properties of the mouse suprachiasmatic nucleus in vitro. Brain Res Bull 65:149–154

Olpe HR, Steinmann (1982) Age-related decline in the activity of noradrenergic neurons of the rat locus coeruleus. Brain Res 151:174–176

Papatheodoropolos C, Kostopoulos G (1996) Age-related changes in excitability and recurrent inhibition in the rat CA1 hippocampal region. Eur J Neuroscience 8:510–520

Pelosi L, Blumbardt LD (1999) Effects of age on working memory: an event- related potential study. Cognitive Brain Res 7:321–334

Pfutz EM, Sommer W, Schweinberger SR (2002) Age-related slowing in face and name recognition: evidence from event-related brain potential. Psychol Aging 17:140–160

Potier BY, Lamour Y, Dutar P (1993) Age-related alterations in the properties of hippocampal pyramidal neurons among rat strains. Neurobiol Aging 14:17–25

Potier B, Rascol O, Jazat F, Lamour Y, Dutar P (1992) Alterations in the properties of hippocampal pyramidal neurons in the aged rat. Neuroscience 48:793–806

Rakhade SN, Loeb JA (2008) Focal reduction of neuronal glutamate transporters in human neocortical epilepsy. Epilepsia 49:226–236

Rapp PR, Bachevalier J (2013) Cognitive development and aging. In: Squire LR, Berg D, Bloom FE, Ghosh A, Spitzer NC (eds) Fundamental Neuroscience. Oxford Academic Press, pp 919–945

Roy D, Singh R (1988) Age-related changes in multiple unit activity in the rat brain parietal cortex and the effect of centrophenoxine. Exp Gerontol 34:161–174

Sharma D, Maurya AK, Singh R (1993) Age-related decline in multiple unit action potentials of CA3 region of rat hippocampus: correlation with lipid peroxidation and lipofuscin concentration and the effect of centrophenoxine. Neurobiol Aging 14:319–330

Sharma V, Prakash-Babu P, Singh A, Singh S, Singh R (2007) Iron-induced experimental cortical seizures: electroencephalographic mapping of seizure spread in the subcortical brain areas. Seizure 16:680–690

Shin E-J, Jeong JH, Chung YH, Kim W-K, Bach J-H, Hong J-S, Yoneda Y, Kim H-C (2011) Role of oxidative stress in epileptic seizures. Neurochem Int 59:122–137

Singh R (1994) Electrophysiological and biochemical features of brain ageing and their pharmacological prevention. In: Mallick BN, Singh R (eds) Environment and Physiology. Narosa Publishing House, New Delhi, pp 227–235

Singh R, Kiloung J, Singh S, Sharma D (2008) Effect of paradoxical sleep deprivation on oxidative stress parameters in brain regions of adult and old rats. Biogerontology 9:153–162

Singh R, Sharma D (2005) Electrophysiological ageing of the brain and pharmacology of ageing. In: Thakur MK, Prasad S (eds) Molecular and Cellular Neurobiology. Narosa Publishing House, New Delhi, pp 135–151

Singh R, Mishra M, Singh S, Sharma D (2012) Effect of L-deprenyl treatment on electrical activity, Na^+, K^+ ATPase, and protein kinase C activities in hippocampal subfields (CA1 and CA3) of aged rat brain. Indian J Exp Biol 50:101–109

Singh R, Sharma D, Singh S, Kaur J (2006) Biochemical correlates of electrophysiological ageing of the brain. In: Mathur R (ed) Pain Updated: Mechanisms and Effects. Anamaya Publishers, New Delhi, pp 116–125

Singh SB, Thakur L, Anand JP, Yadav D, Amitabh Banerjee PK, Selvamurthy W (2004) Changes in visual evoked potentials on acute induction to high altitude. Indian J Med Res 120:472–477
Tanaka Y, Ando S (1990) Synaptic aging as revealed by changes in membrane potential and decreased activity of Na+, K(+)-ATPase. Brain Res 506:46–52
Vaney N, Chouhan S, Bhatia MS, Tandon OP (2002) Event related evoked potentials in dementia: role of vitamin E. Indian J Physiol Pharmacol 46:61–68
Wang M, Gamo NJ, Yang Y, Jin LE, Wang XJ, Laubach M, Arnsten AF (2011) Neuronal basis of age-related working memory decline. Nature 476:210–213
Wilson IA, Ikonen S, Gallagher M, Eichenbaum H, Tanila H (2005) Age-associated alterations of hippocampal place cells are subregion specific. J Neurosci 25:6877–6886
Winterer G, Goldman D (2003) Genetics of human prefrontal function. Brain Res Rev 43:134–163
Yordanova J, Kolev V, Hohnsbein J, Falkenstein M (2004) Sensorimotor slowing with ageing is mediated by a functional dysregulation of motor-generation processes: evidence from high-resolution event-related potentials. Brain 127:351–362
Zhang ZW (2004) Maturation of layer V pyramidal neurons in the rat prefrontal cortex: intrinsic properties and synaptic function. J Neurophysiol 91:1171–1182

[illegible]
Wilson IA, Gallagher M, Eichenbaum H, Tanila H (2006) Neurocognitive aging: prior memories hinder new hippocampal encoding. Trends Neurosci 29:662–670

Antiaging Neurosteroid Dehydroepiandrosterone Counters Epileptiform Activity in Iron-Induced Experimental Epilepsy

Monika Mishra, Rameshwar Singh and Deepak Sharma

Abstract The subject of influence of hormones in epilepsy in the elderly is of great interest. Age-related deficit of certain endogenous neuroactive steroid hormones may result in the loss of physiological antiseizure (antiepileptic) potential in the aged brain. Androgen's antiseizure action is of interest as testosterone showed antiseizure effects in humans and also in experimental models of epilepsy. The androgen steroid dehydrepiandrosterone (DHEA) is of particular interest as its levels decline with aging/senescence, and it has been found to exert antiepileptic effects besides its antiaging actions. This paper focuses on the antiepileptic action of DHEA as an example of androgen hormones' antiseizure action. Experimentally, DHEA treatment of iron-induced epileptic rats (an experimental model of post-traumatic clinical epilepsy) prevented development of electrographic seizure activity, and countered seizure-associated cognitive dysfunctions, oxidative stress alterations, sodium pump activity changes, glutamate levels elevation, and glutamate transporters down regulation. DHEA may thus have a promising therapeutic profile for the treatment of epilepsy in the elderly.

Keywords Dehydroepiandrosterone (DHEA) · Neurosteroids · Post-traumatic epilepsy · Glutamate transporters · Na^+, K^+- ATPase · Hormones and seizures

1 Introduction

There is a greater prevalence and incidence of epileptic seizures in the elderly (Leppik 2007, Gupta and Sanchetee 2014). Experimentally also, aged rats were found to be more susceptible to develop epileptic seizures (Jyoti et al. 2009). Endogenous neuroactive steroid hormones: androgens, estrogens, may influence seizure-generation processes as they affect neurotransmitter receptors involved in

M. Mishra · R. Singh · D. Sharma (✉)
Neurobiology Laboratory, School of Life Sciences, Jawaharlal Nehru University, New Delhi 100067, India
e-mail: deepak57in@yahoo.co.in

© Springer Science+Business Media Singapore 2017

P.C. Rath et al. (eds.), *Topics in Biomedical Gerontology*,
DOI 10.1007/978-981-10-2155-8_17

excitability (Beyenburg et al. 2001; Baulieu and Robel 1996; Frye and Reed 1998). Age-related deficit of certain endogenous neuroactive steroids may result in the loss of physiological antiseizure (antiepileptic) potential. Androgen's antiseizure effects are considered to be of interest (Rhode's and Frye 2004) as testosterone was found to have antiseizure effects in humans and also in experimental models of epilepsy. Antiseizure influences may arise from their agonist-like actions at $GABA_A$ receptors and/or antagonist actions at glutamate receptors (Frye 2008). The androgen neuroactive steroid dehydroepiandrosterone (DHEA) is of particular interest as its levels decline with aging/senescence, and it has been found to have antiepileptic effects besides its antiaging effects (Frye 2009 and Baulieu et al. 2001). This paper describes experimentally observed antiepileptic actions of DHEA as an example of androgen hormones' antiseizure action. DHEA treatment effects were measured in epilepsy-associated alterations in iron-induced experimental epileptic rats. Iron ($FeCl_2$/$FeCl_3$)-induced experimental epilepsy in rodents provides an experimental model of post-traumatic human clinical epilepsy (Willmore et al. 1978; Moriwaki et al. 1990, 1992). Experimental studies on this model have provided information relevant to pharmacology and mechanism of epilepsy (Singh and Pathak 1990; Sharma et al. 2007; Mishra et al. 2010, 2013, Willmore and Rubin 1981; Samuelsson et al. 2003).

2 Materials and Methods

Male Wistar rats (8–10 months of age, and body weight 400–450 g) were housed in pairs in standard laboratory cages and maintained at 23 ± 4 °C, under a 12-h-light/12-h-dark cycle. The chemicals used were procured from Sigma Aldrich. The experimental protocols were approved by the Institutional Animal Ethical Committee (IAEC) of Jawaharlal Nehru University, New Delhi, India. After the completion of the experiments rats were sacrificed as per the guidelines of Committee for the Purpose of Control and Supervision of Experimental Animals (CPCSEA) and the brain tissue was harvested for further experiments (Mishra et al. 2010, 2013).

3 Experimental Observations

DHEA effect on electrographic seizure activity:

Treatment of epileptic rats with DHEA (30 mg/kg) for 7–21 days suppressed both epileptiform activity (paroxysms of spike-wave complexes) on electrocorticograms and hippocampal electroencephalographic (EEG) seizure activity recorded in CA1 and CA3 subfields (Fig. 1a, b). The epileptiform activity quantified in terms of multiple-unit action potentials (MUA) (Sharma et al. 2007, Mishra et al. 2010, 2013) corresponding to electroencephalograms showed that quantitatively the epileptiform activity was completely prevented (Mishra et al. 2010) by DHEA treatments.

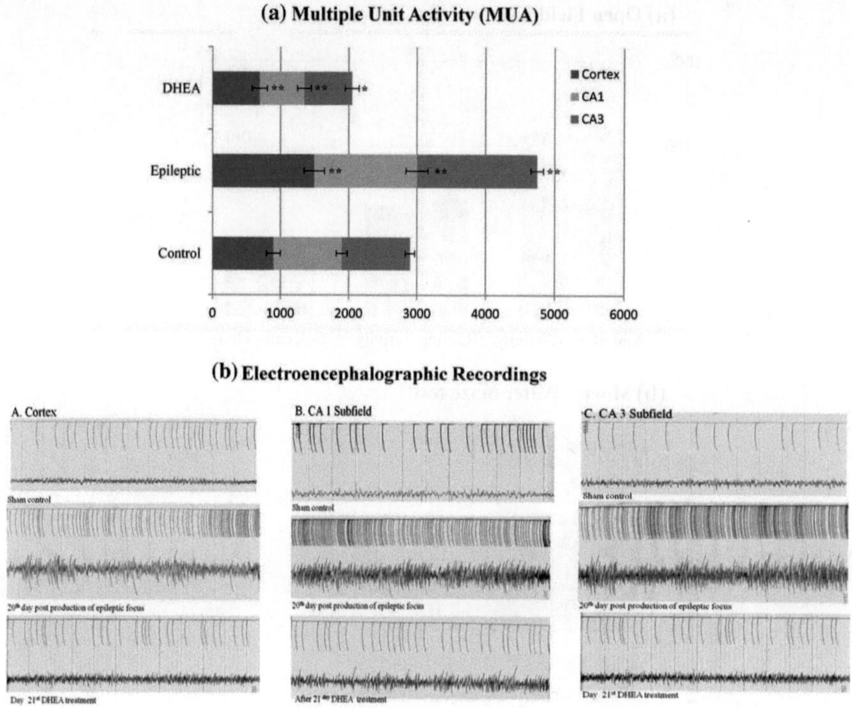

Fig. 1 **a** MUA counts in the cortex, CA1 and CA3 regions of hippocampus in control, epileptic and DHEA treated groups. The labels used are: $^{***}p < 0.001$; $^{**}p < 0.01$; $^{*}p < 0.05$. The significance of the epileptic case in is with respect to the control case and those of the DHEA treatment are with respect to the epileptic case. **b** Representative sample polygraph recordings (EEG and MUA) from the cortex, CA1 and CA3 subfields of hippocampus of the control, epileptic, and DHEA treatment for 21 days specimens. Decreased/suppressed epileptic form activity is evident

DHEA effects on seizure-associated behavioural dysfunctions:

Epileptic rats showed decline in their ambulatory and rearing activities, and elevation in their defecation index (measured by open field test) indicating increased anxiety levels. Treatment of epileptic rats with DHEA restored the behavioural alterations close to the control levels (Fig. 2a).

Epileptic rats showed statistical significant decline in their spatial learning ability (measured by Morris water maze test). DHEA treatment significantly enhanced the learning abilities of these rats indicating restoration of the learning abilities (Fig. 2b).
DHEA effect on epilepsy-associated oxidative stress:
Oxidative stress measured as a change in the levels of lipid peroxidation was found to be significantly elevated in the both cortex and hippocampus (Fig. 3a) of the epileptic rats. DHEA treatment brought down the lipid peroxidation near to the control levels.

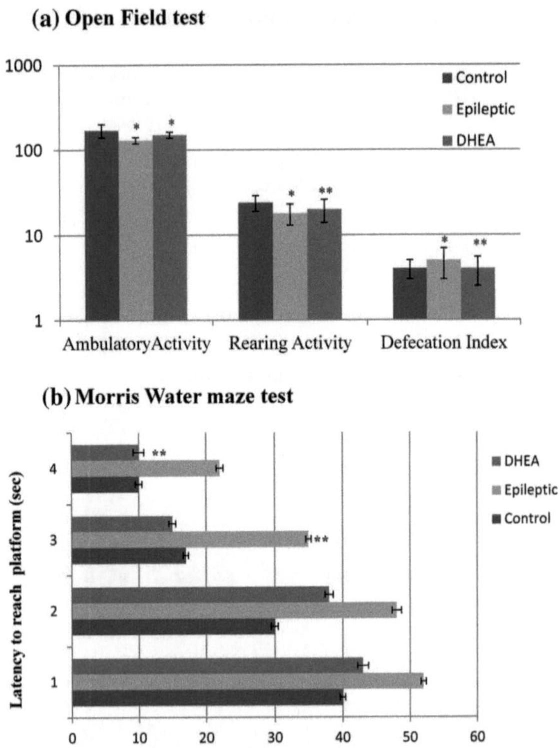

Fig. 2 **a** Comparison of the control, epileptic and DHEA-treated rats with respect to the OFT parameters: ambulatory activity, rearing activity and defecation index. Each data point represents the mean ± SD of n = 5 rats. **b** Comparison of the control, epileptic and DHEA-treated rats with respect to their respective latencies to locate the platform in the MWM test. Each data point represents the mean latency ± SD of n = 5 rats. The significance of the epileptic control case is with respect to the sham control case and that of the 21-day-DHEA-treated case is with respect to the epileptic control case: $^{**}p < 0.001$; $^{*}p < 0.0$

DHEA effect on epilepsy-associated change in Na^+, K^+, ATPase (Sodium pump activity).

Significant decrease in the Na^+, K^+, ATPase activity occurred in the cortex and hippocampus of the epileptic rats. DHEA treatment significantly increased the Na^+, K^+, ATPase (Fig. 3c) indicating that the DHEA treatment normalized the excitability levels.

DHEA effect on epilepsy-associated alteration in glutamate levels and glutamate transporters.

Epileptogenesis-associated elevation in glutamate levels (Fig. 3b), and decrease in glutamate transporters activity (GLT-1, GLAST and EAAC-1) occur in the cortex and hippocampus of iron-induced epileptic rats (Fig. 3d). DHEA treatment upregulated the activities of glutamate transporters, and restored the glutamate levels to near control levels.

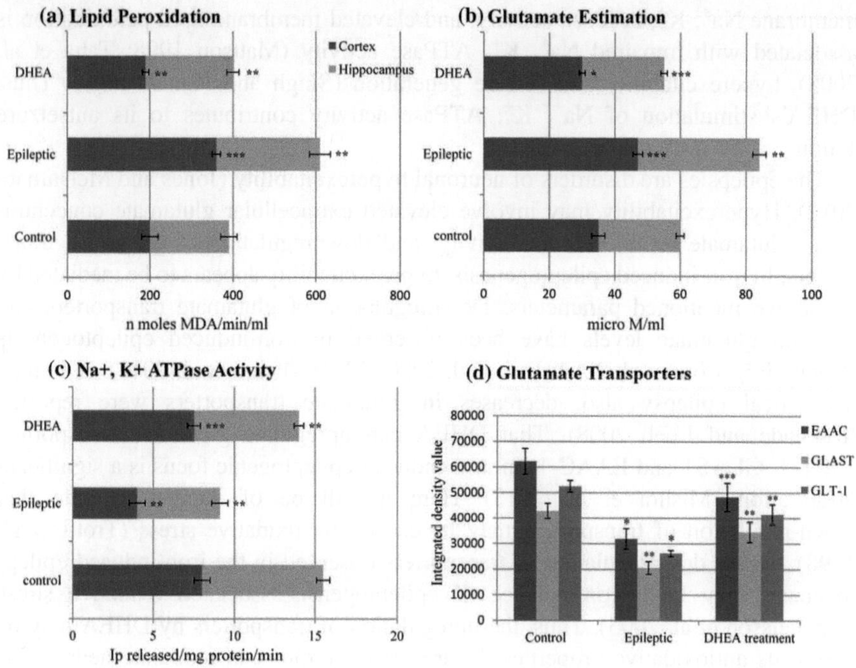

Fig. 3 **a–c** Lipid peroxidation, glutamate levels and Na^+, K^+ ATPase activity in the cortex and hippocampus of epileptic rats. Each bar represents the mean (n = 5) ± SD. The significance of the epileptic case is with respect to the control case and those of the DHEA treatment durations are with respect to the epileptic case: $^{***}p < 0.001$; $^{**}p < 0.01$; $^{*}p < 0.05$. **d** Effect of DHEA treatment on the levels of GLT-1, GLAST and EAAC-1 mRNA. Each bar represents the mean ± SD of n = 5 rats. The significance of the epileptic case is with respect to the control case and those of the DHEA treatment durations are with respect to the epileptic control case: $^{**}p < 0.01$

4 Discussion

DHEA is known to have pharmacological effects in multiple neurological conditions: depression, cognition (learning/memory), anxiety, neuroprotection (Maurice et al. 1999). The experimental data outlined above clearly indicate that DHEA also has the potential to completely prevent the electrographic seizure activity. This action of DHEA is similar to that of ethosuximide which is an antiepileptic drug and was found to prevent electrical seizure activity in iron-induced epilepsy model (Sharma et al. 2007).

DHEA is an effective antioxidant. That is, it counters oxidative stress. In iron-induced epileptogenesis, oxidative stress plays an important role (Singh and Pathak 1990) as membrane lipid peroxidation impacts membrane electrophysiology. Therefore, DHEA's prevention of seizure activity may involve its antioxidative action. DHEA also counters epilepsy-associated decline in Na^+, K^+, ATPase (i.e. sodium pump) activity. Membrane lipid peroxidation is known to influence membrane Na^+, K^+, ATPase activity, and elevated membrane lipid peroxidation is associated with impaired Na^+, K^+, ATPase activity (Mattson 1998; Taha et al. 2008), hyperexcitability and seizure generation (Singh and Pathak 1990). Thus, DHEA's stimulation of Na^+, K^+, ATPase activity contributes to its antseizure action.

The epilepsies are disorders of neuronal hyperexcitability (Jones and McNamara 2012). Hyperexcitability may involve elevated extracellular glutamate concentrations, glutamate receptors hyperactivity, and downregulation of glutamate transporters. In iron-induced epileptogenesis, hyperexcitability appears to be mediated by the above mentioned parameters. Downregulation of glutamate transporters and elevated glutamate levels have been observed in iron-induced epileptogenesis (Ronne-Engström et al. 2001; Doi et al. 2000, 2001; Mishra et al. 2013). In human neocortical epilepsy also, decreases in glutamate transporters were reported (Rakhade and Loeb 2008). That DHEA can upregulate glutamate transporters (GLT-1, GLAST and EAAC-1) in iron-induced epileptogenic focus is a significant observation (Mishra et al. 2013). Here it will be of interest to note that down-regulation of transporters may be caused by oxidative stress (Trotti et al. 1998). So the downregulation of transporters observed in the iron-induced epileptogenesis may be a consequence of epileptogensis-associated oxidative stress (Samuelsson et al. 2003). Thus, the upregulation of transporters by DHEA may be due to its antioxidative properties. Neuroactive steroids can modulate neurotransmitter receptors, excitability channels (Ffrench-Mullen and Spence 1991; Bergeron et al. 1996; Baulieu and Robel 1996; Maurice and Lockhart 1997; Maurice et al. 1999; Beyenburg et al. 2001; Racchi et al. 2001; Wen et al. 2001; Johansson and Greves 2005; Charalampopoulos 2011). Therefore, a direct effect of DHEA on such molecules involved in excitability is also possible.

DHEA treatment also countered epilepsy-associated behavioural dysfunctions (cognition and anxiety) in iron-induced epileptic rats. Epileptic rats showed impaired cognition and increased anxiety as assessed by Morris water Maze test and open field test respectively (Mishra et al. 2010). Epileptic syndromes are known to be associated with cognitive deficits (Gilbert et al. 2000). Therefore, improvement of epilepsy-associated cognitive deficits by DHEA treatment provides an interesting example of pharmacological benefit by an endogenous physiological hormone. Experimentally also, DHEA has, however, been reported to have memory-enhancing properties in rodents (Flood et al. 1992; Melchior and Ritzmann 1996; Frye and Lacey 1999). Overall, it seems possible that DHEA may be used as an antiepileptic substance pharmacologically. Neuroactive steroids in general have been considered valuable against seizure disorders (Budziszewska et al. 1998; Frye 2008; Galimberti et al. 2005). Further studies are thus required to investigate the pharmacological antiepileptic action of DHEA in clinical settings.

Acknowledgments Authors are thankful to UGC (for Dr. D.S. kothari fellowship to MM), ICMR, DST and CSIR for financial supports.

References

Baulieu EE, Robel P (1996) Dehydroepiandrosterone and dehydroepiandrosterone sulfate as neuroactive neurosteroids. J Endocrinol 150:S221–S239

Baulieu EE, Robel P, Schumacher M (2001) Neurosteroids: beginning of the story. Int Rev Neurobiol 46:1–32

Bergeron R, De Montigny C, Debonnel G (1996) Potentiation of neuronal NMDA response induced by dehydroepiandrosterone and its suppression by progesterone: effects mediated via sigma receptors. J Neurosci 16:1193–1202

Beyenburg S, Stoffel-Wagner B, Bauer J, Watzka M, Blümcke I, Bidlingmaier F, Elger CE (2001) Neuroactive steroids and seizure susceptibility. Epilepsy Res 44:141–153

Budziszewska B, Siwanowicz J, Leśkiewicz M, Jaworska-Feil L, Lasoń W (1998) Protective effects of neurosteroids against NMDA-induced seizures and lethality in mice. Eur Neuropsychopharmacol 8:7–12

Charalampopoulos IL (2011) Neuroprotective and neurogenic properties of dehydroepiandrosterone and its synthetic analogs. Horm Neurodegeneration, Neuroprotection, Neurogenesis 137–154

Doi T, Ueda Y, Tokumaru J, Mitsuyama Y, Willmore LJ (2000) Sequential changes in glutamate transporter mRNA levels during Fe^{3+}-induced epileptogenesis. Mol Brain Res 75:105–112

Doi T, Ueda Y, Tokumaru J, Mitsuyama Y, Willmore LJ (2001) Sequential changes in AMPA and NMDA protein levels during Fe^{3+}-induced epileptogenesis. Mol Brain Res 92:107–114

Ffrench-Mullen JMH, Spence KT (1991) Neurosteroids block Ca^{2+} channel current in freshly isolated hippocampal CA1 neurons. Eur J Pharmacol 202:269–272

Flood JF, Morley JE, Roberts E (1992) Memory-enhancing effects in male mice of pregnenolone and steroids metabolically derived from it. Proc Natl Acad Sci USA 89:1567–1571

Frye CA, Lacey EH (1999) The neurosteroids DHEA and DHEAS may influence cognitive performance by altering affective state. Physiol Behav 66:85–92

Frye CA, Reed TA (1998) Androgenic neurosteroids: anti-seizure effects in an animal model of epilepsy. Psychoneuroendocrinology 23:385–399

Frye CA (2008) Hormonal influences on seizures: basic neurobiology. Int Rev Neurobiol 83:27–77

Frye CA (2009) Steroids, reproductive endocrine function, and cognition, a review. Minerva Ginecol 61:563–585

Galimberti CA, Magri F, Copello F, Arbasino C, Cravello L, Casu M, Patrone V, Murialdo G (2005) Seizure frequency and cortisol and dehydroepiandrosterone sulfate (DHEAS) levels in women with epilepsy receiving antiepileptic drug treatment. Epilepsia 46:517–523

Gilbert TH, Hannesson DK, Corcoran ME (2000) Hippocampal kindled seizures impair spatial cognition in the Morris water maze. Epilepsy Res 38:115–125

Gupta S, Sanchetee P (2014) Epilepsy in elderly; in text book of geriatric medicine. In: Sanchetee (ed). Paras medical Publisher, Hyderabad, India, pp 485–489

Johansson T, Greves PL (2005) The effect of dehydroepiandrosterone sulfate and allopregnanolone sulfate on the binding of [(3)H]ifenprodil to the N-methyl-d-aspartate receptor in rat frontal cortex membrane. J Steroid Biochem Mol Biol 94:263–266

Jones DL, McNamara JO (2012) The epilepsies: Phenotypes and mechanisms. In: Basic Neurochemistry. Academic Press Oxford

Jyoti A, Sethi P, Sharma D (2009) Aging accelerates the progression and manifestation of seizures in post-traumatic model of epilepsy. Neurosci Lett 453:86–91

Leppik IE (2007) Epilepsy in the elderly: scope of the problem. Int Rev Neurobiol 81:1–14

Mattson MP (1998) Modification of ion homeostasis by lipid peroxidation: roles in neuronal degeneration and adaptive plasticity. Trends Neurosci 21:53–57

Maurice T, Lockhart BP (1997) Neuroprotective and anti-amnesic potentials of sigma (sigma) receptor ligands. Prog Neuropsychopharmacol Biol Psychiatry 21:69–102

Maurice T, Grégoire C, Espallergues J (1999) Neuroactive neurosteroids as endogenous effectors for the sigma1 (sigma1) receptor: pharmacological evidence and therapeutic opportunities. Jpn J Pharmacol 81:125–155

Melchior CL, Ritzmann RF (1996) Neurosteroids block the memory-impairing effects of ethanol in mice. Pharmacol Biochem Behav 53:51–56

Mishra M, Singh R, Mukherjee S, Sharma D (2013) Dehydroepiandrosterone's antiepileptic action in $FeCl_3$-induced epileptogenesis involves upregulation of glutamate transporters. Epilepsy Res 106:83–91

Mishra M, Singh R, Sharma D (2010) Antiepileptic action of exogenous dehydroepiandrosteron in iron-induced epilepsy in rat brain. Epilepsy Behav 19:264–271

Moriwaki A, Hattori Y, Hayashi Y, Hori Y (1992) Development of epileptic activity induced by iron injection into rat cerebral cortex: electrographic and behavioral characteristics. Electroencephalogr Clin Neurophysiol 83:281–288

Moriwaki A, Hattori Y, Nishida N, Hori Y (1990) Electrocorticographic characterization of chronic iron-induced epilepsy in rats. Neurosci Lett 110:72–76

Racchi M, Govoni S, Solerte SB, Galli CL, Corsini E (2001) Dehydroepiandrosterone and the relationship with aging and memory: a possible link with protein kinase C functional machinery. Brain Res Rev 37:287–293

Rakhade SN, Loeb JA (2008) Focal reduction of neuronal glutamate transporters in human neocortical epilepsy. Epilepsia 49:226–236

Rhodes ME, Frye CA (2004) Progestins in the hippocampus of female rats have antiseizure effects in a pentylenetetrazole seizure model. Epilepsia 45:1531–1538

Ronne-Engström E, Hillered L, Flink R, Kihlström L, Lindquist C, Nie JX, Olsson Y, Silandera HC (2001) Extracellular amino acid levels measured with intracerebral microdialysis in the model of posttraumatic epilepsy induced by intracortical iron injection. Epilepsy Res 43:135–144

Samuelsson C, Kumlien E, Elfving A, Flink R, Lindholm D, Ronne-Engström E (2003) The effects of PBN (phenyl-butyl-nitrone) on GLT-1 levels and on the extracellular levels of amino acids and energy metabolites in a model of iron-induced posttraumatic epilepsy. Epilepsy Res 56:165–173

Sharma V, Babu PP, Singh A, Singh S, Singh R (2007) Iron-induced experimental cortical seizures: electroencephalographic mapping of seizure spread in the subcortical brain areas. Seizure 16:680–690

Singh R, Pathak DN (1990) Lipid peroxidation and glutathione peroxidase, glutathione reductase, superoxide dismutase, catalase, and glucose-6-phosphate dehydrogenase activities in $FeCl_3$-induced epileptogenic foci in the rat brain. Epilepsia 31:15–26

Taha A, Mishra M, Baquer NZ, Sharma D (2008) Na+K(+)-ATPase activity in response to exogenous dehydroepiandrosterone administration in aging rat brain. Indian J Exp Biol 46:852–854

Trotti D, Danbolt NC, Volterra A (1998) Glutamate transporters are oxidant-vulnerable: a molecular link between oxidative and excitotoxic neurodegeneration? Trends Pharmacol Sci 19:328–334

Wen S, Dong K, Onolfo JP, Vincens M (2001) Treatment with dehydroepiandrosterone sulfate increases NMDA receptors in hippocampus and cortex. Eur J Pharmacol 430:373–374

Willmore LJ, Rubin JJ (1981) Antiperoxidant pretreatment and iron-induced epileptiform discharges in the rat: EEG and histopathologic studies. Neurology 31:63–69

Willmore LJ, Sypert GW, Munson JB (1978) Epileptiform activity initiated by pial iontophoresis of ferrous and ferric chloride on rat cerebral cortex. Brain Res 152:406–410

Part IV
Interventions for Healthy Aging

The Holistic Principles of Ayurvedic Geriatrics

Ram H. Singh

Abstract Ayurveda, the ancient Indian traditional system of medicine is essentially the science of life and longevity. It carries a treasure of pro-nature holistic geriatric health care modalities. It deliberates on the science and philosophy of life and longevity with the goal of healthy aging and long life to achieve the *Purusartha catustaya ie* the Four fundamental Instincts of human life viz. *Dharma, Artha, Kama, Moksa*. It considers aging as *Swabhava* or the natural tendency of life and describes in details the pattern of sequential losses of biological strength with advancing age in relation to the doctrine of *Tridosa*. The central focus of strength of Ayurveda in geriatric care swings around the concept of *Rasayana/Rejuvenation* therapy which compensates the age-related biological losses in the mind-body system and affords comprehensive rejuvenative effect. Combining Ayurvedic *Rasayana*, healthy dietetics, positive life style, *yoga* and spirituality it is possible to develop an effective package for geriatric care today for global use. New scientific evidences have been accumulating during last few decades which validate the age-old time tested science of life and health warranting further research and development. This strength of Ayurveda in geriatric health care is becoming more relevant today than ever before because of the rapid rate of population aging world over including India with an obvious shift in the age distribution denoting population aging with increased life expectancy of the people. Such an increase in the number of elderly people in the society is reflecting overtly in rapid rise in the incidence of diseases of old age warranting strategic plans for Geriatric health care and hence Geriatrics is fast emerging as an important medical discipline where the elements of holistic Ayurvedic geriatrics will find an important place.

Keywords *Ayurveda* · Geriatrics · Population aging · *Rasayana* · Rejuvenation

R.H. Singh (✉)
Department of Kayachikitsa, Faculty of Ayurveda, Institute of Medical Sciences, Banaras Hindu University, Varanasi 221005, Uttar Pradesh, India
e-mail: rh_singh2001@yahoo.com

© Springer Science+Business Media Singapore 2017

P.C. Rath et al. (eds.), *Topics in Biomedical Gerontology*,
DOI 10.1007/978-981-10-2155-8_18

1 Introduction—What Is Ayurveda

Ayurveda is the ancient Indian life and health science, its history goes back to 3000–5000 years in the antiquity which has remained in un-broken clinical practice till date and is currently one of the official systems of health care in India and several other South East Asian Countries (Sharma 1992; Singh 2001, 2003; Udupa and Singh 1974). Ayurveda is based on its own fundamental principles, its own biology, diagnostics and therapeutics (Singh 1998, 2005a, b, c, 2008, 2009a, b). Theory of *Panchmahabhuta* and *Tridosha* are the sheet anchors of Ayurvedic biology. Ayurvedic medicine is practiced on three characteristic premises comprising of 1. Pronature approach, 2. Holistic approach and 3. Personalized health care based on the unique *Prakriti/Vikriti* considerations in each individual being.

Ayurveda is essentially a predictive, promotive and preventive health care system. It has eight clinical branches, one of these is called *Rasayana Tantra* which deals with molecular nutrition, immunology, geriatric health care and rejuvenation based on its own principles. The new scientific evidences accumulating beyond the traditional knowledge base reveal that *Rasayana* therapy of Ayurveda has rich scope of revival and development which may strengthen the main stream geriatrics and gerontology.

The original ancient knowledge base of Ayurveda has survived till today through two sets of ancient classical texts initially written in *Samskrit* language now translated in many contemporary languages including English which are easily accessible to the core of the ancient wisdom. These texts are called 1. *Brihattrayi* ie three big books namely *Caraka, Susruta* and *Vagbhatta*, 2. *Laghuttrayi* ie three junior texts namely *Madhavanidana, Sharasngdhara Samhita and Bhawaprakash.* Caraka (700 BC); Madhavkara (900 AD); Susruta (600 BC).

It cannot be overemphasized that Ayurveda is distinctly different from western modern biomedical sciences based on its own fundamental principles, its own biology and therapeutics. It is not easy to draw parallels between these two sciences. However new language is gradually emerging, In spite of such hurdles it is worth pursuing such studies because the ideas contained in ancient traditions seem to be of valuable futuristic significance for reverse innovation and translational studies which may prove to be of bilateral benefit for growth of science.

The present review will deal with the contemporary features of Ayurvedic geriatrics based on the ancient classical knowledge base as well as the recent trends and scientific observations which has been accumulating during last few decades to illustrate the future prospects of this important discipline which is becoming more and more relevant today in the light of rapid population aging world over.

2 Population Aging

Since the turn of the Century there has been a significant rise in the average life expectancy of the people world over including India with an obvious shift in the age distribution denoting population-aging. This trend has been obvious in developed countries for several decades. Most demographers believe that a direct consequence of the ongoing global fertility decline and decline of mortality at elder ages is one of the most prominent global demographic trends of the 21st century which would influence the health status and socioeconomic pattern world-over. This trend is expected to continue over the next few decades. Population aging has a range of socioeconomic and health related consequences including the increase in the old-age dependency ratio presenting challenges for public health concerns as well as for economic development of a Nation due to shrinking and aging of work and labor forces risking the social security systems. The following table provides a reflection of the rate of population aging in terms of the observed and forecasted percentage of elderly (>65 years) in the population of some areas of the world (Kinsella and Victoria 2001; Lutz et al. 2001; Paul and Geoffrey 2003; Preston et al. 1989; Klatz and Goldman 1997).

Countries/regions	Year 1950	Year 2000	Year 2050
World	5.0	6.9	19.3
China	4.5	6.9	22.7
India	3.3	5.0	14.8
Japan	4.9	17.2	36.4
USA	8.3	14.7	29.2

With the increasing number of elderly individuals in today's society the health problems of old age are becoming more and more a matter of concern. Geriatrics is emerging as a major medical specialty all over the world. In India too the last decade has projected significantly rising rate of population-aging and hence a great need is now felt to strengthen the geriatric care system in this fast developing most populous country of the world like India. Conventional modern system of medicine has nothing much to offer in the core area of geriatric care except the medical management of some diseases of the old age with limited success. Geriatrics is still a developing discipline in India.

3 Ayurvedic Geriatrics

On the other hand, Ayurveda is essentially the science of life and longevity. It presents its own concept of aging, its prevention and care. Ayurveda specially incorporates *Rasayana Tantra* as one of its Eight classical specialties, which is exclusively devoted to nutrition, immunology and geriatrics. There is a need to explore the ancient concepts and recent developments in the field of geriatric care in Ayurveda for its appropriate mainstreaming. Practice of Yoga and life style management form an important adjunct in Ayurvedic geriatrics with focus on mental and physical health in old age.

Ayurveda, being the science of life and longevity, offers a treasure of geriatric care from entirely a new angle. It deliberates on the science and philosophy of life and longevity with the goal of healthy aging and long life to achieve the *Purusartha catustaya ie* the four basic instincts of human life viz. *Dharma, Artha, Kama, Moksa*. It considers aging as *Swabhava* or the natural tendency of the life and describes in details the pattern of sequential losses of biological strength with advancing age and advocates the pronature holistic management of aging and diseases of old age on Ayurvedic principles. The central focus of strength of Ayurvedic geriatrics swings around the concept of *Rasayana*/Rejuvenation therapy which compensates the age-related biological losses in the mind-body system and affords rejuvenative effect. The *Medhya Rasayanas* like *Brahmi* and *Aswagandha* are now scientifically known to retard brain-aging and to induce neuro-regeneration. The *Rasayana* remedies being essentially nutraceutical in nature, they work as essential functional food supplements and have high safety margin with benefit of positive healing. Combining Ayurvedic *Rasayana*, rehabilitative Geriatric *Panchakarma* therapy, healthy dietetics, Geriatric *swasthavritta, sadvritta, yoga* and spirituality it is possible to develop an effective package for geriatric care today for global use.

There is a need of awareness among the masses about the consequences of Population-Aging and about the strength of Ayurveda in Geriatric health care. Govt. of India has launched a national campaign to popularize the strength of Ayurveda in geriatric health care. It is advisable to establish Ayurvedic Geriatric Clinics in all hospitals and health centers. There is also a need to setup regional centers of training and CME Cells for value added training in Geriatric care for all practitioners and paramedics. As stated above selective psychophysical practices of Yoga and meditation are useful adjuncts to Ayurvedic geriatric health care.

4 The Biology of Aging

Aging is an inevitable physiological process. Why senile changes occur in the mind-body system, is not yet precisely known. Ayurveda considers aging as the *swabhava* or the very nature of life. In other words senescence is the *swabhava* ie. the very nature of the living being, because the life is time-bound and it is inherently mortal. Ayurveda believes that aging occurs only in the body and mind, not in the real essence of the life, i.e. its conscious component which is eternal and it only changes the body which is created by the divine Nature for one life-span and ends there after through aging and death, while the real being, the *Atman* transmigrates to another body through rebirth.

Conventional biosciences have identified certain probable causes of aging. A number of theories and observations have been projected such as genetic theory of aging, immunological theory, stress theory, free radical injury, hormonal involution theory etc but none of these have yet been proved. However, these researches obviously indicate that aging is a multi factorial phenomenon, the genomic and immune mechanisms seem to be the sheet anchors in biology of aging.

Contemporary Theories of Aging:

1.	The Wear and Tear Theory	10.	Mitochondrial Theory
2.	The Neuro endocrine Theory	11.	Errors and Repairs Theory
3.	The Genetic Control Theory	12.	Redundant DNA Theory
4.	The Free Radical Theory	13.	Cross-Linkage Theory
5.	Waste Accumulation Theory	14.	Autoimmune Theory
6.	Limited no of Cell Division Theory	15.	Caloric Restriction Theory
7.	Hayflick Limit Theory	16.	Gene Mutation Theory
8.	Death Hormone Theory (DECO)	17.	The Rate of Living Theory
9.	Thymic–Stimulating Theory	18.	The Telomerase Theory

5 Process of Aging—Ayurvedic Perspective

Thus there is no fundamental contradiction between east and the west except the philosophy of time-bound life and rebirth and the splendid speculations about life beyond death which is still beyond the grasp of the western world.

Senescence and death are inevitable still a long healthy life has been the cherished wish of all beings, in all cultures and traditions. The *Vedas* also pray—*Jivem sharadah shatam, Pashyem sharadah shatam* ie. wishing for a long life of 100 years with functioning senses and the mind. Hence tremendous attempts have been made since antiquity to prolong life and to sustain healthy aging. Ayurveda also links the process of aging with the doctrine of *Tridosa*. In principles the childhood is dominated with the *Kapha* in the body, adulthood with *Pitta* and old age with *Vata Dosa. Vata Dosa,* by nature, dries and decays the cells and tissues of the body and produces visible senile changes. Thus the aging process can be evaluated in terms of the *Tridosika* physiology of Ayurveda, and aging can be managed in tune with the *Tridosa* theory of Ayurveda through life style care, dietetics and medications like *Rasayana* drugs after periodical bio-purification by *Panchakarma therapy.*

Sarangadhara (1300 AD) describes the process of aging depicting sequence of the involutionary features of specific bio-losses occurring during different decades of life due to aging. These descriptions are relevant even today if the chronology is corrected with present life-span of man. The *Sharangadhar's* scheme of aging may be used as guide line for geriatric health care even today particularly in reference to *Rasayana* therapy. Earlier Vagbhatta (300 AD) also gave similar description of the sequence of aging.

The important features of aging process are weakening of *Agni* ie the digestive and metabolic strength of the body besides, loss of the integrity of *Srotamsi* or the integrity of microcirculation and *ojabala/immune strength*. Hence geriatric care warrants management of *Agni*, *Ama* and *Ojabala* at biological level. Ayurvedic classics propound a unique concept of *Ojas* which is the essence of all *Dhatus ie the* seven primordial tissues as described in Ayurvedic classics. *Ojas* is of two types namely 1. *Para-ojas* which is in minute quantity located in the heart and brain and is vital for life. If *Para-ojas* is exhausted, instantaneous death follows. 2. *Apara-ojas* which is relatively gross, is spread all over the body and is generally responsible for immune strength or *Ojabala*. *Ojabala* is considered to be of three categories,[1] viz. 1. *Sahajabala* or primary natural immunity, 2. *Kalajabala* or acquired natural immunity, 3. *Yuktikrta* or artificially induced immunity. In Ayurveda *Ojabala* forms an important consideration in the process of aging. Besides many other life-style related factors the generic remedy prescribed for promoting *Ojas* status and longevity is *Rasayana* therapy. Practice of Yoga and meditation is advocated for promotion of mental health.

6 The Rasayana Therapy and Rejuvenation

Rasayana (*Rasa* = nutrition + *Ayana* = circulation and promotion) deals with the science of nutrition, geriatric care and rejuvenation. *Rasayana* signifies not a single drug or medication, rather refers to a rejuvenative regimen which uses rejuvenative remedies, healthy dietetics and overall healthy life-style and positive psychosocial conduct. The use of *Rasayana* measures and remedies produces best qualities of *Dhatus*, i.e. bodily cells and tissues by acting through one or all of the following three principal levels of biological system with net result of improved nutritional status leading, in turn, to better qualities of cells and tissues, longevity, immunity, resistance against disease and improved mental faculties. The primary modes of *Rasayana* effect are as below (Singh 2012, 2014, 2015; Singh and Rastogi 2012).

1. At the level of *Rasa* acting as molecular nutrients or functional foods promoting the nutrient value of plasma.
2. At the level of *Agni* promoting bio-fire system of the body responsible for digestion and metabolism acting as metabolic boosters.
3. At the level of *Srotas* promoting microcirculation and tissue perfusion.

Besides the above mentioned three generic modes of action of *Rasayana*, some *Rasayanas* are organ and tissue specific and are used for specific indications like *Medhya Rasayana* as brain tonics, *Hrdya Rasayana* as cardiotonics, *Vrisya Rasayana* as sex tonics, *Twacya Rasayana* as skin tonics, *Stanya Rasayana* as lactogenic tonics, *Kesya Rasayana* as hair tonics, *Caksusya Rasayana* as eye tonics, *Kanthya Rasayana* as tonics for throat and speech and so on. Some *Rasayanas* are also disease-specific and are used in specific disease states as they induce specific

[1]XXX

bio-strength to combat a particular disease. Such *Rasayanas* are called *Naimittika Rasayana*. The classical *Naimittika Rasayanas* are *Silajatu* for diabetes mellitus and *Tubaraka* for skin diseases and leprosy. There can be many other *Naimittika Rasayanas* identifiable by rational planning ie the *yukti*.

As mentioned earlier some *Rasayanas* are also age-specific and can be prescribed for particular age groups to retard aging process. Ayurveda describes the qualities of each decade of the 100 years of estimated life-span. During the process of aging an individual goes on loosing these age-related qualities and if this loss is compensated by age-specific *Rasayana* in specific age groups the rate of aging can be retarded to some extent and one can promote longevity. The decade-wise bio-losses described by Sarangadhara (1300 AD) and Vagbhatta (300 AD) are as mentioned below with some suggested *Rasayana* remedies to compensate the age related bio-losses in different age groups.

Age Specific Biolosses and Suggested Rasayana Remedies:

S. No.	Age in years	Age specific biolosses	Suggested Rasayana remedies
1.	0–10	Declining corpulence	Vaca, Swarna, Kasmari
2.	11–20	Declining growth	Kasmari, Bala, Asvagandha
3.	21–30	Declining lusture	Amalaki, Haritaki, Louha
4.	31–40	Declining intellect	Aindri, Brahmi, Sankhapuspi
5.	41–50	Depleting skin glow	Bhringaraja, Jyotismati
6.	51–60	Declining vision	Jyotismati, Triphala
7.	61–70	Declining virility	Kapikacchu, Aswagandha
8.	71–80	Declining bio-strength	Bala, Amalaki
9.	81–90	Loss of memory and cognition	Sankhapuspi, Brahmi
10.	91–100	Declining locomotion	Bala, Sahachara
11.	101–110	Loss of consciousness	Divya Rasayana
12.	111–120	Loss of life	Divya Rasayana

7 The Scope and Classification of Rasayana Therapy

(1) As per scope of use:

(A) *Kamya Rasayana*—which is used in healthy persons for further promotion of health. It is again of three subtypes:

(i) *Prana kamya* to promote longevity.
(ii) *Srikamya* to promote body lusture.
(iii) *Medha kamya* to promote mental competence

(B) *Naimittika Rasayana* which is used specifically in the treatment of specific diseases viz *Silajatu* in Diabetes mellitus.

(2) As per method of use:

  (i) *Vatatapika Rasayana* i.e. routine outdoor regimen.
  (ii) *Kutipravesika Rasayana* i.e. intensive indoor regimen including biopurification by *Panchakarma* and consumption of selected *Rasayana* in well controlled conditions.

(3) As part of life-style:

  (i) *Ajasrka Rasayana* as content of daily diet.
  (ii) *Acara Rasayana* i.e. Rejuvenative healthy life-style and conduct.

8 The Rasayana Effect and Rejuvenation—Recent Trends

Ayurvedic *Rasayana* remedies possess special nutritional supplement effect. Generally most of the *Rasayanas* are micromolecular nutrients and they act through nutrition dynamics and not really on pharmacodynamics like other drugs. The *Rasayana* drugs are likely to be nutrient tonics, anti-oxidants, anti-stress, adoptogen and immuno-modulators. The net effect of all these attributes is the anti-aging impact. Recent investigations on popular *Rasayana* remedies like *Amalaki, Asvagandha, Guduci, Brahmi* and classical compound *Rasayana Chyavanaprasa* have shown evidence to suggest their efficacy as anti-aging remedies (Dwivedi and Singh 1997; Jayaprakash et al. 2013; Rastogi et al. 2012; Kobayama et al. 2005; Singh 1987; Singh et al. 2008; Singh and Rastogi 2012. The following tables are being reproduced to give a glimpse of such an evidence about the popular *Rasayana* drug *Aswagandha* (*Withania somnifera dunal*).

Anti-stress Activity of *Aswagandha* in terms of Ed 50 in Stressed Rats (Singh 1987):

Drugs Tested	Swimming performance adrenal wt.	Prevention of increase of Ascorbic Acid	Prevention of rise in Adrenal cortisol	Prevention of induced ulcers under stress	Antistress unit/mg/g
Aswangandha	15.0 ± 1.3	13.0 ± 1.4	14.5 ± 1.5	16.0 ± 1.8	14.9 ± 1.5
Tulasi	13.3 ± 1.2	12.0 ± 1.6	13.0 ± 1.5	13.4 ± 2.0	13.7 ± 1.3
P. ginseng	44.0 ± 3.8	15.0 ± 1.8	24.1 ± 2.1	24.07 ± 2.2	25.2 ± 2.3

Biological Age Scale (BAS) and Brief Psychiatric Rating Scale (BPRS) before and after treatment with Aswagandha (Dwivedi and Singh 1997**):**

Observations	Before treatment	After treatment	t	p
Biological age score	18.77 ± 4.55	14.20 ± 4.37	2.92	<0.01
Brief Psych. Score	32.40 ± 5.22	22.93 ± 2.86	6.12	<0.01

The above cited observations indicate that the Indian *Rasayana* drug *Aswagandha* has significant anti-stress effect on notably lower effective dose (Ed^{50}) as compared to popular Chinese herbal tonic *Ginseng*. *Aswagandha* also shows relevant humoral basis for its anti-stress and anti-inflammatory effect (Archana and Namashivayam 1999). The clinical trial of *Aswagandha* in elderly volunteers shows significant reduction in the rate of Biological Aging (BAS) and mental health status as measured by BPRS. Withanolide A isolated from *Aswagandha* has recently been reported to promote regeneration of neurons *in vitro* (Kobayama et al. 2005).

9 Rehabilitative Panchakarma Therapy for Elderly

Panchakarma therapy of Ayurveda is an unique biopurificatory procedure designed to cleanse the micro channels of the body enabling the organism to function normally and also to restore the inner transport system resulting in turn to better nutritional status with adequate chances of repair of wear and tear of the body with rejuvenative activity. In geriatric practice Ayurvedic physicians use selective rehabilitative *Panchakarma* therapy avoiding the drastic evacuatory practices like *Vamana* and strong *Virechana* procedures. The schedule in the elderly should consist of medicated massage, sudation, *Pindasweda, Shirodhara and Brimhana Basti* suitably planned for each individual in consideration of his *Prakriti and Vikriti.*

10 Ayurvedic Management of Diseases of Old Age

In Addition to the rejuvenative approaches Ayurveda has a potential to afford significant complementary therapeutic effect in a range of diseases of the elderly. These procedures can be used by practicing physicians of all streams. *Arjuna, Guggulu* and *Puskarmula* are cardioprotectives for cases of Ischaemic heart disease, *Brahmi* and similar other *Medhya* drugs are used in treatment of senile dementias. *Varuna* and *Shigru are used* in treatment of senile enlargement of Prostate, *Triphala* in senile visual disorders, *Kapikacchu* in treatment of Parkinsons disease, *Amrita* and *Amalaki* in immunodeficiency, *Shirodhara* and *Shirobasti* in Tension headaches and different kinds of CNS neurodegenerative conditions. These are some of the potential areas where Ayurvedic treatment has promise. Similarly the *Pindasweda* procedure of Keraliya *Panchakarma* therapy is known for its rehabilitative effect in many neurodegenerative conditions and myopathies besides chronic arthritis.

11 Yoga for Health and Spiritual Wellbeing

Practice of Yoga is essentially the applied aspect of ancient Indian philosophy which teaches the nature of the ultimate reality as the fundamental entity of consciousness. The scriptures propound that consciousness is the ultimate reality. The basic matrix of the universe is the cosmic *Chetana (consciousness)*, the *Brahman*. The same cosmic consciousness extends itself into each individual being and creates the universe. Thus according to Indian philosophy Consciousness is the creator of the universe. The universal Consciousness, the *Brahman* is all pervasive omnipresent power and its extension in all individual beings is called *Atman*. Thus, the *Atman* and *Brahman* are a continuum. In other words the entire universe is functioning in a unified field of consciousness. This classical ancient Indian concept simulates the new concept of unified field of Energy. As such the unified field theories form the most outstanding advancement in modern physics where the universe is conceived as the ocean of energy in which all other constituents are connectedly embedded. So is the concept in *Vedanta* presenting the entire universe as unified field of Consciousness. Thus ancient Indian *Vedant* is the foundation of modern physics.

Practice of Yoga is essentially the path of spirituality and super consciousness. Super consciousness is the core goal of Yoga. Promotion of health or treatment of a disease is not the core goal of practice of classical Yoga. These are only the side benefits. However now there are new evidences to suggest that there is a good scope of practice of certain forms of Yoga as a heath practice or even as therapy for certain ailments. There is no harm in harnessing such benefits in common man's interest. Some of the psychophysical practices of Yoga including some *Asanas, Pranayama* and meditation are very useful for the elderly people and some such selected practices should form an integral part of the package for geriatric health care today. Lot of new scientific evidence has been accumulating during last 50 years which has validated the health potential of Yoga today (Udupa and Singh 1972; Udupa et al. 1973, 1975; Singh et al. 1982; Singh 2009a, b; Wallace and Benson 1972). It is because of the sound traditional background and emerging scientific evidence behind Yoga which prompted the United Nations to declare 21st June as the International Day of Yoga with effect from 2015. The following tables show some data to illustrate the impact of some yogic practices in apparently normal individuals and in patients suffering from certain diseases common in old age.

Some Psychobiological Changes after practice of Yoga in normal volunteers:

Observations	Initial	3rd month
Vital capacity (ml)	3729 ± 232	4372 ± 336
Memory quotient	89.75 ± 9.15	97.30 ± 13.20
Performance quotient	93.15 ± 12.50	102.00 ± 16.40
Health index (mean CMI score)	192	114

Responses of 3000 ailing users of Yoga to UK Yoga Biomedical Trust Survey

Disorders	No. of cases[a]	% Helped by Yoga[b]
Backache	1142	98
Arthritis	589	90
Anxiety	838	94
Migraine	464	80
Insomnia	542	82
Menopause disorders	247	83
Hypertension, IHD	200	84
Asthma, Bronchitis	226	88
Obesity	240	74
Diabetes	10	80
Cancer	29	90
Tobacco addiction	219	74
Alcoholism	26	100

[a]Those who answered "Yes" to the question "Have you suffered from this condition?"
[b]Those who answered "Yes" to the question "Has Yoga helped this condition?"

12 The Possible Mechanism of Yoga Effect

Yoga as originally conceived in ancient classics is not a therapeutic science. It is essentially concerned with expansion of consciousness and pursuit of spirituality with possibility of wellness. However, postural correction and exercise effect, neuromuscular coordination, vascular and haemodynamic correction, anti-Stress effect, relaxation response, biopurificatory effect, neuro-endocrine effect, energy modulation, immuno stabilization, metabolic correction, improved cognitive ability, emotional stability and longevity seem to be the possible physiological impacts of the practice of Yoga which do have therapeutic potential. There are new experiences and scientific evidences validating the above mentioned presumptions which form the basis of development of Yoga and meditation as health sciences. The practice of Yoga promotes the state of wellness and spiritual wellbeing.

It cannot be overemphasized that Yoga is not a religious ritual. It is a science of health in depth, wellness and spiritual wellbeing. Wellness combines health and wellness together. Spirituality intends to develop an individual's inner life including connectedness with a larger reality. Practice of Yoga integrates all aspects of life, hence is the term Yoga.

References

Archana R, Namashivayam (1999) Antistress effect of Withania somnifera Dunal. J Ethnopharmacology 64:91–93

Caraka (700 BC) Caraka Samhita. In: Sharma PV (ed) Cikitsa Sthana Chap. I. Choukhamba Orientalia, Varanasi

Dwivedi KK, Singh RH (1997) A study on geriatric patients and response of Aswagandha as anti-aging agent. Ph.D. Thesis Kayachikitsa, Banaras Hindu University

Jayaprakash et al (2013) Neuroprotective role of Withania somnifera root extract in mouse model of Parkinsonism. Neurochem Res Feb 27, 2013

Kinsella K, Victoria AV (2001) An aging world: 2001. U.S. Census Bureau, Series P 94/01-1, Washington, DC

Klatz R, Goldman R (1997) Stopping the clock. Keats Publishing House, New Canaan, CT

Kobayama T et al (2005) Neuritic regeneration and synoptic reconstruction induced by Withanolide-A. Br J Pharmacol 144(7):961–971

Lutz W, Warren S, Sergei S (2001) The end of world population growth. Nature 412:543–545

Madhavkara (900 AD) Madhava Nidana. In: Upadhyaya YN (ed). Choukhamba Publications, Varanasi, India

Paul D, Geoffrey M (2003) The Encyclopedia of population. Macmillan reference, USA, New York

Preston SH, Christine H, Mitchell E (1989) Demographic conditions responsible for population aging. Demography 26:291–704

Rastogi S, Chiapelli F, Singh RH (2012) Evidence based practice of CAM, special monograph, Springer Germany

Sarangadhara (1300 AD), Sarangadhara Samhita Part II Chap. 6, vs. 20

Sharma PV (1992) History of medicine in India, Pub. Indian National Science Academy INSA, New Delhi, India

Singh N (1987) Study of antistress effect of plant drugs. Ann Acad Ind Med 1(1)

Singh RH (1998) The holistic principles of ayurvedic medicine, Chap. 8. Choukhamba Surbharati, Varanasi-New Delhi

Singh RH (2001) Ayurveda in India today. Proceedings of international symposium on traditional medicine. World Health Organization, Kobe Centre, Japan, pp 74–85

Singh RH (2003) In India health report. In: Mishra R, Chap. 14, pp 179–187, Oxford University Press

Singh RH (2005a) Advances in ayurvedic medicine, vol 1–5. Choukhamba Vishwabharati, Varanasi

Singh RH (2005b) Pancakarma therapy. Choukhamba Vishwabharati Publications, Varanasi

Singh RH (2005c) Swasthavrtta vijnana, Chaps. 3, 4, 7, 22–23. Choukhamba Surbharati, Varanasi

Singh RH (2008) Strength of ayurveda in geriatric health care. Keynote lecture delivered at state launching of national campaign on geriatric health care through ayurveda, Dept. of AYUSH, Govt. of India, New Delhi

Singh RH (2009a) Body-mind spirit integrative medicine in ayurveda, Yoga and Nature-cure. Choukhamba Surbharati, Varanasi

Singh RH (2009b) Foundations of contemporary yoga and yoga therapy, special monograph, 2nd edn, Pub. Choukhamba Pratishthan, Delhi, India

Singh RH (2012) The contemporary strength of Ayurvedic geriatrics. Ann Ayurvedic Med 1 (1):18–28

Singh RH (2014) Basic tenets of Ayurvedic dietetics and nutrition. In: special volume on Ayurvedic food science and nutrition. Springer, USA

Singh RH (2015) Foundations of immunology in Ayurvedic classics. Indian J Hist Sci 50(1): 83–94. INSA, New Delhi

Singh RH, Rastogi S (2012) Rasayana therapy and Rejuvenation. In Evidenced based practice of CAM, Springer Germany, pp 177–190

Singh RH, Narasimhamurthy K, Singh Girish (2008) Neuronutrient impact of Ayurvedic Rasayana theraopy. Biogerontology 9:350–374

Singh RH, Shettiwar M, Udupa KN (1982) Physiological and Therapeutic studies on Yoga. Yoga Review II/4:185–209

Susruta (600 BC) Susruta Samhita. In: Singhal & Associates, Cikitsa Sthana Chaps. 27–30. Choukhamba Surbharati, Varanasi

Udupa KN, Singh RH (1974) Science and philosophy of Indian medicine. Sri Baidyanath Ayurveda Bhavan, Nagpur, India

Udupa KN, Singh RH, Yadav RA (1973) Certain studies on psychological and biochemical responses to practice of Hatha Yoga in young normal volunteers. Indian J Med Res ICMR 61/2:237–244

Udupa KN, Singh RH (1972) Scientific basis of Yoga. J Am Med Assoc JAMA 220:1365

Udupa KN, Singh RH, Dwivedi KN, Pandey HP, Rai V (1975) Comparative biochemical studies on meditation. Indian J Med Res ICMR 63/12:66–76

Vagbhatta (300 AD) Astanga hridaya uttara tantra, Rasayana Chapter, Choukhamba Publication, Varanasi, India

Wallace RK, Benson H (1972) The physiological basis of meditation. Sci Am 226(2):84–90

Dietary Restriction, an Intervention for Healthy Aging

Ramesh Sharma

I am deeply indebted to Association of Gerontology of India (AGI) for nominating me to Prof. M. S. Kanungo Oration Award Lecture during 17th AGI Meeting at Trivandrum. It is my proud privilege to deliver this lecture in the honour of such an illustrious teacher and a great mentor to many of his students including me. Late Prof. M. S. Kanungo was the first who started pioneering research in the area of Biochemistry of aging in our country during early 1960s and continued to do so till the last breath of his life when we lost him on 26th July 2011 at the age of 84 years. His extraordinary commitment and dedication to the aging research is also exhibited by the fact that he still had couple of Ph.D. students to submit their thesis. I profoundly feel honoured in having got this opportunity to deliver oration lecture in his memory.

Aging is associated with changes at the cellular, tissue, organ and the whole body levels giving rise to decreased functioning and increased susceptibility to diseases, ultimately causing death. The maximum lifespan of any organism is a set feature of speciation to genetic controls influenced by environmental factors. Healthy aging and longevity thus result from the genetic constitution of an organism with a significant influence by epigenetic controls. Many theories have been proposed to explain the mechanisms of aging. They have been grouped as evolutionary, molecular, cellular and systemic theories. Age-dependent changes at molecular levels may lead to cellular alterations contributing to organ and system failures. Based on these theories and their explanations, aging seems to be a multi-factorial process and there is no single theory, at present that explains all the phenotypic changes those occur during aging. However, each one of them attempts to clarify some of the aging characteristics. Aging, thus results from the interplay of

Prof. M. S. Kanungo Oration Award Lecture during 17th AGI Meeting at Trivandrum.

R. Sharma (✉)
Department of Biochemistry, North-Eastern Hill University,
Shillong 793022, Meghalaya, India
e-mail: rsharma@nehu.ac.in

© Springer Science+Business Media Singapore 2017

P.C. Rath et al. (eds.), *Topics in Biomedical Gerontology*,
DOI 10.1007/978-981-10-2155-8_19

parallel and often interacting processes that govern the lifespan of an organism. It is controlled by a combination of genetic (the genes) and the environmental (diet/toxins/bacteria/viruses/radiation/pollution) factors. Hence, aging is dependent on an interplay of both inherent (inborn) and acquired (environmental) factors. It begins with good genes but depends heavily on good habits. A right kind of lifestyle may add significant number of years to one's life.

For more than a decade, I have been concentrating on dietary restriction (calorie restriction without malnutrition) as one of the interventions for achieving healthy aging. It has been found to retard the aging processes in many organisms. Dietary restriction (DR) regimes have been shown to reduce the biomarkers of oxidative damage such as peroxidation of lipids, protein carbonyl formation, DNA breaks in nuclear and mitochondrial genomes. Three months of DR (alternate days of feeding) has been found to elevate the level and activation magnitude of glucocorticoid receptors in mouse model system. It has been attributed to an increased level of stress adaptation in older animals (Dutta and Sharma 2003) (Table 1 and Fig. 1). During aging, the capacity to adjust to internal as well as external variations declines and results in unstable homeostatic balance. Glucocorticoids (GCs) are key regulators of homeostasis. They act through intracellular receptors called the glucocorticoid receptor (GR). The GR comprises an N-terminal immunologic domain carrying the transactivation region, a DNA binding domain (DBD) located centrally with Zn^{2+}-fingers and a domain that binds to hormone towards C-terminal region with ligand- dependent transactivational function (AF-2). AF-1 positioned at the N-terminus is ligand independent and constitutive, whereas AF-2 in the C-terminus is GCs dependent. The unliganded GR is retained in the cytoplasm with attached complex of proteins, that include two sub-units of heat shock proteins 90 (hsp90) acting as molecular chaperones, immunophilins and various other modulatory proteins. This complex is involved in ligand binding and maintaining GR in a state of high affinity for GCs, nucleo-cytoplasmic trafficking and proper folding of the GR. Upon ligand binding, hsp90 dimers dissociate and the hormone-receptor complex (HR-complex) translocates into the nucleus. Nuclear localization of HR-complex is followed by the interaction of the GR with the DNA at specific response elements, called GREs located up-stream of the regulated genes. Even late onset of DR has been shown to reverse the age-related decrease in renal GR in mice (Sharma and Dutta 2006) (Table 2). It may help renal system to control fluid and electrolyte excretion during long-term dietary restriction in mice.

Table 1 Specific binding sites (B_{max}) and dissociation constant (K_d) of glucocorticoid receptors in the liver of dietary restricted (DR) and ad libitum (AL) fed mice

Conditions	B_{max} (fmol/mg protein)	K_d(nM)
AL	122.2 ± 7.3	2.3 ± 0.3
DR	173.0 ± 5.7[a]	2.3 ± 0.2

The results are mean ± standard deviation of four separate experiments for each group

[a]Statistically significant ($p < 0.001$) with respect to AL fed mice

[Adapted from Dutta, D. and Sharma, R. (2003) Horm. Metab. Res. 35: 415–420.]

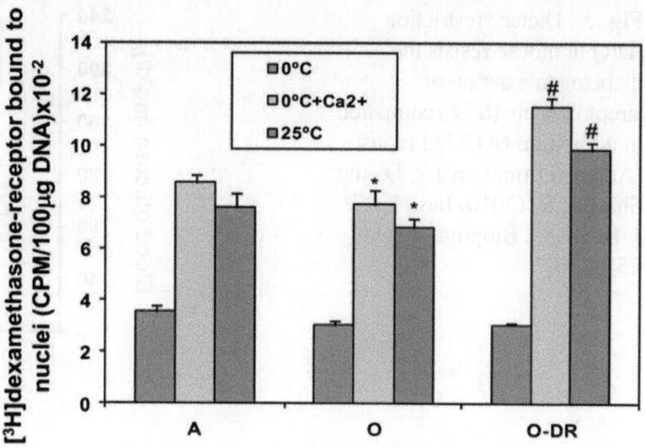

Fig. 1 Specific binding of kidney [^{3}H] dexamethasone-receptor complexes to purified nuclei from the adult (*A*), old (*O*) and old dietary-restricted (*O-DR*) animals. Calcium and temperature (25 °C) were used to activate the receptors [Adapter from Sharma, R. and Dutta, D. (2006) Ann. N.Y. Acad. Sci. USA. 1067: 129–141.]

Table 2 Concentration and affinity of [^{3}H] dexamethasone-receptor in the kidney of adult, old and old-dietary restricted male mice

Animals	Concentration (fmol/mg protein)	Affinity (nM)
Adult	111.2 ± 6.91	2.41 ± 0.03
Old	77.37 ± 6.27[a]	2.36 ± 0.11
Old-DR	99.50 ± 7.08[b]	2.46 ± 0.08

[a]Statistically significant ($p < 0.001$) with respect to adult mice
[b]Statistically significant ($p < 0.001$) with respect to old mice
[Adapted from Sharma, R. and Dutta, D. (2006) Ann. N. Y. Acad. Sci. USA. 1067:129–141.]

DR feeding has been shown to up-regulate malate-aspartate shuttle activity giving-rise to greater transfer of reducing equivalents from cytosol to mitosol used for energetic adaptation of aged animals (Goyary and Sharma 2005) (Fig. 2). The most active NADH shuttle for the movement of reducing equivalents (in the form of NADH) from the cytoplasm to the mitochondria is the malate-aspartate shuttle.

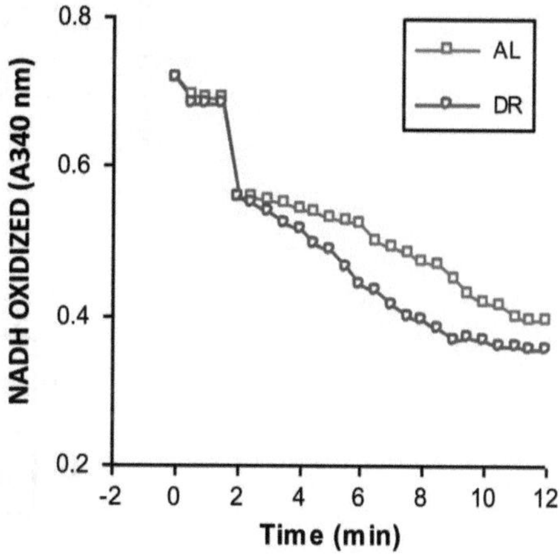

Fig. 2 Effect of dietary restriction (DR) on malate-aspartate shuttle activity as measured by NADH oxidation in the liver os mice compared to ad libitum (AL) fed mice [Adapted from Goyary, D. and Sharma, R. (2005) Ind. J. Biochem. Biophys. 42: 345–349]

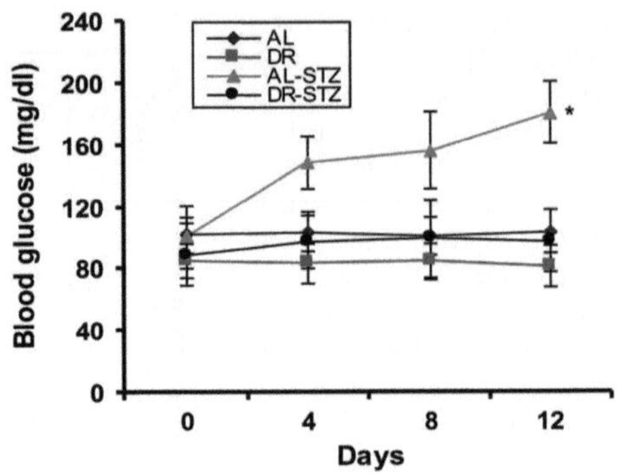

Fig. 3 Dietary restriction (DR) in mouse resists the diabetogenic action of streptozotocin (STZ) compared to ad libitum (AL)-fed mouse [Adapted from Goyary, D. and Sharma, R. (2010) Ind. J. Biochem. Biophys. 47: 254–256]

Since malate-aspartate shuttle plays a pivotal role in the oxidative metabolism, its possible modulation by dietary controls may provide an insight into the regulation of this shuttle during such interventions. These studies also help in understanding regulation of shuttle in age-delaying role of DR.

In addition, DR has been shown to provide protection against streptozotocin-induced diabetes in mouse model (Goyary and Sharma 2010) (Fig. 3). Streptozotocin is a glucosamine-nitrosourea group of compound that

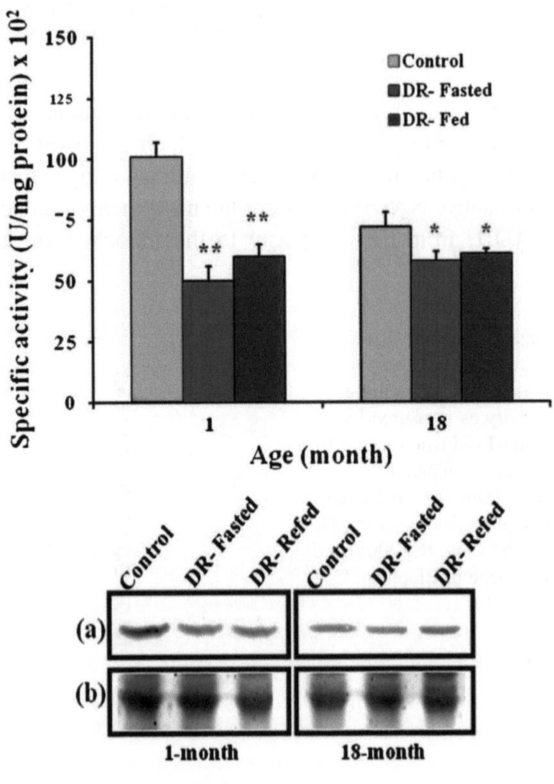

Fig. 4 Long-term dietary restriction (DR) downregulates acetylcholinesterase level in the brain of mice in the both young and old ages [Adapted from Suchiang, K. and Sharma, R. (2011) Biogerontology 12: 581–589.]

produces toxic effect to the beta cells of pancreas in mammals. It enters into β-cells of pancreas by the glucose transporter protein (GLUT2) and causes DNA damage by alkylation reaction. It also generates large amount of superoxide radicals, hydroxyl radicals and hydrogen peroxide giving rise to diabetogenic response due to necrosis of β-cells of pancreas.

In our laboratory, DR has been shown to reduce the activity of acetylcholinesterase (AChE) in the cerebral hemispheres of mice to provide benefits of cholinergic deficits in age-related neurodegenerative diseases like Alzheimer's disease (Suchiang and Sharma 2011) (Fig. 4). It could be a better alternative to pharmaceuticals used as inhibitors of this enzyme in Alzheimer's disease. AChE hydrolyses acetylcholine released from the presynaptic nerve cell in response to electrical signal. It is expressed in a variety of mammalian tissues such as muscle, nerve, placenta, and hematopoietic cells. Besides its localization at neuromuscular junctions, it is also found in cholinergic synapses and cells such as erythrocytes. AChE inhibitors have potential therapeutic roles since its higher expression is linked to increased deterioration in cognitive function in mice. This enzyme is presently the most important molecular target for therapeutic intervention in symptomatic treatment of senile dementia in AD. Many AChE inhibitors such as donepezil, physostigmine, rivastigmine and galantamine are being targeted for their memory-inducing effects.

We have elegantly shown that DR significantly reduces age-dependent increase in the amount of reactive oxygen species (ROS) and protein carbonylation (an oxidative modification of proteins) in brain of older mouse (Figs. 5 and 6). Oxidative damage to proteins is caused by production of ROS, reactive nitrogen species (RNS), superoxide anion, hydroxyl radical, hydrogen peroxide, the

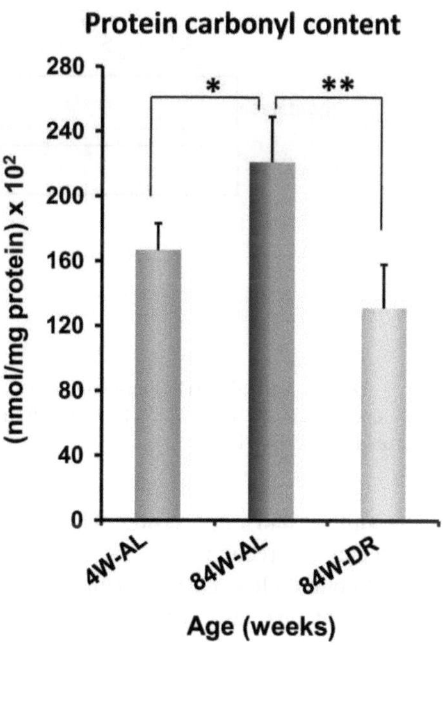

Fig. 5 Dietary restriction (DR) downregulates the level of protein carbonyl content in the brain of mice that otherwise is up-regulated in the older mice [Adapted from Dkhar, P. and Sharma, R. (2014) Cell. Mol. Neurobiol. 34: 307–313.]

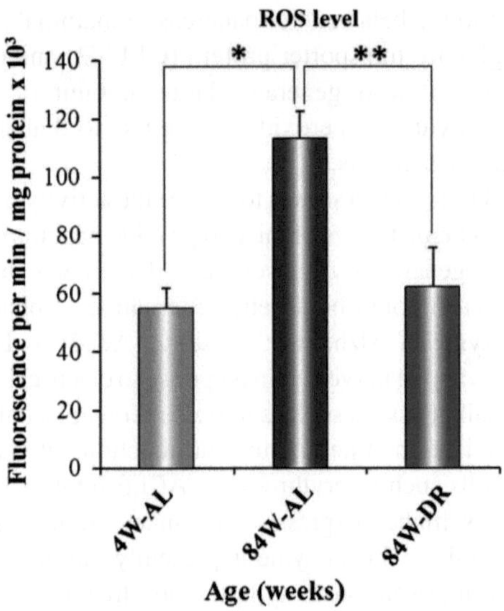

Fig. 6 Dietary restriction (DR) downregulates the level of reactive oxygen species (ROS) in the brain of mice that otherwise is up-regulated in the older mice [Adapted from Dkhar, P. and Sharma, R. (2014) Cell. Mol. Neurobiol. 34: 307–313.]

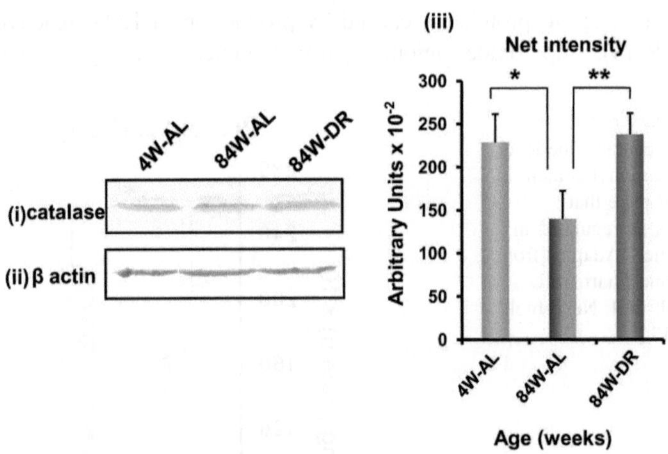

Fig. 7 Dietary restriction (DR) up-regulates catalase level in the brain of mice that otherwise is downregulated in the older mice [Adapted from Dkhar, P. and Sharma, R. (2014) Cell. Mol. Neurobiol. 34: 307–313.]

peroxynitrite and nitric oxide. There is an impaired scavenging of ROS due to age-dependent reduction in the antioxidant enzymes in cerebral hemispheres of mice. In our studies, DR has been shown to elevate the level of catalase, one of the anti-oxidant enzymes, which otherwise gets lowered in aging animals (Fig. 7).

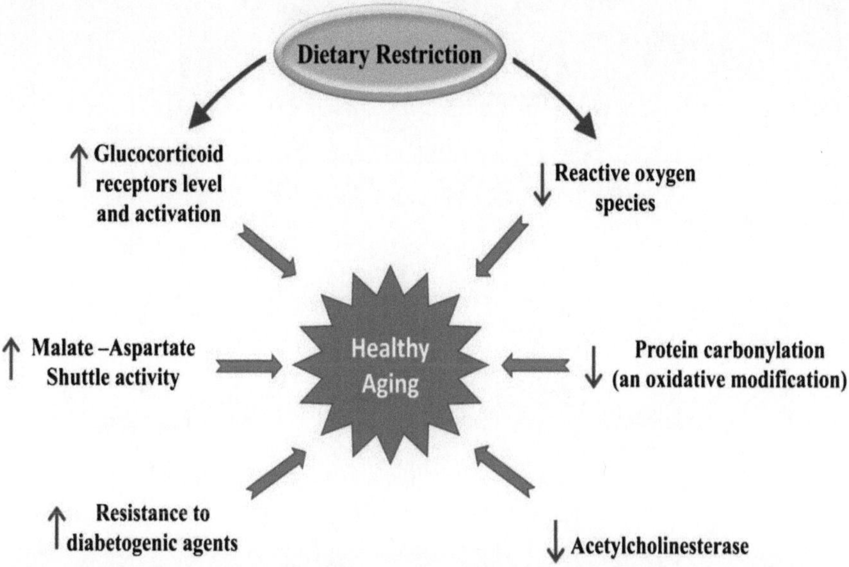

Fig. 8 Dietary restriction up-regulates glucocorticoid receptor level and activation, malate-aspartate shuttle activity and resistance to diabetogenic agents while it down-regulates reactive oxygen species, protein carbonylation (an oxidative modification) and acetylcholinesterase in mice rise to healthy aging

Hence, an increased ROS generation with age can be quenched by DR in mouse models (Dkhar and Sharma 2014).

From the above findings, it seems plausible to realize the immense benefit of dietary restriction in achieving healthy aging (Fig. 8). The practice of such intervention might give-rise to a lesser morbidity during later part of lifespan and may lead to a good quality life with increased health span of aging population.

References

Dkhar P, Sharma R (2014) Late-onset dietary restriction modulates protein carbonylation and catalase in cerebral hemispheres of aged mice. Cell Mol Neurobiol 34:307–313

Dutta D, Sharma R (2003) Regulation of hepatic glucocorticoid receptors in mice during dietary restriction. Horm Metab Res 35:415–420

Goyary D, Sharma R (2005) Dietary restriction and triiodothyronine (T3) regulation of malate-aspartate shuttle enzymes in the liver and kidney of mice. Ind J Biochem Biophys 42:345–349

Goyary D, Sharma R (2010) Dietary restriction prevents diabetogenic effect of streptozotocin in mice. Indian J Biochem Biophys 47:254–256

Sharma R, Dutta D (2006) Age-dependent decrease in renal glucocorticoid receptor function is reversed by dietary restriction in mice. Ann N Y Acad Sci USA 1067:129–141

Suchiang K, Sharma R (2011) Dietary restriction regulates brain acetylcholinesterase in female mice as a function of age. Biogerontology 12:581–589

Diabetes Mellitus Type 2 Induces Brain Aging and Memory Impairment in Mice: Neuroprotective Effects of *Bacopa monnieri* Extract

Surya Prakash Pandey and S. Prasad

Abstract Cognitive impairment is associated with brain aging or age-dependent neurodegenerative disorders. However, experimental evidence for type 2 diabetes mellitus (DM II)-induced aging process, decline in learning and memory and the underlying mechanisms are not known. Present study was aimed to understand the underpinning molecular and neurobiological mechanisms for above. Also, the study includes whether *Bacopa monnieri* extract (CDRI-08), a widely acclaimed nootropic drug, possesses neuroprotective role in DM II-induced brain aging and in learning and memory impairments. To understand above, we developed the DM II mice model by i.p. injection of a selected dose of streptozotocin as mentioned in methodology section and thereafter analyzed the level of neuronal lipofuscin, alterations in working and reference memory, AChE activity and dendritic spine density in pre-frontal cortex as well as in hippocampus. Our data revealed a significant increase in the accumulation of lipofuscin and impairments in working and reference memory in streptozotocin-induced DM II mice compared to normal control. It was correlated with increased AChE activity and decreased neuronal spine density in pre-frontal cortex and hippocampus. Oral administration of a selected dose of CDRI-08 to DM II mice significantly reversed the accumulation of lipofuscin, recovered memory loss, decreased the AChE activity and enhanced dendritic spine density towards their values in normal control mice. Current study clearly demonstrates that DM II induces brain aging and CDRI-08 has neuroprotective role in the recovery of DM II-induced aging and cognitive impairments.

Keywords Brain aging · Diabetes mellitus · Lipofuscin · AChE · Learning and memory · Neuronal spine density · *Bacopa monnieri*

S.P. Pandey · S. Prasad (✉)
Biochemistry & Molecular Biology Lab, Centre of Advanced Study in Zoology, Banaras Hindu University, Varanasi 221005, Uttar Pradesh, India
e-mail: s.sprasadbhu@gmail.com

© Springer Science+Business Media Singapore 2017
P.C. Rath et al. (eds.), *Topics in Biomedical Gerontology*,
DOI 10.1007/978-981-10-2155-8_20

1 Introduction

Recent advancement in medical sciences has increased the average lifespan which has caused phenomenal increase in the life expectancy and thereby increase in the number of global elderly population (Christensen et al. 2009). Whereas the advancement in the technologies has contributed to significant increase in the overall life span on one hand, it has increased the possibilities of the elderly population suffering from age-related diseases on the other hand. Aging, a known cause of cellular stress and various age-dependent neurodegenerative diseases (Beckman and Ames 1998), increases oxidative stress (Simpson et al. 2015) which leads to accumulation of lipofuscin (a protein and lipid aggregate) in tissues including brain and leads to aging (Firlag et al. 2013; Hohn and Grune 2013; Nowotny et al. 2014). Aggregation of the lipofuscin is augmented to alterations in oxidative degradation mechanisms (Mole et al. 2011; Hohn and Grune 2013). Such changes reduce the efficiency of cardiovascular (Chiao and Rabinovitch 2015), nervous (Simpson et al. 2015) and digestive systems and eventually leads to alterations in metabolic processes (Goodpaster et al. 2005). Occurrence of diabetes mellitus has been reported to increase with increasing age (Gunasekaran and Gannon 2011; Liu et al. 2011, 2014). National Diabetes Statistical Survey data indicates that the frequency of the occurrence of diabetes mellitus is about 16 % in age group of 45–65 and 26 % in the age group of 65 years and above (Guariguata et al. 2014). However, its association with accumulation of lipofuscin, the hallmark of aging, is not known.

Recent studies have shown that the diabetes mellitus type 2 (DM II) patients are more prone to develop age-associated neurodegenerative diseases like Alzheimer's disease (AD) which is characterized by dysfunction of cholinergic system (Rosales-Corral et al. 2015; Verdile et al. 2015). The acetylcholine level at synaptic junction is largely regulated by acetylcholinesterase (AChE). Accumulating reports reveal that increased AChE activity underlie the cognitive dysfunction during aging and pathology of Alzheimer's patients (Pasqualetti et al. 2015). Dendritic spines, which help in establishing synaptic connections, achieving synaptic plasticity, acquisition and formation of memory and thereby cognition, are also subject to abnormal alterations during aging and age associated disorder like AD (Zou et al. 2016). However, its association with the AChE remains elusive in DM II-induced brain dysfunction.

Recovery of age-related disease such as diabetes mellitus induced alterations in the brain and thereby cognitive function of the brain has been a major challenge for researchers. In this process, the herbal interventions using ayurvedic medicines having least side effects have attracted scientists all over the world. *Bacopa monnieri* extract (BME), a nootropic drug used for the improvement of dhi (the power of acquisition or learning), dhriti (the power of retention) and smriti (the ability to recall) has been used in the Indian ayurvedic medicine system from ancient time. The major active ingredients in BME have been reported to be Bacoside A and B in its alcoholic extract (Singh and Dhawan 1997). BME has been used frequently to treat many neurological disorders such as amnesia (Uabundit et al. 2009; Rai et al.

2015), hypoxia (Rani and Prasad 2015), diabetic neuropathy (Pandey et al. 2015b), epilepsy (Mathew et al. 2012) and neuropathic pain (Rauf et al. 2013). However, its role in attenuation of diabetes mellitus induced possible aging, alterations in the accumulation of lipofuscin, associated memory deficiency, AChE, dendritic spine density due to DM II is not well understood. Therefore, in the present study we have investigated whether DM II leads to accumulation of lipofuscin and thereby aging of brain which might lead to alterations in the cognition and whether the Bacoside A and B rich fraction of BME called CDRI-08 has advantageous effects in the streptozotocin-induced DM II mouse model.

2 Materials and Methods

2.1 *Chemicals, Drugs and Reagents*

All the chemicals used in experiments were of analytical and molecular biology grade and bought from Sigma, USA or Merck, India. Alcoholic extract of *Bacopa monnieri* (BME) containing 58 % of bacosides A and B (CDRI-08) was obtained from Lumen Research Foundation, Chennai, India, as a gift and suspended in Tween 80 (0.5 % v/v). Streptozotocin was dissolved in citrate buffer (0.01 M, pH 4.5). Freshly prepared drugs solutions were used in all experiments.

2.2 *Animals*

Adult male and female Swiss albino mice were procured from the central animal facility, Banaras Hindu University, Varanasi, India. These mice were inbred and the colony was maintained in the animal house of Department of Zoology, Banaras Hindu University, Varanasi, India, under a 12:12 h light-dark cycle (7:00 am to 7:00 pm) at 25 °C ± 2 °C. One-day male pups were assigned for further experiment and sacrificed after 16 week. Animals were used and handled according to the guidelines of Institutional Animal Care and Use Committee (IACUC) of Banaras Hindu University. Blood glucose content was measured using a Roch Glucometer before radial arm maze test was conducted. After behavioral analysis, mice were sacrificed by cervical dislocation and brain tissues were isolated and used for further experiments directly or stored at −80 °C after processing them in liquid nitrogen.

2.2.1 Group Distribution

In the experiment, twenty-eight male Swiss albino pups were initially divided into two groups (1) Control group, in which one-day old pups received a single i. p. injection of citrate buffer; (2) Streptozotocin-treated group, in which pups received a single i.p. injection of streptozotocin 100 mg/kg dissolved in citrate buffer. After 14-week, glucose content was measured in the blood obtained by cutting the tail vein of mice of each experimental group using a Glucometer (Accu-Check Active, Roche Diagnostics, Mannheim, Germany). Mice having blood glucose content above 250 mg/dl were considered as diabetic. After blood glucose measurement, both groups were further divided in two groups: (1a) Control: Citrate buffer treated group which received Tween 80, (1b) Brahmi: Citrate buffer treated group which received Brahmi extract dissolved in Tween 80 (100 mg/kg BW), and (2a) STZ: STZ dissolved in citrate buffer treated group which received Tween 80 and (2b) STZ-Br: STZ (dissolved in citrate buffer) treated group which received Brahmi extract (dissolved in Tween 80) (100 mg/kg BW).

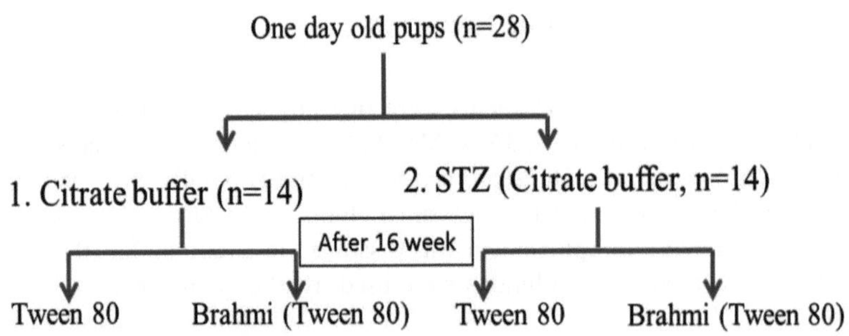

2.3 *Radial Arm Maze (RAM) Test*

To assess the memory (working and reference), eight arm radial maze test paradigm was used. The apparatus was located in the a center of isolated behavior room. The arms (3″ wide × 9″ long × 6″ high) were made of black painted wood. The eight arms were radiated from an octagonal central chamber (8.5″ wide) made up of the same materials as the arms. Food cup was located at the end of all arms below the surface of the floor. Varieties of extra-maze visual cues were fixed on surrounding walls of the room. At approximately 13 weeks of age, all animals were fed on restricted food intake to maintain body weight at 85 % of the free-feeding level. After food deprivation, the food reward of 1–2 mg of chocolate was given to mice in order to acclimatize them with food.

Radial arm maze test was performed when mice were approximately of 14 weeks age. On the very first day, all animals were individually kept for 5 min in the apparatus for acclimatization. On the subsequent day, food reward was scattered on the floor of all four baited arms (1, 3, 5 and 7) and animals were allowed to explore for 5 min. Third-day reward was kept in the corner of baited arms. Finally, reward was kept only in food cups of the reward arms. For each animal, a pattern of four baited and four unbaited arms was fixed for the entire experiment. Test was started after the mouse became comfortable to pick up the food, and when placed alone on the maze, explores the food without hesitation and without excessive defecation or urination. To begin each trial, the mouse was placed into the central chamber of the maze and all eight doors were simultaneously opened. After each trial, the maze was cleaned with alcohol to remove the odors and cues of the animal left before. The animal was allowed to enter the arms until it entered all four baited arms or till 5 min time was elapsed. Entry into an arm was scored when the whole body of an animal traversed into the arm. A working memory error was recorded when a mouse re-entered into a baited arm and the reference memory error recorded when a mouse entered into the un baited arms. Each mouse received two trials per day for 16 days (Rai et al. 2015).

2.4 Tissue Fixation and Cryosectioning

Brain tissue fixation and processing were done by the previously described procedure (Pandey et al. 2015b). Briefly, animals were perfused transcardially with 0.9 % buffered saline followed by cold 2 % buffered paraformaldehyde. After perfusion, brain was carefully dissected out individually from each mouse of various groups and post-fixed in the same fixative overnight at 4 °C. Next day, tissues were washed three times with phosphate buffer for 30 min each and cryoprotected with sucrose gradients of 10, 20 and 30 %, prepared in Phosphate buffer at 4 °C till they finally settled at the bottom. 10 µm thick transverse sections were cut using cryotome (Thermo Scientific, USA) through the frontal cortex and hippocampus regions and collected on poly-L-lysine coated slides. The slides were stored at −20 °C, till further processing (Pandey et al. 2015b).

2.5 Detection of Autofluorescent Lipofuscin and Co-localization with DAPI

To observe the autofluorescence of lipofuscin granules, cryosections were washed in PBS and mounted with aqueous mounting medium (Hardset with DAPI). Thereafter, each slide was directly examined under a fluorescence microscope (Leica DM2000). For visualization of nucleus, UV filter was used and a blue filter was used for observing auto-fluorescent lipofuscin. The integrated density value of lipofuscin granules in the selected area of the images was calculated using Image-J (NIH image analysis software).

2.6 Determination of Acetylcholinesterase (AChE) Activity

AChE activity was measured in pre-frontal cortex and hippocampus of the brain where this enzyme is present in abundant and help in memory formation following the procedure of (Rai et al. 2015). Modified Ellman's colorimetric method (Ellman et al. 1961; Srikumar et al. 2004) was used for acetylcholinesterase (AchE) activity. Enzyme activity was measured by colorimetric reaction between acetylthiocholine (ATC) and di-thio-bis-nitrobenzoate (DTNB). In Brief, hippocampal and pre-frontal cortices were dissected (Pandey et al. 2015b) and quickly homogenized in 0.1 M Phosphate buffer, pH 7.4. The esterase activity was measured by providing an artificial substrate, ATC. Thiocholine, which was released because of the cleavage of ATC by AChE, is allowed to react with the -SH group of the DTNB, which is reduced to thionitrobenzoic acid, an yellow coloured anion with an absorbance maxima at 412 nm. The extinction coefficient of the thionitrobenzoic acid is 1.36 × 104/M/cm. The concentration of thionitro benzoic acid was detected using a UV-Vis Spectrophotometer and was taken as a direct estimate of the AChE activity. The enzyme activity was calculated using the following formula; $R = 5.74 \times 10^{-4} \times A/CO$ Where, R = Rate in moles of substrate hydrolysed/minute/gm tissue, A = Change in absorbance/min and CO = Original weight of the tissue (g).

2.7 Rapid Golgi Staining

Dendritic spine density was studied by rapid Golgi method as described by Patro et al (Patro et al. 2009). In brief, mice were decapitated under deep anesthesia and brain was dissected out quickly. Dissected brains were fixed in Golgi fixative solution containing 5 % Chloral hydrate, 8 % Glutaraldehyde, 6 % formaldehyde, 5 % Potassium dichromate and 1 % Dimethylsulphoxide for 4 days at 4 °C. After fixation, tissue were washed and impregnated with 0.75 % silver nitrate for 48 h in the dark. 100 μm thick sections were prepared by vibratome and collected in chilled 70 % alcohol. The sections were dehydrated in absolute alcohol, cleared in xylene and mounted in DPX. Slides were dried at 37 °C for 3–4 days and then visualized under a light microscope.

2.8 Statistical Analysis

All the experiments were repeated at least three times (n = 7/group/set). The individual data was presented as bar diagram showing Mean ± SEM and the data were examined by One-way ANOVA between treatment groups followed by post hoc Tukey's multiple comparison tests with SPSS 16.0. The p values <0.05 were taken as significant.

3 Results

3.1 BME Has No Effect on the Streptozotocin-Induced Increased Blood Glucose Level

Streptozotocin treatment lead to a significant increase in the blood glucose content in the mice compared to control. After administration of Brahmi extract (CDRI-08), no change in the level of blood glucose content was observed in STZ+BME treated mice. Also, CDRI-08 treatment of the control mice per se did not alter the blood glucose content compared to that in the control mice (Fig. 1).

3.2 BME Reverses Reference and Working Memory Alterations in DM II Mice

Behavioral analysis of memory was performed by radial arm maze (RAM) test to assess errors in reference and working memory. After continuous 16 days of food reward, spatial memory acquisition trial on RAM was given and the last day data was recorded and analyzed by tracking software called Any Maze. Data obtained from the radial arm maze test reveals that wrong entry was significantly higher in diabetic animals compared to control mice. This indicates that the DM II mice developed memory loss. BME treatment of Streptozotocin-treated DM II mice significantly lowered extent of errors which indicates that the BME retrieved the altered memory (Fig. 2a–c).

Tracking report plot (Fig. 2a) demonstrates that the control and CDRI-08 treated mice enters often in baited arms whereas the diabetic mice enter more in unbaited arms. The CDRI-08 treated mice were performing similar the control mice. Analysis of data suggests the diabetic mice were unable to perform on RAM as evident by the significant increase in working and reference memory errors ($p < 0.05$) as compared to control. The significance level was higher in reference memory errors performed by diabetic mice ($p < 0.05$) as compared to working memory errors ($p < 0.05$). Track plot analysis of DM II-induced animals revealed repeated re-entry in baited and

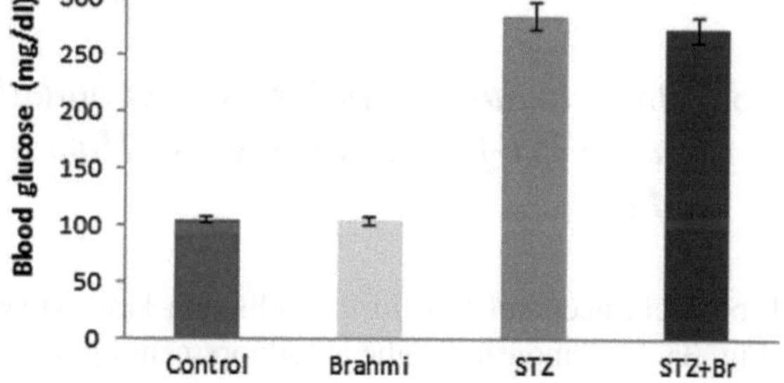

Fig. 1 Blood glucose level in different groups. *Control* vehicles treated; *Brahmi* (*BME*) CDRI-08-treated; *STZ* streptozotocin-treated; *STZ +Br* streptozotocin-treated mice treated with CDRI-08, * denotes comparison with control and other groups (* denote $p < 0.05$)

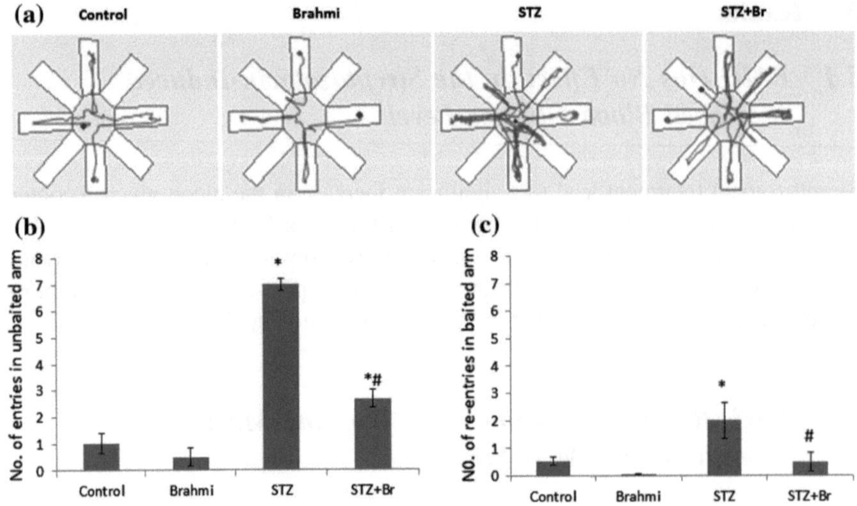

Fig. 2 Radial arm maze analysis of spatial memory of mice. **a** tracking analysis of mouse (b) reference memory errors (c) working memory errors. Data represents Mean ± SEM. *Control* vehicles treated; *Brahmi (BME)* CDRI-08-treated; *STZ* streptozotocin-treated; *STZ+Br*, streptozotocin-treated mice treated with CDRI-08, * denotes comparison between control and other groups, # comparison between STZ and the remaining groups (* and # denote $p < 0.05$)

unbaited arms. Treatment of mice with BME per say resulted into significant decline in the reference and working memory errors during RAM test which suggests improved memory in them compared to control mice. Track plot analysis of BME animals committed fewer re-entries and they were able to locate reward easily. BME treated DM II mice showed significant improvement in memory acquisition and statistically significant reduction in Reference and working memory errors were observed on RAM when compared to DM II mice and control mice ($p < 0.05$).

Figure 2b indicates that working memory error was three times more in diabetic mice as compared to control whereas CDRI-08 treated mice showed fewer errors and were able to track the food more accurately as compared to control mice group. However, diabetic animals committed more reference memory error was (seven times) and less working memory error (three times) as compared to normal mice (Fig. 2c).

3.3 BME Reduces Lipofuscin Accumulation in the Pre-frontal Cortex and Hippocampus of DM II Mice

Lipofuscin accumulation in the cells is a key marker of aging. Photomicrograph (Fig. 3a, b) demonstrates the autofluorescent lipofuscin accumulation in pre-frontal

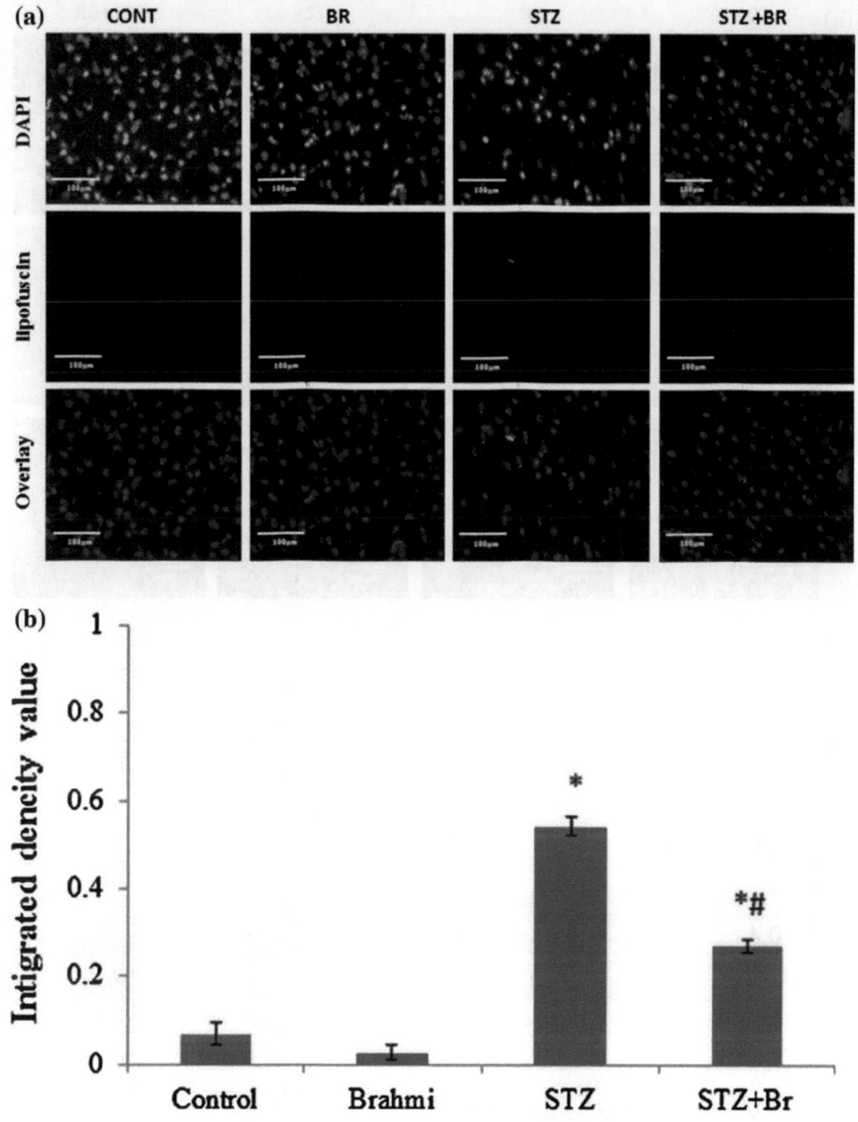

Fig. 3 Autofluorescent pigment lipofuscin accumulation in Pre-frontal cortex. **a** photomicrograph showing autofluorescent pigment (*blue filter*). **b** Densitometry analysis of autofluorescent pigment lipofuscin. Data represents Mean ± SEM. *Control* vehicles treated; *Brahmi* (*BME*) CDRI-08-treated; *STZ* streptozotocin-treated; *STZ+Br* streptozotocin-treated mice treated with CDRI-08, * denotes comparison between control and other groups, # comparison between STZ and the remaining groups (* and # denote $p < 0.05$)

cortex and hippocampus (Fig. 4a, b) along with co localization of DAPI stained nuclei. The concentration of lipofuscin present in photomicrograph was expressed in terms of its intensity in a selected area. Densitometry data of lipofuscin in

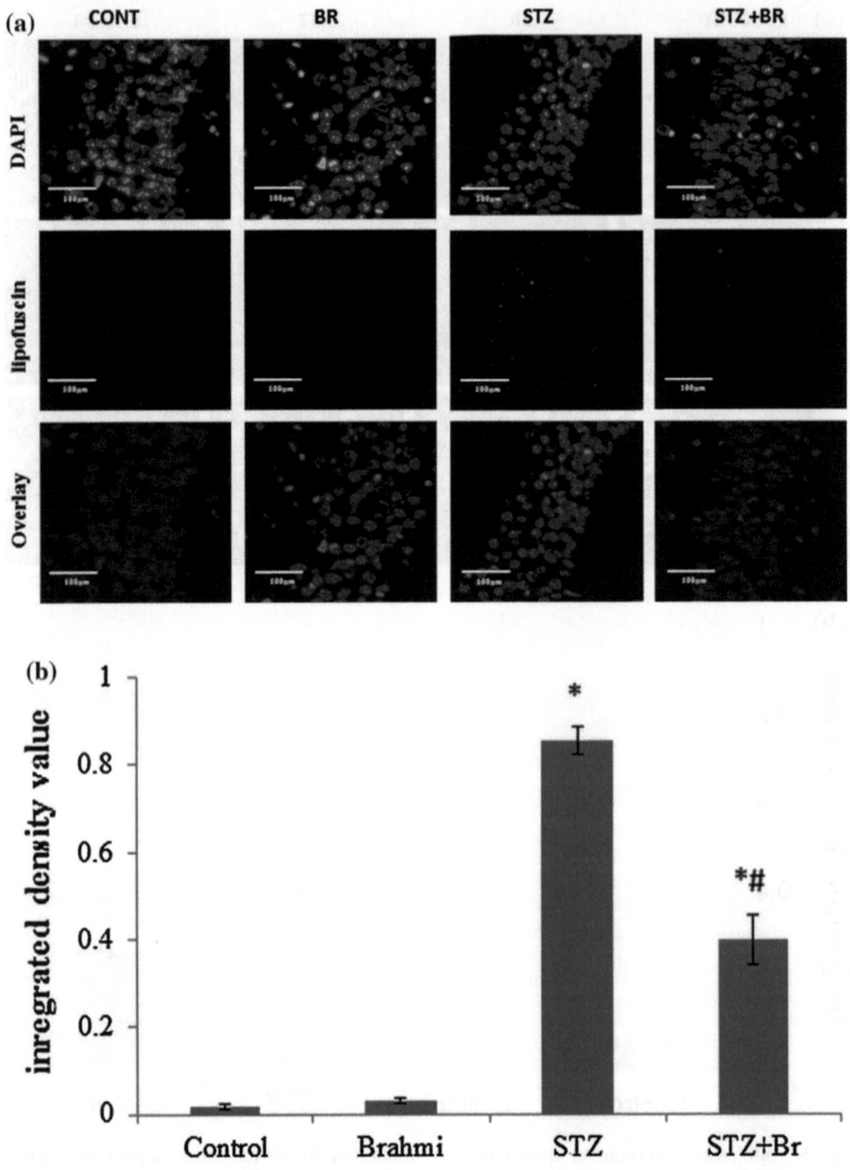

Fig. 4 Autofluorescent pigment lipofuscin accumulation in hippocampus. **a** photomicrograph showing autofluorescent pigment (*blue filter*). **b** Densitometry analysis of autofluorescent pigment lipofuscin. Data represents Mean ± SEM. *Control* vehicles treated; *Brahmi* (*BME*) CDRI-08-treated; *STZ* streptozotocin-treated; *STZ+Br* streptozotocin-treated mice treated with CDRI-08, * denotes comparison between control and other groups, # comparison between STZ and the remaining groups (* and # denote $p < 0.05$)

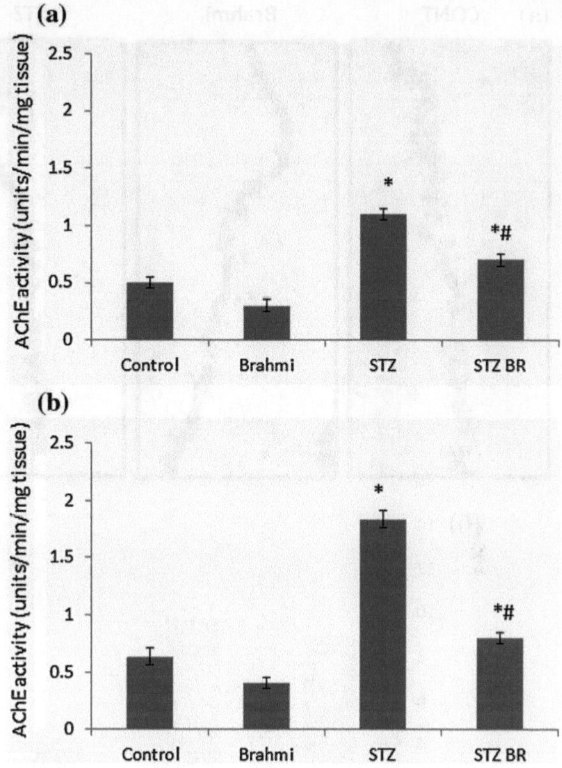

Fig. 5 Colorimetric analysis of acetylcholine esterase activity in pre-frontal cortex (**a**) and hippocampus cortex (**b**). Data represents Mean ± SEM. *Control* vehicles treated; *Brahmi* (*BME*) CDRI-08-treated; *STZ* streptozotocin-treated; *STZ +Br* streptozotocin-treated mice treated with CDRI-08, * denotes comparison between control and other groups, # comparison between STZ and the remaining groups(* and # denote $p < 0.05$)

hippocampus revealed that the autofluorescent lipofuscin accumulation was more in diabetic mice as compared to control ($p < 0.05$). Treatment of CDRI-08 shows significant decline in the accumulation of lipofuscin towards normal. Similar result was observed in the pre-frontal cortex. However, the level of deposition was more in the hippocampus (ten times) as compared to pre-frontal cortex (three times) in diabetic mice.

3.4 BME Reduces AChE Activity in the Pre-frontal Cortex and Hippocampus of DM II Mice

Spectrophotometric assay of AChE in pre-frontal cortex (Fig. 5a) shows no significant change in AChE activity with BME treatment as compared to control. AChE activity was significantly increased in diabetic mice ($p > 0.05$) as compared to that in control. Treatment of diabetic mice with BME significantly reduces the AChE activity in comparison to the diabetic mice group but not with control. Figure 5b shows the activity of AChE in hippocampal lysate in various groups of mice. The CDRI-08 per say showed significant decline in AChE activity as

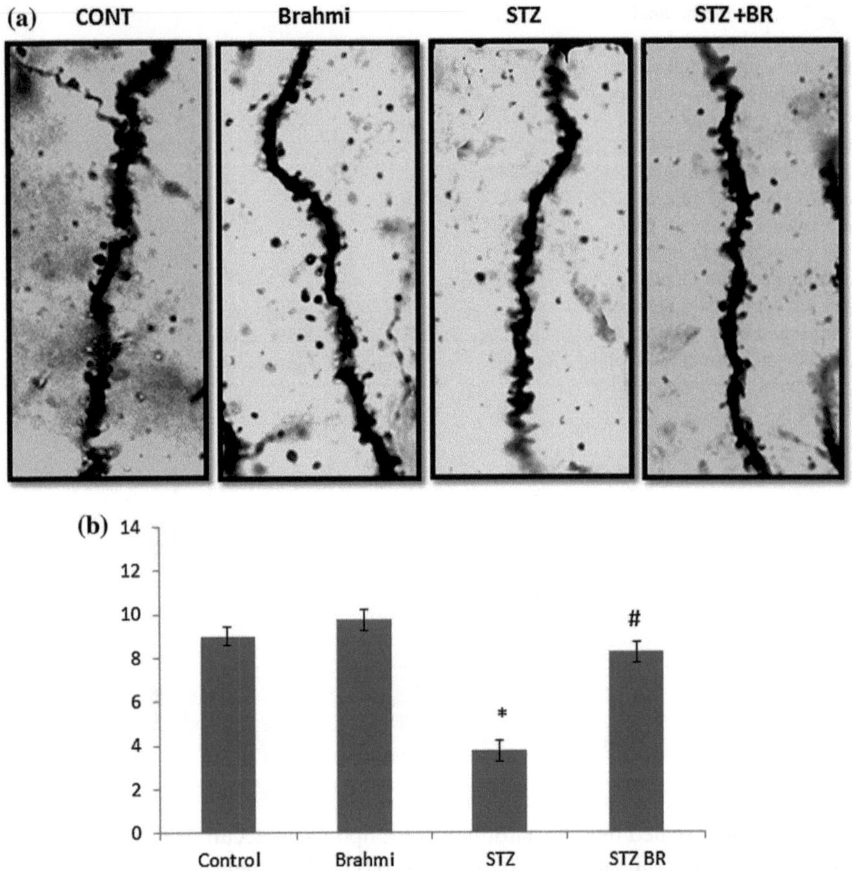

Fig. 6 Neuronal spine density in pre-frontal cortex by rapid Golgi staining of neurons. **a** photomicrograph of the dendritic spine at 100X **b** histograph of the spine. Data represents Mean ± SEM. *Control* vehicles treated; *Brahmi* (*BME*), CDRI-08-treated; *STZ* streptozotocin-treated; *STZ+Br* streptozotocin-treated mice treated with CDRI-08, * denotes comparison between control and other groups, #, comparison between STZ and the remaining groups (* and # denote $p < 0.05$)

compared to control group ($p > 0.05$). Diabetic mice showed significantly elevated value of AChE activity ($p > 0.05$) when compared with control mice. A significant decline in the AChE activity was observed with BME treatment in diabetic mice. Nevertheless, increase in the activity of AchE was more in the hippocampus (four times) as compared to pre-frontal cortex (two times) of diabetic mice.

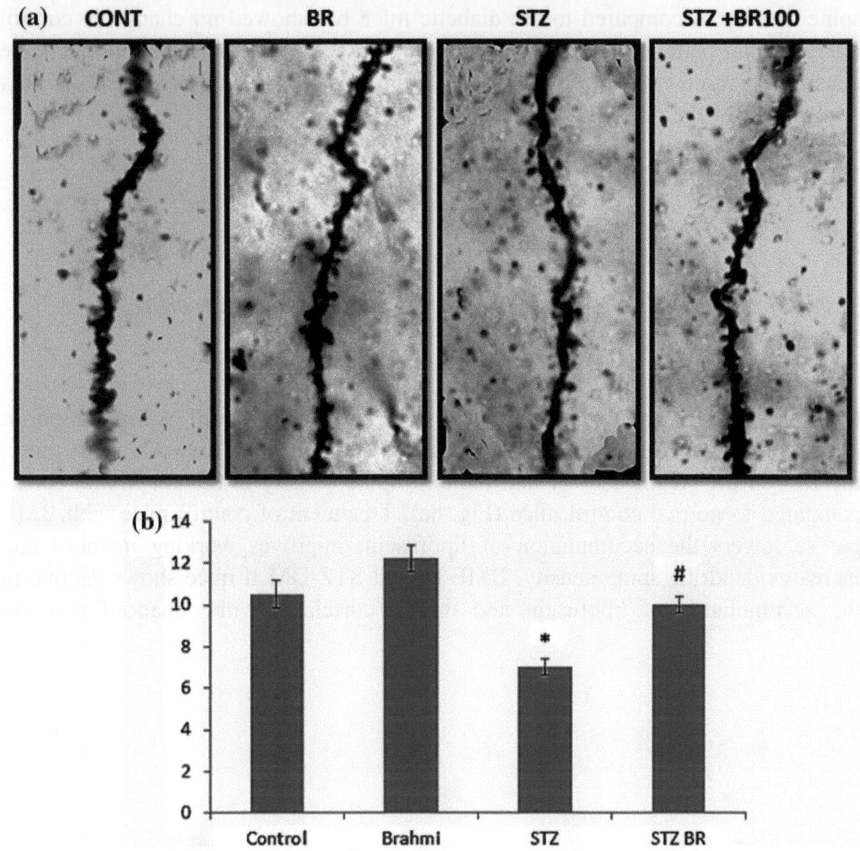

Fig. 7 Neuronal spine density in hippocampal CA1 neurons by rapid Golgi staining of neurons. **a** photomicrograph of the dendritic spine at 100X **b** histograph of the spine. Data represents Mean ± SEM. *Control* vehicles treated; *Brahmi* (*BME*) CDRI-08-treated; *STZ* streptozotocin-treated; *STZ+Br* streptozotocin-treated mice treated with CDRI-08, * denotes comparison between control and other groups, # comparison between STZ and the remaining groups (* and # denote $p < 0.05$)

3.5 *BME Enhances Dendritic Spine Density in Pre-frontal Cortex and Hippocampal Neurons of DM II Mice*

To demonstrate whether the observed memory decline is associated with the morphological changes in neurons, we analyzed the dendritic spine density in the pre-frontal cortex (Fig. 6a, b) and hippocampal CA1 pyramidal neurons (Fig. 7a, b). No significant change was observed in the spine density in above regions with BME administration as compared to control spine density was significantly decreased in above regions in diabetic mice ($p > 0.05$) as compared with control (Figs. 6 and 7). Treatment of diabetic mice with BME increased ($p > 0.05$) the spine density as compared to the diabetic mice but showed no change in control mice. Similar pattern were observed in both regions, however, decline in the spine density was more in the hippocampus (four times) as compared to pre-frontal cortex (two times) of the diabetic mice.

3.6 *BME-Induced Alteration in the Lipofuscin Accumulation Is Correlated with Reversal of STZ-Induced DM II Effects on Memory Decline, Reduced Spine Density and AChE Activity*

Our correlation data suggests that the STZ treatment-caused accumulation of lipofuscin in pre-frontal cortex is correlated with a decline in working memory, increase in the AChE activity and reduced spine density in the pre-frontal cortex compared to normal control mice (Fig. 8a). Treatment of control mice with BME per se lowers the accumulation of lipofuscin, improves working memory and increases dendritic spine density. BME treated STZ-DM II mice shows decline in the accumulation of lipofuscin and this is correlated with enhanced working

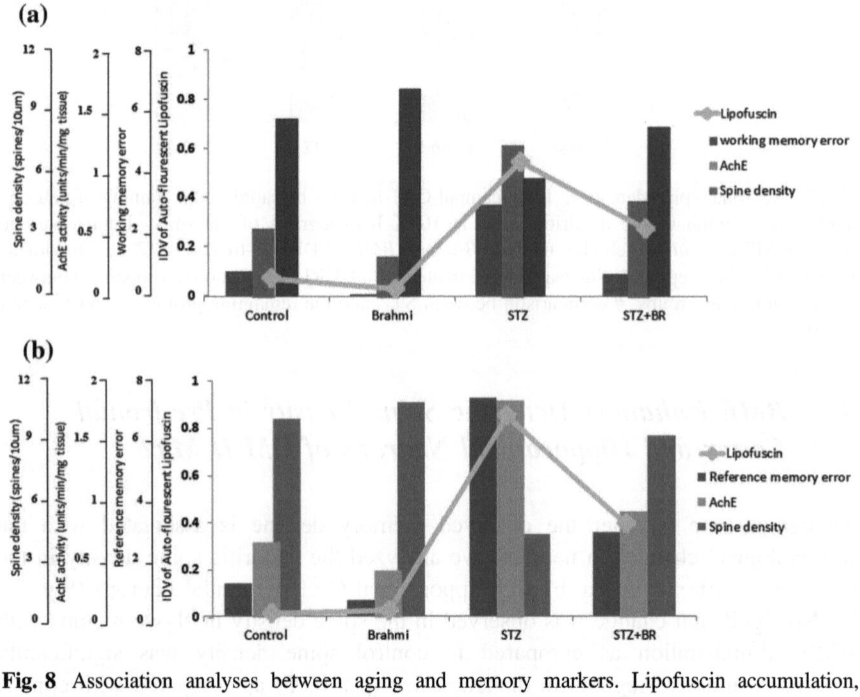

Fig. 8 Association analyses between aging and memory markers. Lipofuscin accumulation, AChE activity and spine density with respect to working memory (pre-frontal cortex function) (**a**), and the same with respect to reference memory (hippocampal function) (**b**)

memory, decreased AChE activity and increased dendritic spine density compared to those in the STZ-treated DM II mice (Fig. 8a). Further, the association between lipofuscin deposition and reference memory and AChE activity in hippocampus suggest that the reference memory decline due to STZ treatment is correlated with increased lipofuscin deposition and AChE activity in hippocampus in comparison to the normal control mice (Fig. 8b). Data also suggests that BME treatment to STZ-treated DM II mice enhanced reference memory and this is also correlated with decline in lipofuscin deposition and AChE activity, and increase in dendritic spine density (Fig. 8b) in comparison to STZ-treated mice. By this correlation data, it is also observed that the neuroprotective effects of BME are more prominent in hippocampus as compared to pre-frontal cortex.

4 Discussion

The present study is focussed on whether the diabetes mellitus, one of the age-related metabolic disorders that changes the milieu of neurons and their metabolic status, which have otherwise effects on the array of brain functions including learning and memory and thereby the cognition. As it is well established that diabetes occurs mainly in older population, however, it is not yet established whether inception of diabetes mellitus per se increases the aging process irrespective of the age of subjects who is inflicted with the disease. Also, the molecular correlation of diabetes-induced aging and whether it is associated with decline in cognition at the level of learning and memory, association with alterations in cholinergic neuronal function and neurobiological parameters is not well understood. However, studies indicate that the diabetes mellitus leads to changes in the brain which are very similar to changes due to Alzheimer's disease. These finding increases the curiosity that despite being an age-related disorder why Alzheimer's appears early in the life of diabetic patients. We, therefore, thought to understand above phenomena in the drug-induced type II diabetes mice model.

In order to study above, we developed DM II diabetic mice model by administering a selected dose of streptozotocin to puff mice and assessed them for the development of diabetes (type II) in their adult age and accumulation of lipofuscin and analysed alterations in AChE activity in pre-frontal cortex and hippocampus of DM II mice as markers of aging as the pre-frontal cortex and hippocampus are associated with learning and memory. Findings on the blood glucose content data validate the diabetic mouse model. This diabetic type II (DM II) has already been reported to be due to increase in the insulin resistance (Pandey et al. 2015b). Our data also clearly suggests that the STZ-induced DM II is associated with increase in the accumulation of lipofuscin and AChE activity in both the memory related regions of the brain. This suggests that neurons undergo stress due to excessive lipofuscin deposition and decline in cholinergic transmission which together indicates the symptoms of brain aging by altering the functions of the pre-frontal cortex in short term memory and hippocampus in formation and retrieval of memory as was earlier reported on H.M. (Henry Molaison), (Scoville and Milner 2000). Working memory is the process for the retrieval and proper utilization of previously acquired knowledge (Goldman-Rakic 1995) via pre-frontal cortex and its channelization to hippocampus plays important role in the initial development of cognition (Goldman-Rakic 1987, 1995). Large body of evidence indicates that lipofuscin accumulation in different tissues is the hallmark of aging (Nowotny et al. 2014), which is also present in large amount in the brain (Firlag et al. 2013). Lipofuscin accumulation is considered as product of "wear-and-tear" process of biomolecules during aging. This is formed due to oxidative damage of many functional proteins specially proteasomal degradation results into accumulation of inclusion bodies and their removal is abnormally affected (Hohn and Grune 2013). In the present study, the large accumulation of lipofuscin in diabetic brain suggests that untreated diabetes accelerate the sign of aging. A group of researchers that believe on the 'lipofuscin hypothesis of AD (Giaccone et al. 2011), have proposed that some type of lipofuscin accumulation have more deleterious effects than others, and have more propensity to induce the formation of senile plaques (Giaccone et al. 2011). Further, our data suggest that DM II increases the activity of AChE which leads to memory decline by reduction in the dendritic spine density in pre-frontal cortex and hippocampus. It is also evident from our data that the hippocampus is more prone to these changes leading to decline in the reference memory severely compared to the pre-frontal cortex. This indicates that long term memory is affected more than the short term memory. Further, data on the treatment of DM II mice with CDRI-08 showed a significant decline in the accumulation of lipofuscin with a corresponding decline the AChE activity in both the domains of brain involved in acquisition and retrieval of memory by increasing the dendritic spine density. During Alzheimer's disease acetylcholine neurotransmitter availability in the synapse get reduced and this might either be due to less secretion of neurotransmitter or its degradation by AChE (acetylcholinesterase) (Pasqualetti et al. 2015). Increased activity of AChE has been found in our study also which might be due to less availability of acetylcholine or enhanced activity of AChE leading to increased break down of the acetylcholine into acetate and choline due to DM II. Thus, reduced level of acetylcholine in synaptic cleft might be associated with deficit in the postsynaptic potential (PSP) development leading to reduced synaptic transmission of signals and thereby the impairment in memory.

We observed in our current study that both forms of spatial memory (working and reference) are compromised in diabetic mice as compared to normal control mice. Learning and memory deficit is a key component of the cognitive impairment in either aging or metabolic disorder (DM II). Report from earlier studies have shown that alterations in different forms of memory i.e. spatial or non-spatial, working and reference memory occurs during aging (Belviranli et al. 2012) and DM II (Pandey et al. 2015b). These memories have been reported also to decline during age-associated neurodegenerative disease (AD) (Uabundit et al. 2009). However, the reference memory is more severally affected (seven-fold more) in the case of DM II mice as compared to the working memory (fourfold more). Thus, the question arises how this memory is altered due to DM II-induced aging as is evident from our data. We, therefore, speculated that the DM II-induced memory impairments might be caused due to alterations in the dendritic spine density in the pre-frontal cortex and hippocampus.

Most excitatory synapses in the brain are small protrusions from dendrites called spines (Nimchinsky et al. 2002), are a major sites of synaptic transmission during brain function, therefore, the morphological changes or altered spine number might be observed in various mental disorders (Lai and Ip 2013). Memory formation occurs at the level of excitatory synapses. Spine density is highly dynamic and is activity dependent that might be altered in pathological conditions. It is reported that lowering of neuronal activity in long term depression (LTD) causes decline in the spine density (Fiala et al. 2002; Chen et al. 2014; Pereira et al. 2014). Memory is directly related to density of dendritic spines (Frankfurt and Luine 2015). The decline in the dendritic spine density might also be correlated with alterations in the glutamatergic synaptic transmission, an important event during development of learning and memory, involving altered activity of ionotropic glutamate receptors such as AMPA (α-amino-3-hydroxy-5-methyl propionic acid) and NMDA (N-methyl-D-aspartate) receptors (Lohmann and Kessels 2014; Pandey et al. 2015a, b; Rai et al. 2015).

Our data on the dendritic spine density shows that number of spine is decreased in both the regions of the brain; however, the extent of decline is more in hippocampus than in the pre-frontal cortex of DM II mice when compared to corresponding control. The decrease in the dendritic spine density may be correlated with decline in the activity of neuronal circuits (Gardoni et al. 2002) or pathological condition created by DM II (Stranahan et al. 2009) or both in these areas of the brain. Our finding on more accumulation of lipofuscin in these regions is correlated with pathological conditions in the pre-frontal cortex and hippocampus whereas increased AChE activity supports the decreased activity in the neurons belonging to these regions. Therefore, our data supports that accelerated accumulation of lipofuscin and increased activity of AChE might be leading to decrease in the dendritic spine density which leads to perturbations in memory.

Our data also reveal that DM II has more negative effects on hippocampus as has been reported also in case of ischemia (Schmidt-Kastner and Freund 1991), oxidative stress (Wilde et al. 1997), epilepsy (Pacheco Otalora et al. 2007), Alzheimer's disease (Detoledo-Morrell et al. 1997) and many neurodegenerative pathological conditions (Hatanpaa et al. 2014). There are three unique features which separate hippocampus from the other brain regions- one, it has more glucocorticoid receptor and so it is more susceptible to stress response (Aronsson et al. 1988); two, it is more sensitive to heat stress as the temperature of the hippocampus is lower than other parts of the brain (Mariak et al. 1999); and three, this is a centre for adult neurogenesis (Curtis et al. 2011) and dividing cell are more susceptible for ROS (Benhar et al. 2002). These finding may be correlated with our data on DM II, which might affect hippocampus and thus the hippocampus-dependent reference memory loss at all above levels. Altogether, we can conclude that DM II induces aging-like symptoms (increased AChE and increased lipofuscin accumulation) in adult animals which lead to decline in memory.

Scientists have been facing a challenge how to prevent or slow these age-related changes in the brain or diabetes mellitus-induced aging or impairments in learning and memory by using alternative medicines. Further, we have reported that BME has dose-dependent antidiabetic and memory enhancing properties in DM II mice (Pandey et al. 2015b). Our present study suggests that hyperglycemia is not reduced by CDRI-08 administration at 100 mg/kg BW. In this regard, our results suggest that changes in cognitive performance of DM II mice are not due to its anti-diabetic effect although it has the neuroprotective effect. The present study suggests the memory enhancing effect of CDRI-08 in diabetic mice. The previous study from our group and elsewhere has shown that BME has a role in the recovery of chemically induced amnesia, neurodegenerative disease and normal aging (Joshi and Parle 2006; Uabundit et al. 2009; Rai et al. 2015). However, the underlying mechanism of BME mediated recovery of amnesia is poorly understood. Oxidative stress causes an adverse effect on the brain and aging by increasing reactive oxygen species (Pandey et al. 2015b). Literatures support that increase in the level of ROS has been associated with Alzheimer's disease and lipofuscin deposition (Hohn and Grune 2013; Rosales-Corral et al. 2015). Therefore, reduction in the ROS level may be associated with these changes as BME has been found to reduce the oxidative stress load (Pandey et al. 2015b; Rai et al. 2015). Therefore, reduction in AChE activity and lowering of the lipofuscin accumulation may be correlated with altered ROS levels. Reduction in AChE activity by administration of AChE inhibitor has already been reported during aging and Alzheimer's disease for the recovery of memory decline (Kar et al. 2002; Jahanshahi et al. 2013; Wang et al. 2014). We have observed that BME treatment reduces the AChE activity in diabetic mice. Similar observation regarding BME dependent inhibition of AChE activity in scopolamine-induced amnesic mice has been reported (Rai et al. 2015). Our correlation study suggest that decrease in the AChE activity and lipofuscin accumulation by BME administration in DM II mice may normalize the dendritic spine density. Thus, it is likely that the BME-induced lowering of AChE activity might underlie the recovery of working and reference memory by way of restoring the dendritic spine density in streptozotocin-induced DM II mice. The BME induced reversal in the DM II-induced alterations in memory thus might be due its anti-oxidative stress activity and AChE inhibitory activity. Our results are novel in identifying the DM II-induced aging process and the associated memory loss due to excessive deposition of lipofuscin, increased AChE activity and altered dendritic spine density. Our findings indicate the possible therapeutic potentials of the CDRI-08 as neuroprotective agent in the treatment of DM II-induced possible brain aging and the associated decline in its cognitive function.

Acknowledgments SPP thanks UGC for providing Senior Research Fellowship. Financial assistance from CAS in Zoology, UGC-UPE Program for the Genomics and Proteomics Thrust Area, DBT-ISLS Program and partial funding from CSIR, BRNS and ICMR to SP is gratefully acknowledged.

References

Aronsson M, Fuxe K, Dong Y, Agnati LF, Okret S, Gustafsson JA (1988) Localization of glucocorticoid receptor mRNA in the male rat brain by in situ hybridization. Proc Natl Acad Sci USA 85:9331–9335

Beckman KB, Ames BN (1998) The free radical theory of aging matures. Physiol Rev 78:547–581

Belviranli M, Atalik KE, Okudan N, Gokbel H (2012) Age and sex affect spatial and emotional behaviors in rats: the role of repeated elevated plus maze test. Neuroscience 227:1–9

Benhar M, Engelberg D, Levitzki A (2002) ROS, stress-activated kinases and stress signaling in cancer. EMBO Rep 3:420–425

Chen X, Hu J, Jiang L, Xu S, Zheng B, Wang C, Zhang J, Wei X, Chang L, Wang Q (2014) Brilliant Blue G improves cognition in an animal model of Alzheimer's disease and inhibits amyloid-beta-induced loss of filopodia and dendrite spines in hippocampal neurons. Neuroscience 279:94–101

Chiao YA, Rabinovitch PS (2015) The aging heart. Cold Spring Harb Perspect Med 5

Christensen K, Doblhammer G, Rau R, Vaupel JW (2009) Ageing populations: the challenges ahead. Lancet 374:1196–1208

Curtis MA, Kam M, Faull RL (2011) Neurogenesis in humans. Eur J Neurosci 33:1170–1174

Detoledo-Morrell L, Sullivan MP, Morrell F, Wilson RS, Bennett DA, Spencer S (1997) Alzheimer's disease: in vivo detection of differential vulnerability of brain regions. Neurobiol Aging 18:463–468

Ellman GL, Courtney KD, Andres V, Feather-Stone RM Jr (1961) A new and rapid colorimetric determination of acetylcholinesterase activity. Biochem Pharmacol 7:88–95

Fiala JC, Spacek J, Harris KM (2002) Dendritic spine pathology: cause or consequence of neurological disorders? Brain Res Brain Res Rev 39:29–54

Firlag M, Kamaszewski M, Gaca K, Balasinska B (2013) Age-related changes in the central nervous system in selected domestic mammals and primates. Postepy Hig Med Dosw (Online) 67:269–275

Frankfurt M, Luine V (2015) The evolving role of dendritic spines and memory: interaction(s) with estradiol. Horm Behav 74:28–36

Gardoni F, Kamal A, Bellone C, Biessels GJ, Ramakers GM, Cattabeni F, Gispent WH, Di Luca M (2002) Effects of streptozotocin-diabetes on the hippocampal NMDA receptor complex in rats. J Neurochem 80:438–447

Giaccone G, Orsi L, Cupidi C, Tagliavini F (2011) Lipofuscin hypothesis of Alzheimer's disease. Dement Geriatr Cogn Dis Extra 1:292–296

Goldman-Rakic PS (1987) Circuitry of the frontal association cortex and its relevance to dementia. Arch Gerontol Geriatr 6:299–309

Goldman-Rakic PS (1995) Cellular basis of working memory. Neuron 14:477–485

Goodpaster BH, Krishnaswami S, Harris TB, Katsiaras A, Kritchevsky SB, Simonsick EM, Nevitt M, Holvoet P, Newman AB (2005) Obesity, regional body fat distribution, and the metabolic syndrome in older men and women. Arch Intern Med 165:777–783

Guariguata L, Whiting DR, Hambleton I, Beagley J, Linnenkamp U, Shaw JE (2014) Global estimates of diabetes prevalence for 2013 and projections for 2035. Diabetes Res Clin Pract 103:137–149

Gunasekaran U, Gannon M (2011) Type 2 diabetes and the aging pancreatic beta cell. Aging (Albany NY) 3:565–575

Hatanpaa KJ, Raisanen JM, Herndon E, Burns DK, Foong C, Habib AA, White CL 3rd (2014) Hippocampal sclerosis in dementia, epilepsy, and ischemic injury: differential vulnerability of hippocampal subfields. J Neuropathol Exp Neurol 73:136–142

Hohn A, Grune T (2013) Lipofuscin: formation, effects and role of macroautophagy. Redox Biol 1:140–144

Jahanshahi M, Nickmahzar EG, Babakordi F (2013) The effect of Ginkgo biloba extract on scopolamine-induced apoptosis in the hippocampus of rats. Anat Sci Int 88:217–222

Joshi H, Parle M (2006) Brahmi rasayana improves learning and memory in mice. Evid Based Complement Altern Med 3:79–85

Kar A, Panda S, Bharti S (2002) Relative efficacy of three medicinal plant extracts in the alteration of thyroid hormone concentrations in male mice. J Ethnopharmacol 81:281–285

Lai KO, Ip NY (2013) Structural plasticity of dendritic spines: the underlying mechanisms and its dysregulation in brain disorders. Biochim Biophys Acta 1832:2257–2263

Liu J, Wu YY, Huang XM, Yang M, Zha BB, Wang F, Zha Y, Sheng L, Chen ZP, Gu Y (2014) Ageing and type 2 diabetes in an elderly Chinese population: the role of insulin resistance and beta cell dysfunction. Eur Rev Med Pharmacol Sci 18:1790–1797

Liu S, Wang W, Zhang J, He Y, Yao C, Zeng Z, Piao J et al (2011) Prevalence of diabetes and impaired fasting glucose in Chinese adults, China National Nutrition and Health Survey, 2002. Prev Chronic Dis 8:A13

Lohmann C, Kessels HW (2014) The developmental stages of synaptic plasticity. J Physiol 592:13–31

Mariak Z, White M, Lewko J, Lyson T, Piekarski P (1999) Direct cooling of the human brain by heat loss from the upper respiratory tract. J Appl Physiol 87:1609–1613

Mathew J, Balakrishnan S, Antony S, Abraham PM, Paulose CS (2012) Decreased GABA receptor in the cerebral cortex of epileptic rats: effect of Bacopa monnieri and Bacoside-A. J Biomed Sci 19:25

Mole S, Williams R, Goebel H (2011) The neuronal ceroid lipofuscinoses (Batten disease). Oxford University Press

Nimchinsky EA, Sabatini BL, Svoboda K (2002) Structure and function of dendritic spines. Annu Rev Physiol 64:313–353

Nowotny K, Jung T, Grune T, Hohn A (2014) Accumulation of modified proteins and aggregate formation in aging. Exp Gerontol 57:122–131

Pacheco Otalora LF, Moffett JR, Garrido-Sanabria ER (2007) Selective vulnerability of hippocampal NAAGergic neurons in experimental temporal lobe epilepsy. Brain Res 1144:219–230

Pandey SP, Rai R, Gaur P, Prasad S (2015a) Development- and age-related alterations in the expression of AMPA receptor subunit GluR2 and its trafficking proteins in the hippocampus of male mouse brain. Biogerontology 16:317–328

Pandey SP, Singh HK, Prasad S (2015b) Alterations in hippocampal oxidative stress, expression of AMPA receptor GluR2 subunit and associated spatial memory loss by Bacopa monnieri extract (CDRI-08) in streptozotocin-induced diabetes mellitus type 2 mice. PLoS ONE 10:e0131862

Pasqualetti G, Tognini S, Calsolaro V, Polini A, Monzani F (2015) Potential drug-drug interactions in Alzheimer patients with behavioral symptoms. Clin Interv Aging 10:1457–1466

Patro N, Shrivastava M, Tripathi S, Patro IK (2009) S100beta upregulation: a possible mechanism of deltamethrin toxicity and motor coordination deficits. Neurotoxicol Teratol 31:169–176

Pereira AC, Lambert HK, Grossman YS, Dumitriu D, Waldman R, Jannetty SK, Calakos K, Janssen WG, McEwen BS, Morrison JH (2014) Glutamatergic regulation prevents hippocampal-dependent age-related cognitive decline through dendritic spine clustering. Proc Natl Acad Sci USA 111:18733–18738

Rai R, Singh HK and Prasad S 2015 a special extract of bacopa monnieri (CDRI-08) restores learning and memory by upregulating expression of the NMDA receptor subunit GluN2B in the brain of scopolamine-induced amnesic mice. Evid Based Complement Altern Med (In press)

Rani A and Prasad S 2015 A Special Extract of Bacopa monnieri (CDRI-08)-Restored Memory in CoCl 2-Hypoxia Mimetic Mice Is Associated with Upregulation of Fmr-1 Gene Expression in Hippocampus. Evid Based Complement Altern Med (In press)

Rauf K, Subhan F, Al-Othman AM, Khan I, Zarrelli A, Shah MR (2013) Preclinical profile of bacopasides from Bacopa monnieri (BM) as an emerging class of therapeutics for management of chronic pains. Curr Med Chem 20:1028–1037

Rosales-Corral S, Tan DX, Manchester L, Reiter RJ (2015) Diabetes and Alzheimer disease, two overlapping pathologies with the same background: oxidative stress. Oxid Med Cell Longev 2015:985845

Schmidt-Kastner R, Freund TF (1991) Selective vulnerability of the hippocampus in brain ischemia. Neuroscience 40:599–636

Scoville WB, Milner B (2000) Loss of recent memory after bilateral hippocampal lesions. 1957. J Neuropsychiatry Clin Neurosci 12:103–113

Simpson T, Pase M, Stough C (2015) Bacopa monnieri as an antioxidant therapy to reduce oxidative stress in the aging brain. Evid Based Complement Altern Med 2015:615384

Singh H, Dhawan B (1997) Neuropsychopharmacological effects of the Ayurvedic nootropic Bacopa monniera Linn. (Brahmi). Indian J Pharmacol 29:359

Srikumar BN, Ramkumar K, Raju TR, Shankaranarayana Rao BS (2004) Assay-of-Acetylcholinesterase activity in the brain. Brain and Behavior 142–144

Stranahan AM, Lee K, Martin B, Maudsley S, Golden E, Cutler RG, Mattson MP (2009) Voluntary exercise and caloric restriction enhance hippocampal dendritic spine density and BDNF levels in diabetic mice. Hippocampus 19:951–961

Uabundit N, Wattanathorn J, Mucimapura S, Ingkaninan K (2009) Cognitive enhancement and neuroprotective effects of Bacopa monnieri in Alzheimer's disease model. J Ethnopharmacol 127:26–31

Verdile G, Fuller SJ, Martins RN (2015) The role of type 2 diabetes in neurodegeneration. Neurobiol Dis

Wang X, Wang LP, Tang H, Shan WY, Wang X, Liu D, Wu YY, Tian Q, Wang JZ, Zhu LQ (2014) Acetyl-l-carnitine rescues scopolamine-induced memory deficits by restoring insulin-like growth factor II via decreasing p 53 oxidation. Neuropharmacology 76 Pt A:80–87

Wilde GJ, Pringle AK, Wright P, Iannotti F (1997) Differential vulnerability of the CA1 and CA3 subfields of the hippocampus to superoxide and hydroxyl radicals in vitro. J Neurochem 69:883–886

Zou C, Shi Y, Ohli J, Schuller U, Dorostkar MM, Herms J (2016) Neuroinflammation impairs adaptive structural plasticity of dendritic spines in a preclinical model of Alzheimer's disease. Acta Neuropathol 131:235–246

FSC
www.fsc.org
MIX
Papier aus verantwortungsvollen Quellen
Paper from responsible sources
FSC® C105338

If you have any concerns about our products,
you can contact us on
ProductSafety@springernature.com

In case Publisher is established outside the EU,
the EU authorized representative is:
Springer Nature Customer Service Center GmbH
Europaplatz 3, 69115 Heidelberg, Germany

Printed by Libri Plureos GmbH
in Hamburg, Germany